Blue Skies, Blue Seas

Blue Skies, Blue Seas

Air Pollution, Marine Plastics, and Coastal Erosion
in the Middle East and North Africa

Martin Philipp Heger, Lukas Vashold, Anabella Palacios,
Mala Alahmadi, Marjory-Anne Bromhead,
and Marcelo Acerbi

 **WORLD BANK GROUP**

Middle East and North Africa Development Report Series

This series features major development reports from the Middle East and North Africa region of the World Bank, based on new research and thoroughly peer-reviewed analysis. Each report aims to enrich the debate on the main development challenges and opportunities the region faces as it strives to meet the evolving needs of its people.

TITLES IN THE MIDDLE EAST AND NORTH AFRICA DEVELOPMENT REPORT SERIES

Blue Skies, Blue Seas: Air Pollution, Marine Plastics, and Coastal Erosion in the Middle East and North Africa (2022) by Martin Philipp Heger, Lukas Vashold, Anabella Palacios, Mala Alahmadi, Marjory-Anne Bromhead, and Marcelo Acerbi

Distributional Impacts of COVID-19 in the Middle East and North Africa Region (2021) edited by Johannes G. Hoogeveen and Gladys Lopez-Acevedo

The Reconstruction of Iraq after 2003: Learning from Its Successes and Failures (2019) by Hideki Matsunaga

Beyond Scarcity: Water Security in the Middle East and North Africa (2018) by World Bank

Eruptions of Popular Anger: The Economics of the Arab Spring and Its Aftermath (2018) by Elena Ianchovichina

Privilege-Resistant Policies in the Middle East and North Africa: Measurement and Operational Implications (2018) by Syed Akhtar Mahmood and Meriem Ait Ali Slimane

Jobs or Privileges: Unleashing the Employment Potential of the Middle East and North Africa (2015) by Marc Schiffbauer, Abdoulaye Sy, Sahar Hussain, Hania Sahnoun, and Philip Keefer

The Road Traveled: Dubai's Journey towards Improving Private Education: A World Bank Review (2014) by Simon Thacker and Ernesto Cuadra

Inclusion and Resilience: The Way Forward for Social Safety Nets in the Middle East and North Africa (2013) by Joana Silva, Victoria Levin, and Matteo Morgandi

Opening Doors: Gender Equality and Development in the Middle East and North Africa (2013) by World Bank

From Political to Economic Awakening in the Arab World: The Path of Economic Integration (2013) by Jean-Pierre Chauffour

Adaptation to a Changing Climate in the Arab Countries: A Case for Adaptation Governance and Leadership in Building Climate Resilience (2012) by Dorte Verner

Renewable Energy Desalination: An Emerging Solution to Close the Water Gap in the Middle East and North Africa (2012) by World Bank

Poor Places, Thriving People: How the Middle East and North Africa Can Rise Above Spatial Disparities (2011) by World Bank

Financial Access and Stability: A Road Map for the Middle East and North Africa (2011) by Roberto R. Rocha, Zsofia Arvai, and Subika Farazi

From Privilege to Competition: Unlocking Private-Led Growth in the Middle East and North Africa (2009) by World Bank

The Road Not Traveled: Education Reform in the Middle East and North Africa (2008) by World Bank

Making the Most of Scarcity: Accountability for Better Water Management Results in the Middle East and North Africa (2007) by World Bank

Gender and Development in the Middle East and North Africa: Women in the Public Sphere (2004) by World Bank

Unlocking the Employment Potential in the Middle East and North Africa: Toward a New Social Contract (2004) by World Bank

Better Governance for Development in the Middle East and North Africa: Enhancing Inclusiveness and Accountability (2003) by World Bank

Trade, Investment, and Development in the Middle East and North Africa: Engaging with the World (2003) by World Bank

All books in the Middle East and North Africa Development Report series are available for free at https://openknowledge.worldbank.org/handle/10986/2168.

Contents

Maps

Photos

Tables

Acknowledgments

This book was prepared by a team led by Martin Philipp Heger (Senior Environmental Economist) and comprising Lukas Vashold (Research Consultant), Anabella Palacios (Urban Planner and Environmental Consultant), Mala Alahmadi (Natural Resources Management Specialist), Marjory-Anne Bromhead (Lead Environmental Consultant), and Marcelo Acerbi (Senior Environmental Specialist). The book has greatly benefited from the strategic guidance of Ayat Soliman (Sustainable Development Regional Director, Middle East and North Africa Region), Karin Kemper (Global Director, Environment, Natural Resources, and Blue Economy Global Practice [ENB GP]), and Lia Sieghart (Manager, ENB GP, Middle East and North Africa Region). In addition, valuable guidance and advice was received from Anna Bjerde (former Strategy and Operations Director, Middle East and North Africa Region) and Stefan Koeberle (Strategy and Operations Director, Middle East and North Africa Region).

The team received technical guidance throughout from Dahlia Lotayef and Frank Van Woerden (Lead Environmental Specialists). The team was supported operationally by Nadege Mertus (Program Assistant) and Marie A. F. How Yew Kin (Senior Program Assistant).

Technical inputs from the Stanford University Center on Food Security and the Environment, the European Space Agency (ESA), and the National Oceanography Centre (NOC) of the United Kingdom were incorporated into this report, including contributions from Sam Heft-Neal and Marshall Burke (both from Stanford University) and Christine Sams and Stephen Carpenter (both from the NOC). The support of Christoph Aubrecht (ESA) is greatly appreciated.

The team greatly benefited from insightful comments and guidance from internal peer reviewers. Peer reviewers for the report were Urvashi Narain (Lead Economist), Nancy Lozano-Gracia (Senior Economist), Delphine Arri (Senior Environmental Engineer), Nicolas Desramaut (Senior Environmental Engineer), Daniel Lederman (Deputy Chief Economist), Asif Islam (Senior Economist), Lili Mottaghi (Senior Economist), and Ruma Tavorath (Senior Environmental Specialist).

Peer reviewers for the concept note were Nancy Lozano-Gracia (Senior Economist), Anjali Acharya (Senior Environmental Specialist), Ernesto Sánchez-Triana (Global Lead, Pollution Management and Circular Economy), and Stephen Dorey (Public Health Doctor).

In addition, the authors received incisive and helpful advice and comments from World Bank colleagues, including Richard Damania (Chief Economist), Roberta Gatti (Chief Economist), Jason Russ (Senior Economist), and Esha Dilip Zaveri (Water Economist).

Excellent publication and editorial support were provided by Stan Wanat (Stanford University), Jewel McFadden (Acquisitions Editor), Mary Anderson (Copyeditor), Mary Fisk (Production Editor), and Yaneisy Martinez (Print And Electronic Conversion Coordinator).

The team thanks the Pollution Management and Environmental Health (PMEH) Trust Fund (https://www.worldbank.org/en /programs/pollution-management-and-environmental-health -program), the Korea Green Growth Trust Fund (KGGTF) (http://www.kgreengrowthpartnership.org), and the PROBLUE Trust Fund (https://www.worldbank.org/en/programs/problue) for supporting specific activities that have informed this report.

Executive Summary

INTRODUCTION

The economies of the Middle East and North Africa[1] have been hit hard by the COVID-19 pandemic, but the recovery brings with it an opportunity—to embark on new development paths that are greener, more resilient, and more inclusive. One crucial lesson from the COVID-19 crisis is that prevention is by far superior to any cure. As the Middle East and North Africa moves from the relief phase (where the focus rightly has been on public health and social protection) to the recovery phase, expansionary fiscal investments will play a critical role. Fiscal stimuli are crucial to kick-start economic growth (Hepburn et al. 2020). Given scarce fiscal resources, it is critical that the region seizes this window of opportunity to shed the old "brown" growth models and switch to a green, resilient, and inclusive development (GRID) path to help prevent the next crisis brought about by unsustainable economic growth.[2] A GRID growth path would have fewer emissions, less environmental degradation, and stronger ecosystems, while at the same time boosting resilience and inclusion, if managed properly (World Bank and IMF 2021).

To commit to a green recovery from the current pandemic crisis would help stem another advancing crisis—that of environmental degradation and climate change. Growing back greener and more resilient is the key for economies to accelerate growth, restore standards of living to precrisis levels, and get on a sustainable growth path, while also preparing for the new normal as opposed to the world of yesterday that locked them into their traditional growth paths. Whether the region's economies make the right type of investments now and in the coming years will determine their trajectories—economically, environmentally, and socially—for decades to come.

Building Back Greener: Returns and Trade-Offs

A green recovery will bring more jobs and growth than a brown recovery, especially in the long term. Besides avoiding the costs of environmental degradation, a green fiscal stimulus will create more jobs and deliver higher short-term returns per US dollar spent than a brown fiscal stimulus. A recent International Monetary Fund analysis showed that the returns of green investments in spurring gross domestic product (GDP) growth are indeed two to three times greater than the returns on comparable brown investments (Batini et al. 2021). Similarly, in a recent survey of more than 200 experts from finance ministries, central banks, and academia from around the world, the collective suggestion was that a green recovery from COVID-19 would be better not only for the environment but also for the economy (Hepburn et al. 2020). The fiscal experts argued that a green recovery strategy has higher economic multipliers, and they highlighted a number of priority investments, including in natural capital for terrestrial, marine, and coastal ecosystem resilience; restoration of carbon-rich habitats; clean mobility; resource efficiency; integrated land management systems; sustainable agriculture; and clean energy production.

Although the positive effects of such efforts are apparent, a green transition also comes with some trade-offs, at least temporarily. The decline of traditional brown industries implies that some people will lose their old jobs and that communities may face a temporary shortfall in tax revenue. Following the principles of a just transition, social protection schemes are important during the transition period to a green growth path, as are training and upskilling opportunities as well as support and active promotion of emerging green industries.

Human, Physical, and Natural Capital: Gains and Losses

In the Middle East and North Africa, residents' living standards (incomes, human capital, and infrastructure) have improved over the past three decades. Despite variations across economies, on average, real incomes have increased by around 40 percent,[3] and the region's people now live longer and are healthier and better educated than 30 years ago (as detailed in chapter 2). They have better access to water, sanitation, electricity, heating and cooling, transportation infrastructure, the internet, and telecommunications.

Not everyone has benefited, however. In countries affected by conflict—such as Libya, the Syrian Arab Republic, and the Republic of Yemen—residents have suffered not only displacement and the tragic loss of family and friends but also a collapse in living standards. Furthermore, challenges to inclusion persist: Youth unemployment is high. For many, work is precarious and informal. And because women in some Middle East and North Africa economies, despite gains in education, lack the same opportunities as men, relatively few women are working outside the home.

Although the region's human and physical capital have improved overall, its natural capital has deteriorated in recent decades. This report reviews the performance of the region's economies on many environmental indicators, most of which show deterioration over the past couple of decades. Emissions increased, terrestrial and marine ecosystems deteriorated, natural habitat was destroyed, marine pollution and ocean acidification increased, and unsustainable water management increased water stress. Some economies and cities have shown positive developments in recent years, but to restore the region's degraded natural capital on a broader scale, more ambitious steps are necessary.

Among other environmental shortfalls, the Middle East and North Africa has been the world's slowest region in decoupling economic growth from air pollutants, and it has yet to decouple economic growth from carbon emissions. It is the only region in the world that has not decoupled economic growth from CO_2, and although the region has decoupled economic growth from some related air pollutants, it did so more slowly than any other region (as discussed extensively in chapter 2). This adverse trend is driven mainly by the region's oil exporting economies; however, non-oil exporting economies are also decoupling rather slowly.

The economic structure of oil exporters is heavily skewed toward the exploitation, processing, and exportation of their natural resources, resulting in high carbon and air pollutant emissions. Past efforts toward economic diversification often targeted sectors whose adverse effects are similar to those of the oil and gas sector (such as metal extraction and processing) or sectors that depend directly on it (such as the petrochemical sector).[4] The abundance of fossil fuels and their subsidized provision (for final consumption as well as feedstock and energy sources for industries) disincentivizes their economical use and impedes the spread of more-sustainable alternatives—for example, public transportation versus personal combustion vehicles or renewable energy sources versus thermal power plants burning fossil fuels—driving up emissions of carbon as well as air pollutants.

"Blue Capital": Threats to Skies and Seas

This report focuses on the Middle East and North Africa's "blue" natural assets—its skies and seas—which are under severe threat. Specifically, it addresses three of the most significant threats to blue natural capital:

• *Air pollution* levels in the region's cities are second only to those in South Asia. The average urban resident in the Middle East and North Africa breathes in air that exceeds by more than 10 times the level of pollutants considered safe (figure ES.1, panel a).

FIGURE ES.1

Urban Air Pollution, Marine-Plastic Pollution, and Net Coastal Erosion, by Region

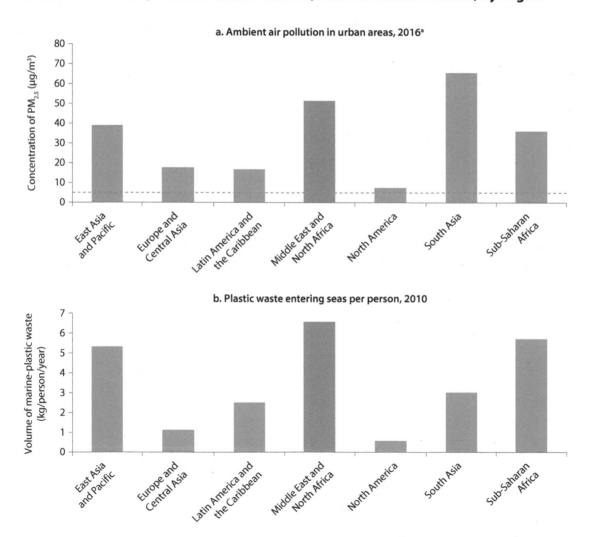

a. Ambient air pollution in urban areas, 2016[a]

b. Plastic waste entering seas per person, 2010

(continued on next page)

FIGURE ES.1

Urban Air Pollution, Marine-Plastic Pollution, and Net Coastal Erosion, by Region (*continued*)

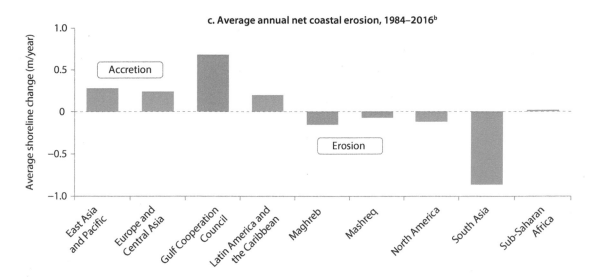

Sources: Based on Jambeck et al. 2015; Luijendijk et al. 2018; and 2016 data from the World Health Organization's Global Health Observatory (https://apps.who.int/gho/data/view.main).

Note: "North America" includes Bermuda, Canada, and the United States. Orange bars designate the Middle East and North Africa region or its subregions.

a. Particulate matter (PM) is made up of solid or liquid matter associated with Earth's atmosphere and suspended as atmospheric aerosol (the particulate/air mixture). $PM_{2.5}$ is a fine particle of 2.5 micrometers. The orange line denotes the World Health Organization (WHO) $PM_{2.5}$ threshold of 5 μg/m³ (micrograms per cubic meter of air).

b. In panel c, positive values represent net coastal accretion, and negative values, net coastal erosion. Middle East and North Africa subregions are as follows: (a) Maghreb, including Algeria, Libya, Malta, Morocco, and Tunisia; (b) Mashreq, including Djibouti, the Arab Republic of Egypt, the Islamic Republic of Iran, Iraq, Jordan, Lebanon, the Syrian Arab Republic, West Bank and Gaza, and the Republic of Yemen; and (c) Gulf Cooperation Council, including Bahrain, Kuwait, Oman, Qatar, Saudi Arabia, and the United Arab Emirates.

- *Marine-plastic pollution* is a severe and growing problem in the Middle East and North Africa. While there are regions that produce and leak more plastics in total into the seas, such as South and East Asia, MENA has the highest per capita footprint of plastics used and leaking into the region's seas and oceans. The Mediterranean is among the world's most plastic-polluted seas, with as much plastic flowing into it each year as the volume of fish taken out from the two most commonly caught species.[5]

- *Coastal erosion* rates of the Maghreb's coasts are among the fastest in the world—second, again, only to South Asian coasts (figure ES.1, panel c). Coastal erosion rates in some Middle East and North Africa economies exceed the global average (7 millimeters per year) by almost 10 times.

WHAT IS AT STAKE?

Health and Livelihood

The depletion of blue assets has reversed some of the region's improvements in human and economic development, threatening lives and livelihoods. In the Middle East and North Africa, ambient air pollution (AAP) costs the average resident at least 60 days of illness over their lifespan and caused 270,000 premature deaths in 2019. And the exposure of pregnant women to periods of elevated AAP has been shown to increase the risk of stunted growth in children. In the Mediterranean, one of the world's most marine-plastic-polluted seas, plastic debris damages marine life (plants and animals) as well as human life, since microplastics (very small fragments of plastics) have been found in several human organs and are suspected to lead to significant health problems.[6] (The health effects of air and water pollution are further detailed in chapters 3 and 4.)

Coastal erosion meanwhile relentlessly threatens to devour entire beaches and thereby the livelihoods of millions of people dependent on coastal tourism, fisheries, and related activities. The tourism industry, often focused on beach tourism and its connected value chains, is a major source of employment and accounts for 10 percent or more of GDP in several of the region's economies, for example, in Morocco and Tunisia (as discussed in chapter 5).

Economic Growth

The environmental degradation of skies and seas is estimated to cost more than 3 percent of GDP in some of the region's economies. A review in this report (chapter 3) shows that the annual cost of AAP in Middle East and North Africa economies averages 2 percent of GDP—ranging from around 0.4 percent of GDP in Qatar to more than 3 percent in the Arab Republic of Egypt, Lebanon, and the Republic of Yemen. Productivity falls if residents cannot work after they or their family members fall ill from air pollution, and health care costs can be a substantial burden on both individuals and governments. In addition, coastal erosion destroys sectors such as tourism and fishing that rely on intact beaches and clean seas. The average annual costs of coastal erosion are estimated to be 0.6 percent of GDP in the Maghreb (the subregion most afflicted by coastal erosion), ranging from 0.2 percent of GDP in Algeria to 2.8 percent in Tunisia (see chapter 5). Finally, the annual costs of marine-plastic pollution amount to around 0.8 percent of GDP on average, reaching more than 2 percent of GDP in countries such as Djibouti, Tunisia, and the Republic of Yemen (see chapter 4).

Polluted air, marine spaces full of plastics, and changing coastal landscapes all decrease the efficiency of important economic sectors and make the region less attractive. AAP has been shown to decrease the energy yields of photovoltaic solar panels (for which the region holds great prospects) and also decreases agricultural yields, both by reducing workers' productivity and, through its impact on radiation, by affecting temperature or precipitation. Air pollution can also decrease a city's attractiveness to tourists and its competitiveness by hemming its growth in population and value added. Marine-plastic pollution reduces fishery yields; damages ships and their equipment, increasing operational delays; and contributes to blocked drainages, thus increasing inundation risks and, potentially, disease outbreaks. Changes in the coastal landscape—mainly in the form of erosion but also accretion—impair the functioning of coastal infrastructure (for example, by damaging or blocking navigation paths near ports), and their impacts on biodiversity also have repercussions on coastal activities. The gradual disappearance, and the ever-higher pollution, of the region's beaches reduces their attractiveness, lowers tourism revenues, and decreases the competitiveness of coastal cities.

Trade and Competitiveness

In a world striving for net-zero emissions and phasing out fossil fuels, banking on green investments is critical for the Middle East and North Africa. It would allow the region to transition gradually to the "new normal" climate economy of the future instead of being stuck in the unsustainable economy of the past. Major economies like China, the European Union (EU), and the United States are pushing for substantial cuts in carbon emissions all along their value chains, with the EU and the US striving to be carbon neutral by 2050 and China by 2060. Overreliance on carbon-intensive fossil fuel industries is in direct contrast to global efforts toward phasing out fossil fuels and achieving net-zero emissions growth and can hence restrain the future economic performance of the Middle East and North Africa's economies. Even among the region's oil exporting countries—some of which are more exposed than others to the direct effects of trade measures such as carbon border adjustment mechanisms, as envisaged in the European Green Deal—the indirect effects of decreasing oil prices resulting from lower global demand can be substantial. Green growth is hence imperative, not only to reduce the current social and economic costs of burning fossil fuels—costs that air pollution is already imposing—but also to prepare the region's economies for the future.

Sticking to the traditional "brown growth" paths will leave the Middle East and North Africa at risk to end up with stranded assets and outdated business models. Where fiscal space allows it, countries globally are ramping up investments as stimulus to overcome the economic crisis caused by the COVID-19 pandemic. The Middle East and North Africa is no exception. The investments made now and in the coming years in response to this crisis will shape the trajectories of the region's economies for decades in terms of both economic advancement and environmental sustainability. Directing these investments toward sectors that will likely be at odds with global decarbonization trends is not prudent. The EU's recovery program plans to allocate 37 percent of its €800 billion stimulus for climate-friendly investments (IMF 2021). Similarly, the United States plans for large green-infrastructure investments as a response to the pandemic.[7] With the global push to decarbonize value chains and trade—for example, through the European Commission's proposed introduction of a carbon border adjustment mechanism—it becomes ever more important for Middle East and North Africa economies to increase efforts to avoid ending up with stranded assets and outdated business models. This is especially true for the economies strongly relying on fossil fuel exports, considering that large shares of fossil fuels must remain unextracted to meet the 2015 Paris Agreement targets for limiting global warming to well below 2 degrees Celsius, preferably to 1.5 degrees Celsius (Welsby et al. 2021).

Tackling air pollution, marine-plastic pollution, and coastal erosion will deliver considerable benefits, whether from an environmental, social, or economic perspective. Given the costs that these issues impose and the fact that climate change will exacerbate many of their adverse effects, setting changes in motion toward more sustainable management of blue assets is paramount. The restoration of the region's skies and seas will bring benefits on many fronts but also require strong policy responses to the various factors driving their degradation. Therefore, it is necessary to identify those factors and formulate appropriate strategies to get a grip on them.

WHY ARE THE REGION'S BLUE ASSETS DETERIORATING?

The Middle East and North Africa's blue natural capital is deteriorating for numerous reasons, many of which necessitate a collective answer from public authorities. Among these reasons, the region's economies lag in a range of areas when benchmarked against international best practices. Problematic areas include sending the wrong price signals by subsidizing polluting behavior; setting weak rules for limiting polluting activities as well as for enforcement of those rules; and lacking comprehensive management plans, whether for waste treatment or coastal development. These weaknesses both perpetuate and exacerbate the degradation of the region's blue assets.

Low environmental standards in the transportation and industry sectors as well as inefficient use of resources and burning of waste contribute to the continued pollution of the region's air. Outdated vehicle fleets, often lenient emission standards,[8] low-quality fuel that is often heavily subsidized and the cheapest globally, and inadequate public transportation all increase the transportation sector's contribution to lower air quality. Industrial emissions are often not well regulated, with the region's economies lagging in air quality laws and regulations (UNEP 2017).[9] The Middle East and North Africa uses a large amount of energy, of which more than 95 percent is derived from fossil fuels (Menichetti et al. 2019), to produce a given amount of economic output. As for energy intensity, the region is moving in the wrong direction: the Middle East and North Africa is the world's only region where energy use per output (BTUs per ton of output) has increased in the past three decades. Furthermore, clean-production incentives are largely absent, with the Middle East and North Africa the only region that has neither put nor initiated a price on carbon in the form of a carbon tax or an emissions trading system (World Bank 2021). Additionally, regulations regarding waste burning (both municipal and agricultural) are often poorly enforced, and the practice is still common in some Middle East and North Africa economies, contributing to the deterioration of the region's skies.

Weak solid waste management (SWM) in the Middle East and North Africa is a major reason why so much plastic is flowing into the region's seas. Especially in the Maghreb and the Mashreq subregions, a large share of waste (including plastics) is mismanaged. These deficiencies not only have an adverse impact on marine plastics[10] but also have important ramifications for air quality because of uncontrolled waste burning. Low recycling rates and few alternatives for reusing plastics help exacerbate the plastic tide. Price discrepancies between plastics and their greener alternatives as well as between virgin and recycled plastics are major reasons for the low adoption of these environmentally less harmful options. These discrepancies are driven largely by heavy subsidization of feedstock and necessary energy (also derived mostly from fossil fuels) for petrochemicals, first and foremost in the Gulf Cooperation Council (GCC) countries.

Inadequate management of coastal assets, exacerbated by rapid expansion of development along the coasts, has increased erosion

of the region's shorelines. Pressures from natural forces contributing to erosion, such as the frequency and intensity of coastal floods, will increase as a consequence of climate change. Ill-conceived adaptations have also created certain coastal erosion hot spots in the Maghreb. For example, although coastal protection infrastructure may protect a specific beach, it may obstruct the sedimentation flow down current and causes coastal erosion there. At the same time, the current absence of knowledge about the state and evolution of the region's coasts impedes proper management. Consequently, despite some progress, most Middle East and North Africa economies lack comprehensive frameworks for coastal development. Strong urbanization pressures in coastal cities have contributed to the fragmentation of coastal areas and their management in an unintegrated fashion. Furthermore, watershed and river management (including dam construction without sufficiently considering its impact on sediment discharge) has reduced sediment transport to the coastline, exacerbating coastal erosion.[11]

Thus, numerous environmental challenges persist, and addressing them is imperative to conserve and restore the Middle East and North Africa's blue natural assets. Understanding these challenges and providing a way forward to build and strengthen the contribution of blue capital to the economy and human well-being will play a key role in the region's transformation toward a greener, more resilient, and inclusive development path. Although this path may be different for individual economies given their heterogeneous starting points and levels of development, it is also important to note that to be successful in restoring the blue assets, regional cooperation among economies is vital. The transboundary nature of air pollution, marine-plastics pollution, and coastal erosion highlights the need for regional cooperation on these issues (not to mention the positive knowledge spillovers from such cooperation).

WHAT SHOULD MIDDLE EAST AND NORTH AFRICA ECONOMIES DO?

The multifaceted problems affecting the region's skies and seas require integrated solutions that this report identifies as the "4 I's": *Inform stakeholders, provide Incentives, strengthen Institutions,* **and** *Invest in abatement options.* Each of these objectives is crucial for successfully tackling air pollution, marine-plastic pollution, and coastal erosion, as follows:

- *Informing the policy discussion* across stakeholders (such as the private sector, nongovernmental organizations, and civil society organizations as well as across government ministries) with evidence about the sources of negative externalities is imperative while also helping to avoid frictions between them where possible. Similarly, the broad and frequent dissemination of information to the public is important to increase awareness and nurture demand for change.

- *Providing incentives* to the private and public sectors as well as households—whether by increasing prices for polluting activities or by providing subsidies for greener alternatives—is a viable way of nudging actors to change their behavior and switch to more sustainable patterns of production, consumption, and disposal.

- *Strengthening institutions* is important to limit and lower air and plastic pollution as well as to manage and mitigate the uncontrolled development and erosion of the Middle East and North Africa coasts. This effort includes the development and implementation of legally backed, clearly communicated regulations and mandates; clearly defined competencies for the ministries and authorities enforcing them; and provision of a transparent legal framework for some of the incentive schemes.

- *Making sizable investments* can tackle the respective degradation of the region's skies and seas in certain areas, including improvement of SWM, expansion of renewable energy production and public transportation infrastructure, and promotion of sustainable options (such as nature-based solutions) to combat coastal erosion.

If residents of the Middle East and North Africa have access to good-quality information about degradation, it will raise public awareness and build stakeholder ownership for policy change—both key elements in successful solutions. More broadly, ensuring open access to information is important in building public trust, making it a fundamental element of a functioning social contract between government and people (World Bank 2019). Appropriate collection of environmental data requires investments not only in physical infrastructure (for example, ground monitoring stations for air pollution) but also in the *use* of technological advancements. The latter means training staff in the necessary skills, such as working with remote sensing data to detect coastal erosion patterns or conducting life-cycle analyses of plastic products.

Informing about air pollution. Even though many Middle East and North Africa economies monitor at least some air pollutants, many do not make this information publicly available in an easily comprehensible way. However, some have made progress in this regard. The United Arab Emirates, for example, now provides real-time information on air quality, with guidance on how to minimize pollution on the worst days. Public awareness campaigns should include messaging on health and other negative economic outcomes—thus conveying the importance of behavioral change; providing a rationale for new regulatory requirements; and also addressing younger residents, as the Qualit'Air program in Morocco did with a dedicated online learning platform.

Raising awareness about marine plastics. Residents need to understand the consequences of plastic pollution, not only for beaches, fisheries, and marine wildlife but also on drainage systems and public health. This will help build consensus for change, including for restriction of plastics use and the broader adoption of recycling. To that end, the Tunisian National Waste Management Agency, in cooperation with the Sweepnet network, set up a dedicated communications and awareness office together with awareness-raising programs. It is also necessary to work with the plastics industry to jointly develop solutions and reach the youth to educate them about the consequences of excessive plastic consumption and inadequate disposal. This can be achieved through specialized campaigns on social media. One such effort is Jordan's "One Dead Sea Is Enough" initiative under the EU-funded SwitchMed project, which aims to induce a switch to a circular economy all along the southern and eastern Mediterranean.

Transparency about coastal erosion. In coastal areas, residents, municipalities, and affected industries need information on the processes behind erosion. This is key to building acceptance for policy changes that may involve restrictions on future development. Integrated coastal zone management (ICZM) processes also require open and transparent discussions of the impacts on different stakeholders.

Strengthening information about the sources of all three environmental issues is an important precursor to effective policy responses. High uncertainty remains about various drivers' contributions to the degradation of the skies and seas on a local scale. Which sector contributes how much to the pollution of a city's skies? Which cities, industries, and types of plastic products are the major culprits for the continuing flow of plastics into a subregion's seas? And which drivers—marine or terrestrial—cause the most coastal erosion? Obtaining clarity about these matters requires deep analyses in the form of source-apportionment studies for air pollution, life-cycle and flow analyses for plastic items, and geomorphological as well as wave-dynamics and sediment-transport studies for coastal erosion. Based on the information derived from these analyses and studies, locally specific policies can be tailored and adopted. Given the transboundary nature of cause-and-effect in the degradation of skies and seas, increasing regional cooperation is paramount. The sharing of knowledge and data across countries as well as across agencies *within* countries is important to guide policy makers in selecting the most effective policies.

"NO-REGRETS" POLICIES: A PRIORITY LIST

Even though source information is a crucial prerequisite for choosing the most effective policy mix, governments can readily take several priority measures to improve the management of their blue assets—their skies and seas. Just as the issues have multisectoral causes, the solutions must be multisectoral as well. Even as many questions persist about the drivers of negative externalities in many cities and countries—and related analytical work must be continuously supported—a set of critical, no-regrets policies can be implemented now to hit the ground running. This Executive Summary briefly describes these priority policies, but the report's main chapters provide more information, including discussions of the distributional implications of certain measures and detailed reviews of a plethora of additional policies. Table ES.1 summarizes some of the most critical measures, highlighting their main objectives and time horizons in the respective problem areas. As it shows, some of these priority recommendations apply to more than one sector, implying possible cross-benefits or similarities between them. Given these synergies, cooperation across sectors is highly desirable to increase the measures' efficiency and effectiveness.

Air Pollution

Regional modeling shows that, for the Middle East and North Africa's residents, the largest contributors to ambient air pollution are (a) road vehicles, (b) municipal waste burning, and (c) industrial processes. In addition, agricultural waste burning is a key source (especially in North Africa) as well as power plant emissions (especially in the Middle East), as further discussed in chapter 3.

Urban Transportation

In cities throughout the Middle East and North Africa, urban transportation is a significant contributor to air pollution. Improving urban planning and traffic management and supporting modal shifts from motorized personal transportation to public transportation (also supporting the greening of public transportation) and to nonmotorized personal transportation are key steps to take. Increasing fuel prices, especially in countries where the prices are extremely low (because of existing subsidies), is another critical step because it incentivizes people to use fuel-efficient or noncombustion cars and also to switch to public transportation and nonmotorized options. Several economies in the Middle East and North Africa have initiated such reforms even though some of them

TABLE ES.1

Priority Recommendations for Tackling Air Pollution, Marine-Plastics Pollution, and Coastal Erosion in the Middle East and North Africa

OBJECTIVE (4 I'S)	AIR POLLUTION	MARINE PLASTICS	COASTAL EROSION
Inform stakeholders	• Create public awareness to incentivize behavior with fewer negative environmental effects and to foster demand for interventions • Strengthen source information to help design appropriate interventions • Consult and plan jointly with the private sector, NGOs, civil society organizations, and across ministries to develop solutions		
	• Disseminate information frequently, better enabling individuals to change behavior so they can avoid exposure		
Provide incentives	• Reduce fuel subsidies, while at the same time making provisions for compensation mechanisms, such as reducing income taxes or social transfers		
	• Create markets for emissions and/or pollution (through ETS), and strengthen recylable-plastic markets (to make prices competitive with virgin plastics)		
	• Support greener alternatives (for example, through subsidies for cleaner technologies and alternatives, support of nature-based solutions)		
Strengthen institutions	• Manage and control polluting practices (for example, usage of certain SUPs, fuels, and technologies) • Mandate targets (for example, recycling targets and emission thresholds)		• Mandate coastal zoning and integrated coastal zone management
Invest in abatement options	• Install emission control technology (for example, fume scrubbers) • Facilitate uptake of clean production and resource efficiency technology	• Support R&D for new plastic alternatives • Implement marine cleanup technologies	• Implement nature-based solutions (for example, construction of wind fences, restoration of dune vegetation, and cultivation of seagrass and mangroves)
	• Strengthen SWM infrastructure (for collection, treatment, and disposal) • Switch to renewable energy production • Expand and "green" the public transportation infrastructure		

Timeline for implementation ● short term (0–2 years) ● medium term (2–5 years) ● long term (5–10 years)

Source: World Bank.
Note: Green shading indicates short-term measures with an expected implementation horizon of up to 2 years; orange shading, medium-term measures with an implementation horizon of 2–5 years; and blue shading, longer-term measures with an implementation horizon of 5–10 years. ETS = emissions trading system; NGOs = nongovernmental organizations; R&D = research and development; SUPs = single-use plastics; SWM = solid waste management.

unfortunately backtracked partially during the COVID-19 pandemic. However, it is also important to consider the adverse impacts of such reforms on low-income households and critical to make timely provisions for compensatory measures, such as reducing income taxes or increasing social transfers to those affected most.

Expanding public transportation and raising fuel prices (by removing fossil fuel subsidies) have both proven effective in reducing air pollution levels. Additional key measures are (improved) monitoring and inspection schemes for combustion vehicles, low-emission zones, and fuel-efficiency and emission-control mandates—as demonstrated in Tehran with successes in reducing the concentrations of harmful air pollutants. Additionally, reducing the number of internal combustion engine vehicles in cities should be advanced and the switch to low-emission alternatives supported. An important first step is to induce such a switch for public transportation fleets like the adoption of electric buses, as was done in some of the region's cities such as Doha, Marrakesh, or Tunis.

Municipal and Agricultural Waste Burning

Municipal waste burning, still practiced in and around many of the region's cities, must be tackled. The priority measures center around strengthening municipal SWM, reducing waste generation, and moving to a "circular economy"—an approach in which products are sustainably managed throughout their life cycles, from production to disposal or reuse. The region has made great progress in curbing agricultural waste burning. For example, over the past couple of years, strengthening regulation, enforcing penalties, and creating prices and markets for agricultural residue have proven to be key in reducing agricultural waste burning in Egypt.

Industrial and Energy Emissions

The regionwide adoption of best-in-class emissions control technology is crucial. There is large scope for the expansion of end-of-pipe emissions reduction programs paired with continuous monitoring schemes and regulations mandating emission caps. Reducing emissions is done most cost effectively by adopting an emissions trading system (ETS), paired with a regulatory cap as implemented in the EU ETS. Such cap-and-trade programs will create important incentives for resource efficiency and switching away from fossil fuels. Although no examples of such a system currently exist in the Middle East and North Africa, there are promising international examples such as the recent introduction of a cap-and-trade system directly targeting air pollutants in Gujarat, India. Switching to renewable energy sources is a crucial prerequisite for a transition to a less carbon-intensive energy sector and requires investments in energy generation, storage, and transmission infrastructure. The Middle East and North Africa region is highly suited for the adoption of renewable energy technologies, and projects are under way from Morocco all the way to the GCC countries.

Marine-Plastic Pollution

For reducing marine-plastic debris, improving SWM, including collection and safe disposal, is a key step. This will require adequate financing mechanisms for public utilities and building capacity in local utilities management. In parallel, work on reducing the generation of waste—and moving toward a circular economy with less waste and keeping resources in continuous use—is the end goal. Switching to a circular economy will require a bouquet of policies ranging from charging consumer fees for single-use plastics (SUPs), to bans on extremely harmful types of plastics, to working with producers on reuse options and subsidizing alternatives (such as bioplastics). Morocco introduced an eco-tax for producers of plastic products. And producers of plastic alternatives are gaining some foothold in the United Arab Emirates, where Abu Dhabi is also moving forward strongly in banning SUP items. All these measures need support from carefully managed information campaigns.

The price of fossil fuels that are not only the feedstocks of plastics but are also burned to create the energy to make plastics must be increased. Otherwise, environmentally friendly alternatives and recycling options cannot compete in the market. Here, cross-benefits for air pollution control could arise from reforming fossil fuel subsidies. Finally, beach cleanups may appear to be only a drop in the bucket, but they are an extremely effective approach to mitigation because plastics that accrue on shores often are dragged into the ocean by waves, where they last for decades or end up in the bellies of animals that in turn are eaten by humans. Along with directly reducing the amount of plastic ending up in marine spaces, such cleanups also raise awareness for the issue, as recognized by groups such as the Ervis Foundation, which organizes such events, as well as with the help of a mobile app to better reach the youth in the United Arab Emirates.

Coastal Erosion

For reducing coastal erosion, development of multistakeholder mechanisms for ICZM is important because there are many competing interests for using the coast. Setting up such collaborative ICZM processes—with a focus on land-use planning—is a particularly critical step for Middle East and North Africa economies, which lag in this respect, especially for parts of the coast that have not yet been developed. The recently introduced coastal management plan in the Rabat-Salé-Kénitra region in northern Morocco provides a regional example of how such schemes can benefit coastal areas.

For parts of the coasts that have already been developed, measures must be taken to mitigate further losses and, in some cases, to restore beaches. Ecosystem restoration and nature-based solutions (NBS) using locally adapted species (of seagrass, seaweed, mangroves, corals, or dune grasses) are often no-regret solutions. In addition to regulating coastal erosion by controlling floods and storm surges, NBS such as coral reef or seagrass rehabilitation have significant co-benefits—for example, by capturing "blue carbon"[12] and offering a habitat for fish or bird species. Egypt and Saudi Arabia are implementing large-scale reforestation programs for mangrove woods along the Red Sea, while artificial reefs have been used in Morocco. Their multiple benefits make these solutions particularly interesting for combating coastal erosion.

Time Horizons for Change

The time horizon for implementing priority measures varies: some measures are realizable in the short term, while others will take longer to unfold. Recognizing these differences is important when selecting policies to tackle a particular issue in a timely manner. However, kicking off the process for implementing measures whose results reveal themselves only with a certain lag is important. In this sense, information measures can be approached right away, together with certain regulations and closer cooperation across ministries and the private sector. Other regulations such as the creation of markets for emissions or the strengthening of recyclable-plastic markets as well as the introduction of ICZM may need more time. This longer time frame stems from the necessity for stocktaking before their introduction and the need for some form of transition period during which affected parties can adjust. Hence, realizing large-scale infrastructure projects—such as strengthening SWM, switching to renewable energy sources, and launching public transportation schemes—have a longer-term horizon. But the preparations should start immediately.

CONCLUSION

Restoring the Middle East and North Africa's blue skies and blue seas will benefit not only the environment but also the health, livelihoods, and incomes of residents. This report identifies and discusses the human and economic impact of blue-asset degradation and proposes solutions to support the transition to greener, more inclusive, more resilient growth paths. In addition to reducing the cost of environmental degradation, green growth paths would have higher economic multipliers in job creation and economic development. Swift action is imperative in the face of the current challenges posed by the COVID-19 crisis, but the region's economies should not lose sight of the much greater challenge posed by climate change and environmental degradation. In a world that painfully starts to feel the consequences of a warming planet, moving toward less-harmful economic models, including less reliance on fossil fuels such as oil and gas, becomes crucial.

Directing investments at carbon-intensive activities that will face ever-growing pressure in the coming decades is unsustainable from both the economic and environmental perspectives. Hence, even though setting the stage for a green transition that is just and that prepares the Middle East and North Africa for the challenges ahead is demanding and will not come without some adjustment costs, *not* doing so will likely have an even larger price tag. The region's leaders have the opportunity now to create jobs and growth with green investments, diversify their economies, and thereby make the region a more attractive place to live and work for today's residents and for future generations. Just as past decisions have shaped the region's current development, so will the actions taken by policy makers today shape these economies' trajectories for the coming decades. Laying the groundwork to address future challenges posed by environmental degradation, climate change, and a world striving to mitigate it is imperative.

NOTES

1. In this report, 20 economies are considered to be part of the Middle East and North Africa region, following the definition of the World Bank Group (except for Israel, which is excluded for the purposes of this report).

Because of the region's heterogeneity, the report sometimes clusters these economies into three subregions: (a) *Maghreb*, comprising Algeria, Libya, Malta, Morocco, and Tunisia; (b) *Gulf Cooperation Council (GCC)*, comprising Bahrain, Kuwait, Oman, Qatar, Saudi Arabia, and the United Arab Emirates; and (c) *Mashreq*, comprising Djibouti, the Arab Republic of Egypt, the Islamic Republic of Iran, Iraq, Jordan, Lebanon, the Syrian Arab Republic, West Bank and Gaza, and the Republic of Yemen.

2. Policies referred to as being "green" in this report are those with the potential to reduce long-run greenhouse gas (GHG) emissions, while "brown" policies are likely to increase net GHG emissions. Similarly, "green growth," "blue growth," or similar terms refer to increases in output and incomes that are accompanied by reductions in emissions and environmental degradation of blue assets—the region's skies and seas. Conversely, "brown growth" or "brown recovery" refer to activities that foster economic growth at the likely expense of increased GHG emissions and intensified degradation of the region's natural capital.

3. Growth rates differed between Middle East and North Africa subregions. On average, real incomes rose by around 40 percent—by more than 50 percent in the Maghreb and Mashreq subregions and by about 11 percent in the GCC countries, albeit from a much higher starting point (Human Development Data Center, United Nations Development Programme, http://hdr.undp.org/en/data).

4. These sectors are directly dependent on low input prices, both in the form of feedstock (for example, oil and gas for petrochemicals) and energy (predominantly derived from fossil fuels); are high-emitting sectors given their high energy intensity; and also contribute to environmental degradation (for example, by contributing to plastics pollution in the region's seas), as examined in chapter 4.

5. These fish species are the European pilchard (*Sardina pilchardus*) and the European anchovy (*Engraulis encrasicolus*).

6. Research on the public health effects of microplastics is emerging quickly but is still in its infancy, and although many worrisome findings are beginning to emerge—such as discovery of microplastics in human placentas and an array of other human organs—microplastics have yet to be conclusively linked to diseases.

7. The US green-infrastructure investments recently enacted include the modernization of bus and rail fleets (including, for example, replacement of school buses with zero- and low-emission alternatives); large-scale expansions of clean energy transmission networks, including half a million electric vehicle chargers; and environmental remediation measures such as cleaning up pollution from former industrial and energy sites and capping orphaned gas wells. The plans also include substantial support for restoring, monitoring, and researching forests—recognizing them as important infrastructure and endangered by increasingly widespread forest fires.

8. A global analysis of vehicle emission standards showed that not a single Middle East and North Africa economy requires new vehicles to adhere to international best practices regarding European emission standards, with some countries such as Algeria or Tunisia lacking any regulations in this respect at all as of February 2019 (Abdoun 2019).

9. A 2017 United Nations Environment Programme report finds that only 2 out of the 18 Middle East and North Africa economies surveyed have specific air quality laws and regulations in place (UNEP 2017). However, several have at least defined ambient air quality standards, which is a sign of progress. Furthermore, the report notes that only four of the region's economies have implemented clean production incentives for industries.

10. Recent studies have shown that the vast majority of plastic that ends up in the seas is from land-based activities as opposed to marine sources such as fishing equipment. These studies also highlight the high share of single-use plastics (SUPs) that end up in the world's oceans and seas (Morales-Caselles et al. 2021).

11. This has been especially the case in North African countries, as in Egypt's Nile delta or the Medjerda River flowing into the Gulf of Tunis (Hzami et al. 2021).

12. The oceans are major sinks of carbon dioxide and annually store amounts of carbon comparable to those stored by terrestrial ecosystems.

REFERENCES

Abdoun, A. 2019. "Global Fuel Quality Developments." Presentation to the 12th Global Partners Meeting of the Partnership for Clean Fuels and Vehicles (PCFV), Paris, March 5–6.

Batini, N., M. Di Serio, M. Fragetta, G. Melina, and A. Waldron. 2021. "Building Back Better: How Big Are Green Spending Multipliers?" Working Paper 2021/087, International Monetary Fund, Washington, DC.

Hepburn, C., B. O'Callaghan, N. Stern, J. Stiglitz, and D. Zenghelis. 2020. "Will COVID-19 Fiscal Recovery Packages Accelerate or Retard Progress on Climate Change?" *Oxford Review of Economic Policy* 36 (Suppl 1): S359–S381.

Hzami, A., E. Heggy, O. Amrouni, G. Mahé, M. Maanan, and S. Abdeljaouad. 2021. "Alarming Coastal Vulnerability of the Deltaic and Sandy Beaches of North Africa." *Scientific Reports* 11 (1): 1–15.

IMF (International Monetary Fund). 2021. "Reaching Net Zero Emissions." G-20 Background Note, IMF, Washington, DC.

Jambeck, J. R., R. Geyer, C. Wilcox, T. R. Siegler, M. Perryman, A. Andrady, R. Narayan, and K. Lavender Law. 2015. "Plastic Waste Inputs from Land into the Ocean." *Science* 347 (6223): 768–71.

Luijendijk, A., G. Hagenaars, R. Ranasinghe, F. Baart, G. Donchyts, and S. Aarninkhof. 2018. "The State of the World's Beaches." *Scientific Reports* 8 (1): 1–11.

Menichetti, E., A. El Gharras, B. Duhamel, and S. Karbuz. 2019. "The MENA Region in the Global Energy Markets." In *Foreign Policy Review* Special Issue, "MENARA: Middle East and North Africa Regional Architecture": 75–119. Institute for Foreign Affairs and Trade, Budapest.

Morales-Caselles, C., J. Viejo, E. Martí, D. González-Fernández, H. Pragnell-Raasch, J. I. González-Gordillo, E. Montero, et al. 2021. "An Inshore–Offshore Sorting System Revealed from Global Classification of Ocean Litter." *Nature Sustainability* 4 (6): 484–93.

UNEP (United Nations Environment Programme). 2017. "Middle East & North Africa: Actions Taken by Governments to Improve Air Quality." Report, UNEP, Nairobi, Kenya.

Welsby, D., J. Price, S. Pye, and P. Ekins. 2021. "Unextractable Fossil Fuels in a 1.5°C World." *Nature* 597 (7875): 230–34.

World Bank. 2019. "Our Expanded Strategy." MENA Region Brief, October 1. World Bank, Washington, DC. https://www.worldbank.org/en/region/mena/brief/our-new-strategy.

World Bank. 2021. *State and Trends of Carbon Pricing 2021*. Washington, DC: World Bank.

World Bank and IMF (International Monetary Fund). 2021. "From COVID-19 Crisis Response to Resilient Recovery: Saving Lives and Livelihoods while Supporting Green, Resilient, and Inclusive Development (GRID)." Document No. DC2021-0004 for the April 9, 2021, Meeting of the Development Committee (Joint Ministerial Committee of the Boards of Governors of the Bank and the Fund on the Transfer of Real Resources to Developing Countries), Washington, DC. https://www.devcommittee.org/sites/dc/files/download/Documents/2021-03/DC2021-0004%20Green%20Resilient%20final.pdf.

Abbreviations

AAP	ambient air pollution
ANGed	National Agency for Waste Management (Tunisia)
APAL	Agency for Coastal Protection and Planning (Tunisia)
AQM	air quality management
BTU	British thermal unit
CO	carbon monoxide
COPD	chronic obstructive pulmonary disease
CO_2	carbon dioxide
DRS	deposit-refund scheme
EIA	environmental impact assessment
EPA	Environmental Protection Agency (US)
EPAP	Egyptian Pollution Abatement Programme
EPR	extended producer responsibility
ESA	European Space Agency
ESIA	environmental and social impact assessment
ETS	emissions trading system
EU	European Union
GCC	Gulf Cooperation Council (Bahrain, Kuwait, Oman, Qatar, Saudi Arabia, and United Arab Emirates)
GDP	gross domestic product
GHG	greenhouse gas
GRID	green, resilient, and inclusive development
GVC	global value chain
HERRCO	Hellenic Recovery Recycling Corporation (Greece)
ICZM	integrated coastal zone management
IHME	Institute for Health Metrics and Evaluation
IPCC	Intergovernmental Panel on Climate Change (UN)
LEPAP	Lebanon Environmental Pollution Abatement Project
LEZ	low emission zone
$\mu g/m^3$	micrograms per cubic meter of air
MW	megawatts
NBS	nature-based solution(s)

NEEP	National Energy Efficiency Program (Saudi Arabia)
NGO	nongovernmental organization
NO_2	nitrogen dioxide
NO_x	nitrogen oxide
PET	polyethylene terephthalate
PM	particulate matter
$PM_{2.5}$	fine particulate matter (2.5 microns or less in diameter)
PM_{10}	fine particulate matter (10 microns or less in diameter)
PNL	National Coastal Plan (Morocco)
ppm	parts per million
R&D	research and development
RCP	Representative Concentration Pathway
ROPME	Regional Organization for the Protection of the Marine Environment
RSA	ROPME Sea Area
SABIC	Saudi Basic Industries Corporation
SEEP	Saudi Energy Efficiency Center
SLR	sea level rise
SMEs	small and medium enterprises
SOE	state-owned enterprise
SO_2	sulfur dioxide
SRL	regional coast management plan (Morocco)
SUP	single-use plastic
SWM	solid waste management
UN	United Nations
UNDP	United Nations Development Programme
UNEP	United Nations Environment Programme
WHO	World Health Organization
YLD	years lived with disability

Country abbreviations

ARE	United Arab Emirates
BHR	Bahrain
DJI	Djibouti
DZA	Algeria
EGY	Egypt, Arab Rep.
IRN	Iran, Islamic Rep.
IRQ	Iraq
JOR	Jordan
KWT	Kuwait
LBN	Lebanon
LBY	Libya
MAR	Morocco
MLT	Malta

OMN	Oman
PSE	West Bank and Gaza
QAT	Qatar
SAU	Saudi Arabia
SYR	Syrian Arab Republic
TUN	Tunisia
YEM	Yemen

CHAPTER 1

Introduction

OVERVIEW

Economies in the Middle East and North Africa have a window of opportunity to make their economic recovery from the COVID-19 pandemic a green one while tackling two major challenges ahead: environmental degradation and climate change. This chapter briefly lays out the structure of this report on the issues that the region's "blue" assets—namely, its skies and seas—face. The region's various forms of natural capital are under many pressures, but the report focuses on three particularly urgent ones: air pollution, marine-plastic pollution, and coastal erosion.

The pandemic and the economic crisis it induced have shown the world, and also the Middle East and North Africa, that these varying shocks can hit in ways both unexpected and unprecedented. Although the impacts and consequences of these shocks are still playing out, it is important to recognize the other crisis that has been unfolding more quietly but steadily over recent decades: the region's environmental degradation. Additionally, the looming threat of climate change is becoming clearer as it exacerbates some of pollution's adverse effects on the region's natural assets and threatens widespread social, health, and economic devastation.

The recovery from COVID-19 provides the Middle East and North Africa with a unique set of circumstances to mount a concomitant response to these next looming crises—and brings the region to a crossroads where some important decisions must be made: Will the region's economies continue down the "brown growth" path that has led to the degradation of most of their natural assets, a deterioration in public health, and unsustainable fiscal budgets? Or will they reset their growth strategies and move toward greener, more resilient, and more inclusive development?

This report argues that rebuilding better now can lay the foundation for the region to abandon its unsustainable growth paths from the past and emerge as a more livable region that offers its residents brighter prospects for the future. Highlighting the already advanced degradation in some areas, it also provides recommendations on how to tackle and even reverse some of the harms from these issues.

"BLUE" CAPITAL AND THREE CORE ISSUES

This report focuses on "blue" forms of natural capital—skies and seas—hence, the title of this *Blue Skies, Blue Seas* report. Blue assets are essential for the development of this region's economies, and the degradation of these assets presents a major challenge, one that the looming threat of climate change can only exacerbate.

Three chapters of this report analyze, in turn, three core issues that worsen this deterioration: air pollution, marine-plastic pollution, and coastal erosion. Degradation of these assets is at an advanced stage and has severe adverse impacts on ecosystems and biodiversity—as well as on human health—while also bearing substantial economic costs.

The three core topics have cross-cutting impacts on several other environmental issues affecting the region's blue assets. These include ocean acidification, water pollution more generally, overexploitation of fish stocks, and sea level rise caused by climate change. As such, this report complements a 2019 World Bank report—"Sustainable Land Management and Restoration in the Middle East and North Africa Region"—that explores some of the region's issues regarding its "green" assets. It also complements a forthcoming World Bank report on the economics of water that will address the crucial issue of sustainable water management throughout the Middle East and North Africa.

ROAD MAP TO THE REPORT

Chapter 2: "Human Advancement and Sustainable Natural Capital Use in the Middle East and North Africa"

Before diving into the core "Blue Skies, Blue Seas" issues, chapter 2 examines the evolution of national income and three types of capital (human, physical, and natural capital) over the past 30 years in the region in the context of a capital accounting framework. The analysis concludes that, by and large, there have been significant improvements in national incomes, human capital, and physical capital, regardless of which indicator one chooses. But the opposite holds for natural capital, which deteriorated substantially in recent decades, no matter which aspect one examines.[1] With the exception of economies affected by war and conflict—notably Libya, the Syrian Arab Republic, and the Republic of Yemen—these opposing trends regarding incomes and human and physical capital versus natural capital are apparent.

As further described below, the chapters on air pollution (chapter 3), marine plastics (chapter 4), and coastal erosion (chapter 5) show how polluted or degraded the region's air and seas already are and how these developments degrade its ecosystems, harm the health of its residents, and weaken the productivity of its economies. Each of these chapters analyzes one of the core issues, then presents priority recommendations for policies suitable in the Middle East and North Africa context. All chapters found that there is a critical dearth of source information on the core issues—whether pertaining to the sources of air pollution in a given city, the sources and reasons for marine-plastic leakage into the seas, or the sources and reasons for coastal erosion in a particular stretch of the coast. Each chapter concludes with a broad menu of possible actions for tackling the various issues.

Chapter 3: "Blue Skies for Healthy and Prosperous Cities"

Chapter 3 analyzes the current state of and trends in air pollution of the Middle East and North Africa, presents estimates of the human and economic toll, examines the sources (or lack of knowledge thereof), and proposes measures to reduce the region's air pollution. Its core analysis section focuses on fine particulate matter (since it is the most consequential pollutant for human health), showing that the Middle East and North Africa region has the second most polluted urban air in the world, trailing only South Asia. It also shows that residents of the region's major cities are subjected to air pollution levels considerably higher (by about 10 times) than the level considered safe by the World Health Organization.

The next major section estimates the toll of air pollution on human health and quantifies the morbidities and premature mortalities that can be attributed to low air quality in the Middle East and North Africa. The final major section proposes policy actions to be taken by governments of the region's countries.

This chapter of the report highlights the importance of improving the current understanding of the source contribution in the region's countries and cities (as a first step) and stresses the need for suitable source-apportionment studies and dissemination of those results. Despite the scarcity of evidence on sources of air pollution in these countries and cities, the chapter highlights some priority recommendations that many Middle East and North Africa countries can implement nonetheless. Furthermore, the policy section discusses ways to tackle air pollution in the region, presenting options for each main source—mobile transportation, industries and energy, waste burning, and other sources—focusing on market-based incentives, regulations, and technology.

Chapter 4: "Blue Seas: Freeing the Seas from Plastics"

Chapter 4 discusses the state of marine-plastic pollution in the region's seas, focusing on the Mediterranean Sea. Its core analysis section shows that residents of the Middle East and North Africa's economies are among the world's top polluters and that large volumes of plastics are entering the Mediterranean from their coasts. It investigates the reasons behind these economies' relatively high contributions to marine-plastic pollution, pointing largely to inadequate solid waste management but also to unsustainable production and consumption patterns of plastics. The next major section then provides an overview of the economic impacts of marine-plastic pollution, its increasing threat to ecosystems and public health, and the recent impacts of COVID-19 on the surge of plastic pollution.

After this initial diagnostic, the policy review section discusses the way forward for stemming the plastics tide. A current lack of understanding about the sources of plastic leakage again presents itself as a bottleneck for the formulation of specific policies in many Middle East and North Africa economies and cities. Nonetheless, the section proposes policies that aim to reduce the production and consumption of plastics, increase the reuse of plastics to improve solid waste management, and enhance the region's capabilities for recycling of plastics.

Chapter 5: "Blue Seas: Fighting Coastal Erosion"

Chapter 5 assesses erosion along the coasts of the Middle East and North Africa, focusing on the Maghreb subregion.[2] The coasts are home to large shares of the population and many economic hot spots. This chapter's core analysis section, using a global dataset, shows that especially the coasts in the Maghreb are retreating rapidly, and many beaches are set to shrink or even disappear in the future. Drawing on a novel dataset for Morocco and Tunisia (created in cooperation with the National Oceanography Centre in the United Kingdom), the analysis highlights specific examples of such coastal regions.

The next major section then quantifies the direct economic costs of coastal erosion in four North African countries. It shows that coastal erosion carries substantial costs, even when not considering forgone revenues from tourism. The policy review section highlights, once again, the need to understand specific sources of coastal erosion and dynamics at the shorelines of the Middle East and North Africa in order to formulate suitable policies. It provides a broad set of such policies, emphasizing the coordination of protective measures and economic activities under the tenets of a comprehensive integrated coastal zone management (ICZM) scheme. These policies are supplemented by a host of regional examples and international best practices.

NOTES

1. National incomes; physical capital (access to energy, access to water and sanitation, road density, infrastructure quality, digital infrastructure, to name but a few); and human capital indicators (infant and maternal mortality, life expectancy, and years of school, among others) show positive trends over the past 30 years. In contrast, natural capital has suffered as various indicators of ecosystem and environmental health have continuously worsened. It could be argued whether human and economic advancements could have been faster (other regions, for example, have had faster progress in selected areas) or whether these aggregate indicators mask inequalities (there are lagging regions, and progress has not affected all strata of society equitably). However, the importance of these issues necessitates analyses that go far beyond the scope of this report, which is narrowly focused on some of the most pressing environmental issues.

2. Among the data and findings presented throughout the report, several subregions of the Middle East and North Africa will be specified. The *Maghreb* (in western North Africa) comprises Algeria, Libya, Malta, Morocco, and Tunisia. The *Mashreq* (in eastern North Africa and western Asia) comprises Djibouti, the Arab Republic of Egypt, the Islamic Republic of Iran, Iraq, Jordan, Lebanon, the Syrian Arab Republic, West Bank and Gaza, and the Republic of Yemen. In addition, the Gulf Cooperation Council (GCC) countries—Bahrain, Kuwait, Oman, Qatar, Saudi Arabia, and the United Arab Emirates—are considered as a separate group given their shared characteristics, although they are also part of the Mashreq subregion geographically.

Human Advancement and Sustainable Natural Capital Use in the Middle East and North Africa

OVERVIEW

Even though the Middle East and North Africa is a region of diverse economies, it has on aggregate achieved impressive gains in human and economic development. This chapter adopts a capital accounting framework whose results show that most of the region's economies have enhanced both produced capital and human capital in recent decades but have seen deterioration of their natural capital.

Except in economies affected by conflict, the Middle East and North Africa's advances in human capital over the past 30 years have included increased years of children's formal education and overall life expectancy as well as reductions in young child and maternal mortality. Many challenges persist, including high levels of inequality and economic and social vulnerability, low rates of female participation in the workforce, high youth unemployment, and public health issues related to unhealthy lifestyles. Nonetheless, overall human capital has improved.

Likewise, the region has made advances in produced capital, including in access to water and sanitation, access to electricity, transportation infrastructure, and digitalization. Urbanization has facilitated structural changes in the economy in a number of the region's economies. And there has been income growth, even if—within the World Bank's green, resilient, and inclusive development (GRID) framework (World Bank and IMF 2021)—it has not always been particularly inclusive.

Opposed to this human and economic progress, the region's natural capital has deteriorated. Poorly planned and executed urban development and high dependence on fossil fuels—especially for transportation but also for heating, cooling, and industry—have increased air pollution, with impacts on human health, productivity, and broader urban livability. Coastlines are eroding in some key areas whose economies depend on beach tourism, and plastic is increasingly flowing into the region's seas.

These stresses are interdependent in several respects and are exacerbated by climate change. Periods of extreme heat are becoming more frequent, increasing the vulnerability of those exposed to air pollution. Rainfall is becoming sparser and less predictable. Global warming is contributing to sea level rise, making the region's coastlines more vulnerable to erosion. And even though the region is water-stressed, it does not manage its resources sustainably: Riverine and coastal ecosystems are threatened by poorly planned urban development and pollution, including plastic pollution. Poor land and watershed management contribute to loss of productive agriculture, to downstream riverine and coastal degradation, and to outdoor air pollution. The degradation has spread throughout terrestrial, coastal, and marine ecosystems, resulting in substantial biodiversity loss.

Insofar as both COVID-19 and climate change stem from inappropriate interaction with nature, they are both symptoms of inadequate management of natural capital, and both have economic and social consequences. Tackling climate change and ensuring inclusive and resilient growth will require restoration of this natural capital.

In the Middle East and North Africa, development overall has not been green or sustainable. A past pattern of "brown growth" threatens the longer-term regional goals of lasting prosperity and well-being. During the COVID-19 recovery period, economies have an opportunity to make a transformational change toward a GRID trajectory, which will improve their residents' quality of life. Such a transformation would, at the same time, address the challenge of climate change and conserve and restore the natural capital that is the foundation of longer-term prosperity and resilience.

This premise of transformation and improving resilience forms the foundation of this report, which focuses on three key challenges: improving air quality (blue skies) and addressing coastal and marine degradation (blue seas) stemming from marine plastic pollution and coastal erosion. This chapter sets these three key challenges in context by summarizing some of the region's broader economic, human development, and environmental trends.

IMPROVEMENT IN INCOMES, PRODUCED CAPITAL, AND HUMAN CAPITAL

This section looks at incomes and different forms of capital (human, produced, and natural) to convey how economies in the Middle East and North Africa have been faring in recent decades. In judging the fundamentals of an economy, it is important to look beyond standard macroeconomic aggregates, taking into consideration a country's entire "balance sheet" for a comprehensive picture of assets. A country's assets are its infrastructure (produced capital), its people and their skills (human capital), and its nature (natural capital). This chapter uses a capital accounting framework (figure 2.1) to demonstrate that most of the region's economies have made progress in enhancing both produced capital and human capital in recent decades and correspondingly also raising average incomes.[1]

Income Growth

Despite variations across countries, real incomes per capita grew from 1990 to 2018 in all subregions of the Middle East and North Africa (figure 2.2). Even though the region's overall income growth has been sluggish compared with other regions,[2] average national incomes grew by around 40 percent and 50 percent for the Maghreb and the Mashreq subregions, respectively, and for the Gulf Cooperation Council (GCC) countries by a little more than 11 percent (a region that had a relatively high starting point).[3]

FIGURE 2.1

Capital Accounting Framework

Source: Adapted from WAVES, n.d. ©World Bank.
Note: In this framework, as adapted, "net foreign assets" (as part of "Total Wealth") has been omitted.

FIGURE 2.2

Gross National Income Per Capita, by Subregion, Middle East and North Africa, 1990–2018

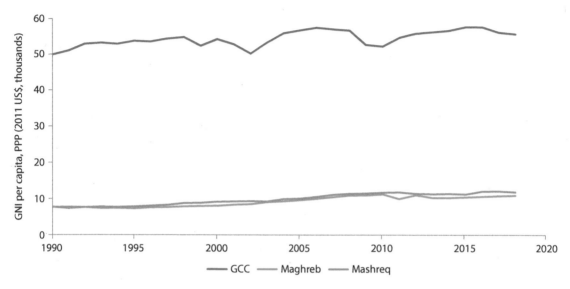

Source: Based on Human Development Data Center, United Nations Development Programme, http://hdr.undp.org/en/data.
Note: Gulf Cooperation Council (GCC) countries include Bahrain, Kuwait, Oman, Qatar, Saudi Arabia, and the United Arab Emirates. The Maghreb subregion comprises Algeria, Libya, Malta, Morocco, and Tunisia. The Mashreq subregion comprises Djibouti, the Arab Republic of Egypt, Iraq, the Islamic Republic of Iran, Jordan, Lebanon, the Syrian Arab Republic, West Bank and Gaza, and the Republic of Yemen. GNI = gross national income; PPP = purchasing power parity.

On a national level, real incomes in Morocco and Tunisia increased by almost 70 percent, and in Djibouti and in the Arab Republic of Egypt by more than 60 percent. However, income growth in Libya, the Syrian Arab Republic, and the Republic of Yemen has been disrupted because of conflict, and the current situation is fragile in Iraq, Lebanon, and West Bank and Gaza. GCC countries that recorded strong increases included Bahrain (30.3 percent), Kuwait (30.8 percent), and Qatar (41.2 percent), whereas average income grew more slowly in Saudi Arabia and Oman (11 percent and 5.8 percent, respectively). In the United Arab Emirates, average real incomes decreased because of large influxes of expatriates starting in the early 2000s, which drove up population figures.

Produced Capital Improvements

Improvements in the development of key infrastructure services have allowed more people to live healthier, more productive lives. Proper access to sanitation and basic drinking water services, for example, is important to reduce the risk for diseases that would decrease people's ability to work productively. Similarly, access to transportation infrastructure, electricity, or more recently to the internet enables businesses to operate more efficiently. Such access is a building block for technological advances. These services are also key to improved connectivity and quality of life.

Water and sanitation. Improvements in delivering clean drinking water and basic sanitation have been among the governments' greatest services to populations throughout the region. As a result, standards of hygiene have improved, and the incidence of waterborne diseases has fallen.

Increases in the share of population using these services in the past two decades were particularly pronounced in the Maghreb subregion, with over 90 percent of the population having access to clean drinking water and almost 90 percent to basic sanitation in recent years, compared with under 80 percent in 2000 (figure 2.3). The Mashreq economies, except for Djibouti and the Republic of Yemen, started from already higher levels in 2000 but have also made great strides in this respect. The GCC countries achieved a coverage rate of almost 100 percent in 2017.

FIGURE 2.3

Trends in Access and Use of Basic Sanitation and Drinking Water Services in the Middle East and North Africa, by Subregion, 2000–17

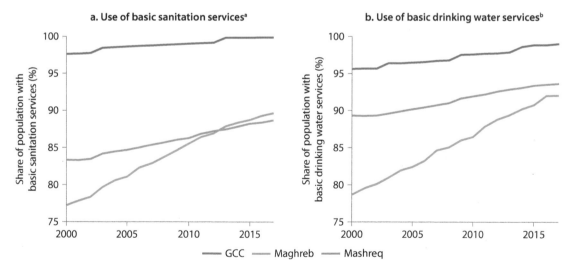

a. Use of basic sanitation services[a]

b. Use of basic drinking water services[b]

GCC — Maghreb — Mashreq

Source: Based on Global Health Observatory database, World Health Organization (WHO), https://apps.who.int/gho/data/view.main.
Note: Gulf Cooperation Council (GCC) countries include Bahrain, Kuwait, Oman, Qatar, Saudi Arabia, and the United Arab Emirates. The Maghreb subregion comprises Algeria, Libya, Malta, Morocco, and Tunisia. The Mashreq subregion comprises Djibouti, the Arab Republic of Egypt, Iraq, the Islamic Republic of Iran, Jordan, Lebanon, the Syrian Arab Republic, West Bank and Gaza, and the Republic of Yemen. Data for West Bank and Gaza are unavailable.
a. Under WHO definitions, access to "basic" sanitation refers to (a) "an improved sanitation facility that is not shared with other households and where excreta are safely disposed of in situ or treated off-site"; and (b) "a handwashing facility with soap and water at home."
b. Under WHO and United Nations Children's Fund (UNICEF) definitions, access to "basic" drinking water services refers to "an improved water source that is accessible on premises, available when needed, and free from fecal and priority chemical contamination."

However, access to water supply and sanitation has been disrupted in conflict-affected economies, with key infrastructure services destroyed. For example, residents of war-torn countries such as Libya, Syria, and the Republic of Yemen often lack access to safe sanitation services that were destroyed as the conflicts raged on (Nonay 2020; World Bank 2017a). Lack of access to parts for maintenance of sewage treatment systems is also referred to as an environmental crisis in the most recent Assistance Strategy for West Bank and Gaza (World Bank 2021a).

Electricity, transportation, and internet. As for expanding access to electricity, transportation infrastructure, and the internet, the region has also made great progress. Within the past 25 years, electricity has become available for almost the entire rural population, with Maghreb countries rapidly catching up with the other subregions and even outpacing Mashreq economies in the past few years (figure 2.4, panel a). The share of the total population with internet access has also increased rapidly (figure 2.4, panel b).

These improvements are key to economic diversification and the development of digitally connected economies. However, the gains have

FIGURE 2.4

Trends in Rural Access to Electricity and Total Population's Internet Use in the Middle East and North Africa, by Subregion

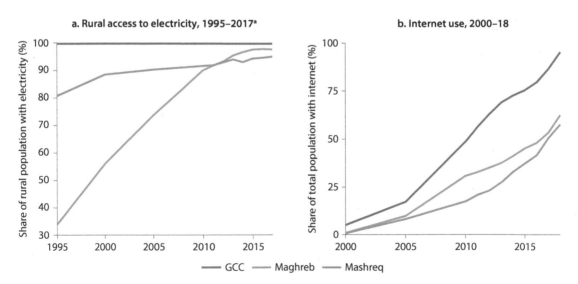

Sources: Based on International Telecommunication Union 2020 data and the World Development Indicators database.
Note: Gulf Cooperation Council (GCC) countries include Bahrain, Kuwait, Oman, Qatar, Saudi Arabia, and the United Arab Emirates. The Maghreb subregion comprises Algeria, Libya, Malta, Morocco, and Tunisia. The Mashreq subregion comprises Djibouti, the Arab Republic of Egypt, Iraq, the Islamic Republic of Iran, Jordan, Lebanon, the Syrian Arab Republic, West Bank and Gaza, and the Republic of Yemen.
a. For individual years at the beginning of the sample, data on the rural population with access to electricity are unavailable for several economies.

not been sustained in the conflict-affected economies, and some economies face issues with the reliability and financial sustainability of services. Despite the improvements over recent decades, there is scope for further improvement in both "hard" and "soft" infrastructure.[4]

Improvements in these core infrastructure services are key building blocks of GRID but must be developed to maximize opportunities for resilience and long-term sustainability. There are opportunities for transforming energy, transportation, and urban development trajectories in the Middle East and North Africa, for example, in ways that use the region's ample renewable energy resources, reduce the use of fossil fuels, improve urban air quality and urban livability, and reduce greenhouse gas (GHG) emissions. There is scope for improving the quality of infrastructure services, including sewage treatment—a core element in improving water quality—and solid waste management (see next section, on natural capital). The rapid increases in digitalization also offer opportunities for technological improvements across a range of sectors. Accompanying these investments with improved policies, institutions, and governance frameworks as well as enabling policies for private sector development provide the basis for sustainable business growth and job creation.

Human Capital Improvements

Reductions in infant, child, and maternal mortality. Infant, young child (under five years), and maternal mortality rates have declined substantially over the past three decades. Mortality rates for infants under one year have dropped by more than half since 1990 throughout the Middle East and North Africa (figure 2.5, panel a). Progress has been especially impressive in the Maghreb and Mashreq economies. Improved standards of hygiene, linked with better access to water and sanitation, have contributed in large part to these advancements (Alemu 2017). A similarly positive pattern emerges for maternal mortality, although these rates are still higher than in most other world regions, especially in the Maghreb countries (figure 2.5, panel b).

Regarding maternal and neonatal health in these areas, most of the long-run decline in mortality can be attributed to (a) dramatic improvements in economic well-being (that is, increased incomes, as shown in figure 2.2); (b) associated improvements in health care services; and (c) substantial declines in the fertility rates of the region's women (Sagynbekov 2018).

Increased education and literacy. Education levels in the Middle East and North Africa have also advanced significantly in recent decades. The region's residents now spend, on average, twice as many years in

FIGURE 2.5

Selected Human Capital Indicators in the Middle East and North Africa, by Subregion

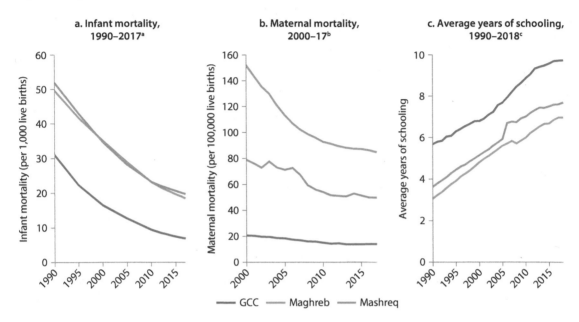

a. Infant mortality, 1990–2017[a]

b. Maternal mortality, 2000–17[b]

c. Average years of schooling, 1990–2018[c]

GCC — Maghreb — Mashreq

Sources: Based on the United Nations Development Programme's Human Development Center (http://hdr.undp.org/en/data) and World Health Organization's Global Health Observatory (https://apps.who.int/gho/data/view.main) databases.
Note: Gulf Cooperation Council (GCC) countries include Bahrain, Kuwait, Oman, Qatar, Saudi Arabia, and the United Arab Emirates. The Maghreb subregion comprises Algeria, Libya, Malta, Morocco, and Tunisia. The Mashreq subregion comprises Djibouti, the Arab Republic of Egypt, Iraq, the Islamic Republic of Iran, Jordan, Lebanon, the Syrian Arab Republic, West Bank and Gaza, and the Republic of Yemen.
a. Infancy is less than one year of age.
b. Maternal mortality data for West Bank and Gaza are unavailable and therefore not included in the Mashreq calculations.
c. Data on average years of schooling in Oman are unavailable before 2000 and therefore not included in the 1990–99 GCC calculations.

school as they did 30 years ago (figure 2.5, panel c). Nonetheless, the average years of schooling in the Mashreq and the Maghreb subregions (7.6 and 7.0 years, respectively) are still lower than the global average of about 8.5 years. Notable exceptions include the Islamic Republic of Iran and Jordan, where young people enjoyed, on average, more than 10 years of education in 2017.[5]

Another positive development is that, in many of the region's economies, the gender gap between boys and girls in secondary schooling and university education has disappeared (Belhaj 2018). In addition, the region's literacy rates have also improved greatly, rising from around 50 percent in 1990 to almost 80 percent in 2019, mostly because of strong increases in women's literacy rates.[6]

The importance of environmental education is also gradually being recognized in the region, with environmental topics like pollution, climate change, biodiversity, and sustainable development gaining ground in schools' curricula. However, the economies vary widely regarding the depth of the content covered and methods of delivery (Saab, Badran, and Sadik 2019). And challenges persist regarding the quality of education and training in general. General concerns are that higher education is geared more toward "providing credentials" for public sector employment than toward building the skill sets and culture of autonomy and inquiry necessary for a diversifying and increasingly digitalized modern economy with a dynamic private sector (El-Kogali and Krafft 2019).

Employment and inclusion challenges. Inclusion remains a challenge. One of the most pressing issues is pervasive gender inequality. In most Middle East and North Africa economies, female labor force participation remains low, social barriers for women remain high, and in some economies, women do not have the same legal rights as men (OECD 2017). On average, only a quarter of the region's full-time workers are female, and women are rarely top managers in businesses (6.5 percent compared with an international average of 18 percent), with less than 5 percent of all firms having majority female ownership.[7]

In addition, young people face chronically high, and rising, rates of unemployment and underemployment. Two-thirds of the region's population is under the age of 35, and over one-quarter of young people (ages 15–24) are unemployed (Bjerde 2020). The World Bank's 2019 strategy to fight youth unemployment in the Middle East and North Africa highlights the need for skills development in areas such as digital technologies and business skills for young entrepreneurs in addition to improving human capital outcomes more generally (Bjerde 2020). Recent World Bank initiatives in the region include a US$55 million program (approved in May 2019) to support youth employment in Morocco (World Bank 2019a).

Tackling these and other issues, including those related to income and wealth inequalities, is imperative to reduce exclusion and the risk of further political turmoil. These challenges require policy actions across a broad range of areas in addition to education and training. Some of these areas are addressed in other recent publications, including a report focused on overcoming spatial differences in opportunity (World Bank 2020a).

COVID-19 impacts. The COVID-19 pandemic has had a severe economic impact on Middle East and North Africa economies, exacerbating existing vulnerabilities. Uncertainty remains high regarding the speed and direction of recovery. Expected macroeconomic losses reached almost 7.2 percent of the region's 2019 GDP as of mid-2021 relative to the counterfactual scenario of no crisis (World Bank 2021d). The expected GDP losses are highest for Lebanon, with an expected accumulated loss in 2021 equivalent to almost 26.1 percent of its 2019 GDP.

This economic fallout has made it much harder for people to pursue their jobs and generate steady incomes, especially among those employed in the informal sector (World Bank 2020b), who make up a large proportion of workers in non-GCC Middle East and North Africa economies (Gatti et al. 2014). As a result, poverty rates have increased. In some countries (Djibouti, Egypt, the Islamic Republic of Iran, Kuwait, Lebanon, Morocco, Qatar, Saudi Arabia, Syria, and the Republic of Yemen), interruptions in supply chains caused food prices to increase by 20 percent or more in 2020.[8] Tourist arrivals have plummeted in countries such as Egypt, Lebanon, Morocco, Tunisia, and the United Arab Emirates, where the tourism sector accounts for a substantial share of GDP and employment.

Most governments, often with the aid of multilateral organizations, responded to the COVID-19 crisis with programs to cushion the impact on vulnerable groups. Nonetheless, the pandemic's impact poses a huge challenge, particularly for the less advantaged. It also exacerbates gender inequalities, with more women than men leaving the workforce—including in the informal sectors—to care for children and other family members if they are sick (OECD 2020). Poverty and vulnerability have increased in countries such as Egypt and Lebanon. Moreover, COVID-19 has added to the obstacles faced by conflict-affected economies such as Libya, Syria, West Bank and Gaza, and the Republic of Yemen.

In summary, the Middle East and North Africa region can claim improvements in incomes, physical capital (infrastructure), and human capital over the past three decades, but it also faces outstanding challenges regarding the quality and inclusiveness of growth. The degree of progress has also varied widely between countries, as illustrated in more detailed country-specific analyses.[9] Notably, residents of conflict-affected countries have faced displacement and tragic losses of life.

Furthermore, the pattern of growth in some instances has not been consistent with longer-term sustainability, and the foundations of a GRID path are at risk. Unfortunately, the region's growth over recent decades has come at the cost of degradation of its natural capital—a key building block for its future long-term development and well-being.

NATURAL CAPITAL DETERIORATES AS ENVIRONMENTAL DEGRADATION ACCELERATES

Even as the Middle East and North Africa's produced capital and human capital have improved over the past three decades, its natural capital has deteriorated. The region's environmental resources are depleted in all dimensions—including air quality; coastal and marine ecosystems; land

and freshwater resources; and biodiversity within both terrestrial and marine natural habitats. Although this report focuses on air quality (and climate change) and coastal and marine resources, this section reviews the trends in all forms of natural capital to provide an overview of how the region's environment has fared in recent decades.

Air and Climate Pollution

Emissions that lead to local air pollution and contribute to global climate change are rising in the Middle East and North Africa. These translate into poor air quality for the region's residents, especially in cities. Since the early 2000s, carbon dioxide (CO_2) emissions have more than doubled in the GCC and Mashreq subregions and have grown by around two-thirds in the Maghreb subregion (figure 2.6, panel a). Similarly, nitrogen oxide (NO_x) and sulfur dioxide (SO_2) emissions have increased throughout the region, with the exception of relatively stable SO_2 emissions in the Maghreb (figure 2.6, panels b and c).[10] These increases—linked mostly to

FIGURE 2.6

Recent Trends in Selected Gas Emissions in the Middle East and North Africa, by Subregion

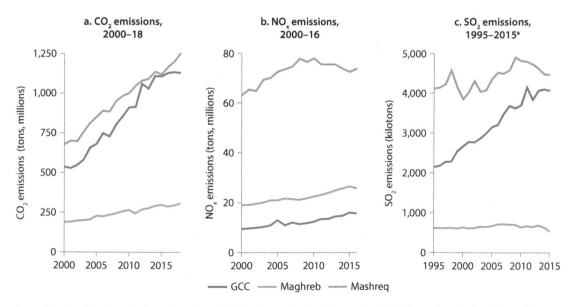

Sources: Based on GCP 2019; Hoesly et al. 2018; and the World Resources Institute's Climate Analysis Indicators Tool (CAIT) 2.0 Climate Data Explorer (http://cait.wri.org/).
Note: Gulf Cooperation Council (GCC) countries include Bahrain, Kuwait, Oman, Qatar, Saudi Arabia, and the United Arab Emirates. The Maghreb subregion comprises Algeria, Libya, Malta, Morocco, and Tunisia. The Mashreq subregion comprises Djibouti, the Arab Republic of Egypt, Iraq, the Islamic Republic of Iran, Jordan, Lebanon, the Syrian Arab Republic, West Bank and Gaza, and the Republic of Yemen. CO_2 = carbon dioxide; NO_x = nitrogen oxide; SO_2 = sulfur dioxide.
a. Data on SO_2 emissions are unavailable for West Bank and Gaza and therefore not included in the Mashreq calculation.

increased use of fossil fuels for power generation, transportation, industry, and the building sector—are the main contributors to climate change.

Poor air quality makes cities less attractive places to live and work. All who live in cities with high levels of air pollution are exposed to it, but low-income residents are the most vulnerable. They often have jobs that require them to work outside, increasing their exposure. They also have worse access to good health care and defensive technologies (such as air purifiers). Air pollution in its various forms also exacerbates the vulnerability of those with underlying respiratory diseases to severe illness and death and is linked to increased severity of COVID-19 disease. Improving air quality will bring multiple benefits across sectors. Air pollution is discussed in more detail in chapter 3, and climate change and its impacts are discussed later in this chapter.

Marine-Plastic Pollution

Increasing quantities of plastic debris from Middle East and North Africa economies are entering the seas and polluting marine ecosystems. Waste generation of all kinds, including plastics, has increased rapidly, but solid waste management remains inadequate, with widespread littering, little recycling, and poorly managed landfills. As a result, plastic pollution of the region's shorelines and seas has escalated dramatically. In 2010, Algeria, Egypt, and Morocco were among the top 20 marine-polluting countries worldwide (Jambeck et al. 2015), with much of their waste entering the Mediterranean Sea. Furthermore, total marine debris is expected to grow significantly, with plastic flows to marine spaces of some countries projected to double their 2010 levels by 2025 (figure 2.7).

Discarded plastics affect other sectors by polluting beaches, degrading marine ecosystems, entering the food chain, and damaging fisheries. Plastics can also block storm drainage channels, increasing the risk of urban flooding after heavy rainfall and, with other discarded waste, can attract insects and rodents that carry health risks. Plastic pollution is discussed in more detail in chapter 4.

Coastal Erosion

The coasts of the Middle East and North Africa have changed in recent decades. Figure 2.8 shows net coastal erosion rates from 1984 to 2016 for the three subregions compared with other major regions worldwide. Whereas GCC coasts have been accreting, mainly because of land reclamation and coastal developments by some countries, shorelines have been on the retreat in both the Maghreb and the Mashreq subregions.

FIGURE 2.7

Volume of Plastic Debris Entering the Seas from the Middle East and North Africa, by Economy, 2010 and 2025

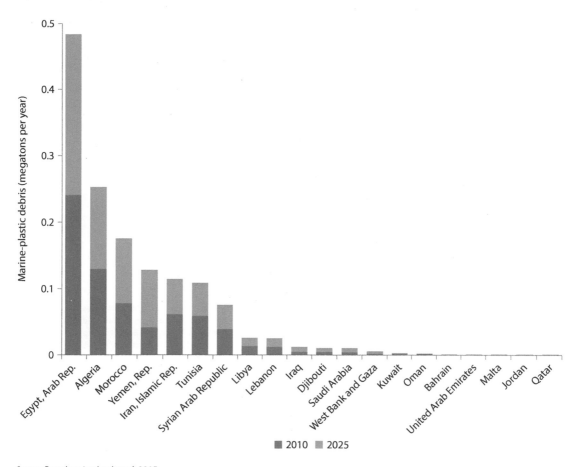

Source: Based on Jambeck et al. 2015.
Note: Figures for 2025 are projections.

The figures mask heterogeneity between and within countries, but the overall speed of coastal erosion in the Maghreb is second only to that of South Asia globally. Mashreq coasts have been eroding more slowly; nevertheless, erosion rates there have also led to large losses of land area. This shoreline retreat has adversely affected key economic sectors including beach tourism, which accounts for a significant share of GDP in several of the region's economies.

Coastal erosion is caused partly by natural processes but also by poorly planned coastal urban development and poor river basin management, including (a) construction of upstream dams that can block silt flows; (b) urban riverside development, which can impede natural

FIGURE 2.8

Average Annual Net Coastal Accretion and Erosion, Global Regions and Middle East and North Africa Subregions, 1984–2016

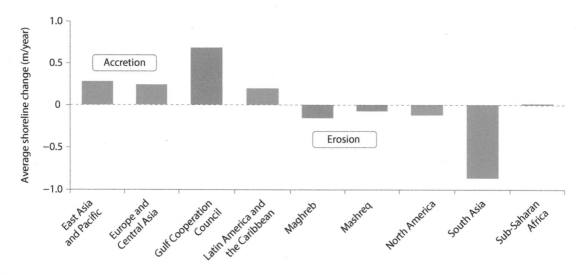

Source: Based on Luijendijk et al. 2018.

Note: Positive values indicate accretion, and negative values, erosion. "North America" includes Canada and the United States. Orange bars designate Middle East and North Africa subregions, as follows: Gulf Cooperation Council (Bahrain, Kuwait, Oman, Qatar, Saudi Arabia, and the United Arab Emirates); Maghreb (Algeria, Libya, Malta, Morocco, and Tunisia); and Mashreq (Djibouti, the Arab Republic of Egypt, Iraq, the Islamic Republic of Iran, Jordan, Lebanon, the Syrian Arab Republic, West Bank and Gaza, and the Republic of Yemen).

drainage channels and interfere with the natural functioning of flood-plains; and (c) gravel and sand extraction from riverbeds and beaches. Coastal erosion is exacerbated by sea level rise linked to global warming and climate change.

Restoring coastal resilience requires an integrated and cross-sectoral approach involving urban planners, local communities, nature and environmental protection agencies, water resource planners, and ports authorities. Chapter 5 discusses coastal erosion in more detail.

Land Degradation

More than half of all land and one-fourth of arable land in the Middle East and North Africa region is degraded, and land degradation affects both urban and rural areas.[11] The economic cost of this reduced productivity of arable land and rangelands is estimated to be about 1 percent of GDP annually, but this estimate does not consider broader ecosystem service values (Larsen 2011).

Overall, the Mashreq subregion is more severely affected than the Maghreb or the GCC (World Bank 2019c). The drivers include poverty and increased population pressure in the poorest regions as well as poor watershed and agricultural water management. Unsustainable land and water management are interlinked: Poor land management contributes to erosion from water runoff and vegetation loss as well as to water pollution from agricultural runoff. Poor water management contributes to erosion on sloping lands and to increased salinity of irrigated land (World Bank 2019c). Land degradation and erosion have also added to the severity of the region's sandstorms and dust storms, which have interacted with local pollutants and exacerbated the impact of air pollution, especially in urban areas. In the region's dust storms, 85 percent of the particles are smaller than 10 micrometers in diameter and are respirable (World Bank 2019b). Climate change exacerbates the impacts.

Interventions to reduce land degradation can restore productivity and livelihoods in rural areas and lead to reduced air pollution. One recent example is the ongoing US$132 million Integrated Landscapes Management in Lagging Regions Project, targeting selected regions of Tunisia (World Bank 2017b). Investments such as these to support climate-smart agriculture, reforestation, and value chains are intended not only to increase productivity and incomes but also to provide broader ecosystem services including reduced dam sedimentation, increased watershed protection, and reduced erosion—leading in turn to cleaner air, greater biodiversity, expanded recreational services, and more carbon sequestration. Because poverty levels are higher in rural areas than in urban areas, interventions of this kind can both increase inclusion and lead to more resilient green development.

Threatened Terrestrial and Marine Ecosystems

The Middle East and North Africa region has important terrestrial and marine ecosystems that are under threat. Terrestrial ecosystems are largely arid, semiarid, and Mediterranean biomes (Tolba and Saab 2009) but also include forests (especially in mountainous areas) and flooded grasslands and wetlands.[12] The tidal flats of the Gulfs are among the world's most important overwintering areas, annually hosting 1–2 million waders from 125 species (IBP 2016; Scott 1989). The proportion of each country under formal national protection varies widely—ranging from only 0.21 percent of total land area in Libya, to 13.1 percent in Egypt, 17.5 percent in Kuwait, and over 30 percent in Morocco.[13]

Terrestrial Ecosystems

More progress has been made on terrestrial protection than on coastal and marine protection. Overall, however, species habitats have been deteriorating throughout the Middle East and North Africa in recent decades (figure 2.9, panel a). In many protected areas, difficulties stem from inadequate management and funding, insufficient community involvement, competing demands for natural resources, or poor tourism management, as in the following examples:

- *In Lebanon,* the Al-Shouf Biosphere Reserve is threatened by uncontrolled quarrying for urban development (SPNL 2018).

- *In Iraq,* full restoration of the southeastern wetlands and delta ecosystems would require use of 11 percent of the Tigris-Euphrates river system (Alwash et al. 2018).

FIGURE 2.9

Natural Habitat Index Trends in the Middle East and North Africa, by Subregion

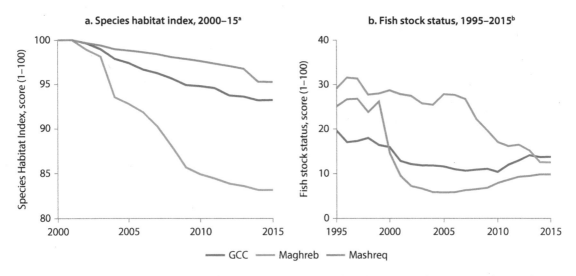

Source: Based on the Environmental Performance Index (EPI) 2020 (https://epi.yale.edu/) of the Yale Center for Environmental Law & Policy and Columbia University's Center for International Earth Science Information Network.
Note: The Species Habitat Index (SHI) and fish stock status (FSS) are indicators ranging from 0 to 100, with lower values denoting worse status. The Gulf Cooperation Council (GCC) includes Bahrain, Kuwait, Oman, Qatar, Saudi Arabia, and the United Arab Emirates. The Maghreb subregion includes Algeria, Libya, Malta, Morocco, and Tunisia. The Mashreq subregion includes Djibouti, the Arab Republic of Egypt, Iraq, the Islamic Republic of Iran, Jordan, Lebanon, the Syrian Arab Republic, West Bank and Gaza, and the Republic of Yemen.
a. The SHI "measures the average proportion of species' suitable habitat remaining within a country relative to the baseline year 2001" (Wendling et al. 2020). It is one of six indicators used to calculate the EPI's Biodiversity & Habitat issue category. Data on this indicator are unavailable for Bahrain, Malta, and West Bank and Gaza.
b. FSS "measures the percentage of a country's total catch that comes from overexploited or collapsed fish stocks, based on an assessment of all fish stocks caught within a country's exclusive economic zone" (Wendling et al. 2020). It is one of three indicators used to calculate the EPI's Fisheries issue category. Data on this indicator are unavailable for Bahrain, Djibouti, Iraq, Jordan, Kuwait, Lebanon, Qatar, Syrian Arab Republic, and West Bank and Gaza.

- *In Saudi Arabia*, illegal hunting of endangered ungulates persists in protected desert landscapes (Al-Tokhais and Thapa 2019).

Improved protected area management, forest and watershed restoration, and support for agrobiodiversity and ecotourism are all strategies for ecosystem restoration. For example, Morocco's Vision 2020 strategy for sustainable tourism aims to ensure the conservation of Morocco's natural resources and its residents' well-being as well as to respond to tourists' evolving sensitivities and make sustainability one of the country's distinguishing features (Roudies 2013).

Marine Ecosystems

Overfishing. Marine ecosystems are under particular threat, with the Mediterranean being one of the most overfished seas in the world. And in the Mediterranean, Red Seas, and Atlantic areas of the Middle East and North Africa, coastal and marine tourism are of growing importance to the region's economies. However, although more than 7 percent of the Mediterranean's coastal and marine areas are currently protected, over 90 percent of those areas are within the northern Mediterranean. In the territorial waters of the Middle East and North Africa, fish stocks have often been overexploited (figure 2.9, panel b). Together with the Black Sea, the Mediterranean has the highest rate of overfishing across all oceans and seas, with more than 60 percent of fish stocks being fished at unsustainable levels (FAO 2020).

Protection of marine areas, including no-take zones in sensitive areas, seasonal and depth restrictions on catch, and appropriate fishing gear can allow fisheries and marine resources to regenerate (MedPAN, UNEP/MAP, and RAC/SPA 2016). Carefully managed recreational fishing and diving can increase local incomes, especially if governments set commercial marine species catch limits.

Water pollutants. The water quality of marine ecosystems has been degrading. Polluted marine waters with high levels of nutrients from fertilizers, septic systems, sewage treatment plants, and urban runoff often have high concentrations of chlorophyll-a and excess amounts of algae. Chlorophyll-a levels (an indicator of such effluent discharge) in the Mediterranean have increased, especially along the Moroccan and Tunisian coasts but also in the Nile delta and along the eastern Mediterranean coast (map 2.1). Overall, chlorophyll-a in the Mediterranean has been increasing by around 0.9 percent per year in the past two decades (CMEMS 2021).

Contributing to this problem is an overall lack of adequate waste management and wastewater treatment facilities. Furthermore, marine resources are threatened by marine-plastic pollution, which will increase in the coming decades if no action is taken.

MAP 2.1

Annual Percentage Change of Chlorophyll-a Levels in the Mediterranean Sea, 1997–2019

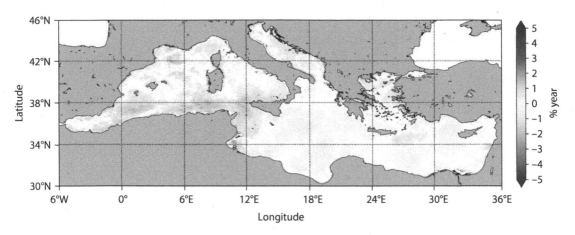

Source: EU Copernicus Marine Service Information (CMEMS 2021). © European Union.
Note: Multisensor satellite observations show the chlorophyll trend from 1997 to 2019, expressed as the average percentage change per year, with positive (increasing) trends in red and negative (decreasing) trends in blue.

Water Scarcity and Poor Quality

The Middle East and North Africa is a water-scarce region that does not manage its water resources sustainably. Over 60 percent of the region's residents live in areas with high or very high surface water stress, compared with the global average of 35 percent, making the Middle East and North Africa the world's most water-scarce region (World Bank 2018).[14]

Figure 2.10 illustrates the sustainability of water resource management, by water source, in the region's economies and shows that groundwater in particular, and a substantial proportion of surface water, are unsustainably managed. Poor water resource management—especially of agricultural water, which accounts for over 80 percent of the region's water use—contributes to water resource depletion and water quality degradation (World Bank 2018). In GCC countries, desalinated water accounts for an increasing share of water supply, which has a significant environmental footprint because desalination is highly energy intensive. Decreasing water flow in rivers, linked to increasing upstream water extraction, also leads to insufficient flows entering the sea, hence contributing to saline intrusion. And as mentioned earlier, poor upstream water management can contribute in turn to coastal erosion.

FIGURE 2.10

Sustainability of Water Withdrawals, by Source, as a Share of Total Withdrawals in Middle East and North Africa Economies, 2010s

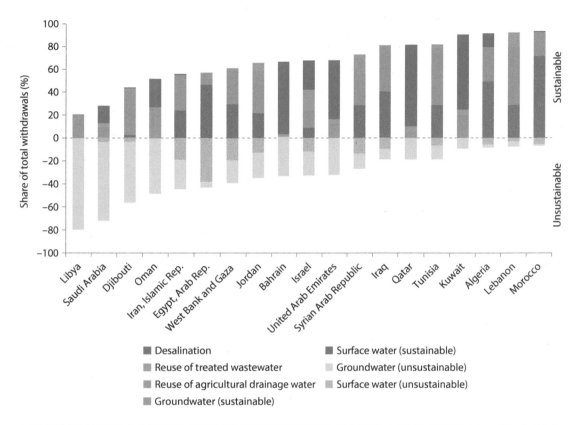

Source: World Bank 2018 using desalination capacity data from Global Water Intelligence (2016) and the AQUASTAT database of the Food and Agriculture Organization of the United Nations.
Note: The Middle East and North Africa economies are shown in order of least to most sustainable water withdrawal sources. No data are available for the Republic of Yemen on sustainability of water use. The percentage of "unsustainable" groundwater and surface water withdrawals was estimated using the Blue Water Sustainability Index (Wada and Bierkens 2014).

    Water quality management is also inadequate in most Middle East and North Africa economies, resulting in pollution, health impacts on the population, and loss of ecosystem services and fisheries. The lack of adequate water quality management practices in the region lead to unsustainable water consumption; untreated municipal and industrial wastewater; and the hazardous or harmful by-products of intense desalination, poor solid waste management, and poor agricultural practices that pollute the region's rivers and seas (World Bank 2018). This pollution contributes to the loss of riverine, marine, and coastal ecosystems and fisheries (Argimon 2019).

    In the Middle East and North Africa, only half of discharged wastewater is collected, and of this, less than half is treated (Damania et al. 2019).

Surface water salinization is pervasive, resulting mostly from agriculture but also from urban wastewater discharge. Partly because of this pollution, an estimated 17 percent of freshwater species are threatened with extinction (García et al. 2015). Moreover, coastal industrial expansion in Tunisia's Gulf of Gabès has contributed to a high discharge of industrial effluents into seawater with a range of contaminants, including metals (Naifar et al. 2018). Poor agricultural land and water management contributes increasingly to water pollution and salinization from discharges of agrochemicals, organic matter, drug residues, sediments, and saline that drain into water bodies and affect human health and productivity (FAO 2014).

Programs are under way to improve wastewater treatment and agricultural water management in several countries including Egypt (*Egypt Today* 2021; World Bank 2010). Better water management requires policies and financial instruments to create incentives for resource-use efficiency and waste reduction. The challenge of water resource and water quality management illustrates the links between land and water management and the impact of water pollution on coastal and marine pollution.

Climate Change

Climate change is exacerbating the environmental challenges the region already faces. The Middle East and North Africa is already hot and water-stressed, and periods of extreme heat and drought will become more prolonged and severe. Much of the region's population and economy is concentrated along coastlines, and even with modest sea level rise, several large cities will be become more vulnerable to coastal flooding. At the same time, several of the region's economies, for their size, are significant contributors to GHG emissions. There is significant scope for these countries to reduce resource-use intensity and to manage land and water resources sustainably to contribute to both climate change adaptation and mitigation.

Other reports—for example, *Turn Down the Heat: Confronting the New Climate Normal* (World Bank 2014)—have addressed these intensifying challenges in detail, so they are discussed only briefly here, drawing on this existing documentation.

Rising Temperatures and Desertification

The Middle East and North Africa region will experience increasing temperatures and more frequent periods of extreme heat. Warming of about 0.2 degrees Celsius per decade has been observed in the region from 1961 to 1990 and has increased even faster since then (World Bank 2014).

Geographically, the strongest warming is projected to take place close to the Mediterranean coast but also in the inland areas of Algeria, Libya, and large parts of Egypt, Iraq, and Saudi Arabia. The effect is likely to be felt most in large cities.

Extreme heat brings health impacts, especially on those with underlying health conditions and those who have to work outside, who are generally lower-income workers. The better-off residents will have access to air conditioning, but without a significant shift to renewable energy, this will only further increase GHG emissions. Desertification processes will interact with dust storms to increase air pollution in both rural and urban areas, including in the wealthy GCC cities. Climate change highlights the importance of addressing urban air pollution and greening cities.

Increasing Aridity and Declining Crop Yields

An already dry region will become even more arid, putting pressure on already scarce water resources and reducing agricultural productivity. Rainfed crop yields are expected to decline by 30 percent with warming of 1.5–2 degrees Celsius (2.7–3.6 degrees Fahrenheit) without considering adaptation (World Bank 2014). The Middle East and North Africa region already imports 50 percent of its wheat, and this share will likely increase. Livestock will be subject to increasing heat stress, with consequent health and productivity impacts. With agriculture already accounting for over 80 percent of the region's water use, existing tensions concerning competing demands for water are likely to increase, and aquifer depletion will also increase.

Reductions in water quantity are likely to also reduce water quality while increasing salinity as well. Although the GCC countries can create desalinated water for municipal purposes, these processes are energy intensive, and technology shifts are needed. As mentioned earlier, climate-smart agriculture has great potential, and some countries are investing heavily in this area to secure the "triple win" of increased productivity, adaptation, and mitigation. (Box 2.1 refers to such a project in Morocco and other countries.)

Rising Seas and Coastal Floods

Middle East and North Africa countries are vulnerable to the impacts of rising seas. Even a modest sea level rise of only 0.35 meters, consistent with a global temperature rise of 1.5 degrees Celsius (the most optimistic scenario), would have substantial impacts. The Mediterranean's jewel, Alexandria; the Nile delta coastal cities; and the Mediterranean metropolises of Port Said, Egypt; Benghazi, Libya; and Algiers, Algeria, are all particularly vulnerable to flooding (Elsharkay, Rashed, and Rached 2009). A sea level rise of only 0.3 meters would flood 30 percent of metropolitan

FIGURE 2.11

Mean Sea Level Rise of the Mediterranean Sea, 1993–2020

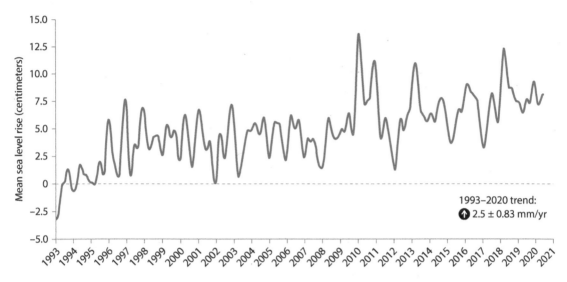

Sources: EU Copernicus Marine Service Information (CMEMS 2021) and Copernicus Climate Change Service (https://climate.copernicus.eu/).
Note: mm/yr = millimeters per year.

Alexandria, forcing about 545,000 people to abandon their homes and land and leading to the loss of 70,500 jobs. Hence, it is worrisome that the Mediterranean's mean sea level has risen steadily over the past three decades (figure 2.11). Other vulnerable cities, although not on the Mediterranean coast, include Muscat, Oman; Dubai, United Arab Emirates; Aden, Republic of Yemen; and Basra, Iraq.

Coastal tourism—an important component of GDP in several Middle East and North Africa economies—will be affected by beach erosion, and some key port and industrial facilities will be vulnerable. Coastal flooding will cause saline intrusion in low-lying areas, affecting agriculture, especially in Egypt, Tunisia, and the Republic of Yemen (World Bank 2014). The vulnerability of the region's coastlines and coastal cities highlights the urgency of scaling up resilience measures as part of a broad green recovery plan.

Leading in GHG Emissions and Other Air Pollutants

The Middle East and North Africa also includes economies with high GHG emissions, and there is significant scope in all of them for climate change mitigation. Five of the world's top 10 countries in per capita GHG emissions are in the Middle East and North Africa, all in the GCC.[15] Overall, GHG emissions have been increasing in the region over the past several decades. As the next subsection shows, the Middle East and North

Africa has also been the only region worldwide that did not manage to decouple average income growth from per capita carbon emissions and has been the least successful in doing so for other air pollutants.

As for trends in energy intensity (per unit of GDP), whereas Organisation for Economic Co-operation and Development (OECD) countries have seen a significant decline on average, the Middle East and North Africa has achieved significant declines only in Bahrain, Jordan, Qatar (which has largely eliminated gas flaring), and Tunisia, as well as slight declines in Algeria, Egypt, Iraq, Lebanon, and Morocco. Despite economic diversification measures, hydrocarbons remain by far the principal export and source of government revenue for 10 Middle East and North Africa countries—and for many of the others, remittances from residents working in these countries are significant.

A LAG IN DECOUPLING GROWTH FROM AIR POLLUTION AND GHGS

Unlike other regions of the world, the Middle East and North Africa has not decoupled its income growth from carbon emissions.[16] As the previous sections noted, the development path of the region as a whole, its subregions, and individual economies has harmed the environment. What would it mean to decouple growth from emissions? Absolute decoupling occurs when emissions per capita decrease from year to year while income per capita continues to grow, whereas relative decoupling means that emissions per capita still increase but more slowly than average income growth.

Decoupling Carbon Emissions from Growth

Most of the world's regions have been decoupling their economic growth from their negative environmental externalities such as carbon emissions, either absolutely or relatively. In contrast, carbon emissions in the Middle East and North Africa's Mashreq and GCC subregions have grown faster than incomes, and carbon emissions in the Maghreb have grown as fast as incomes (figure 2.12).

However, such an aggregate view masks some of the heterogeneity across countries. Some of the region's economies have indeed been able to decouple income growth from carbon emission growth (figure 2.13). Although Iraq, the Islamic Republic of Iran, Oman, and Saudi Arabia have not decoupled at all, other countries such as Lebanon and Tunisia have achieved relative decoupling. Bahrain and Jordan have even lowered their emissions per capita, albeit only slightly.

FIGURE 2.12

Trends in Growth of GNI Per Capita in Relation to CO_2 Emissions Per Capita in Middle East and North Africa Subregions and Other Global Regions, 1990–2018

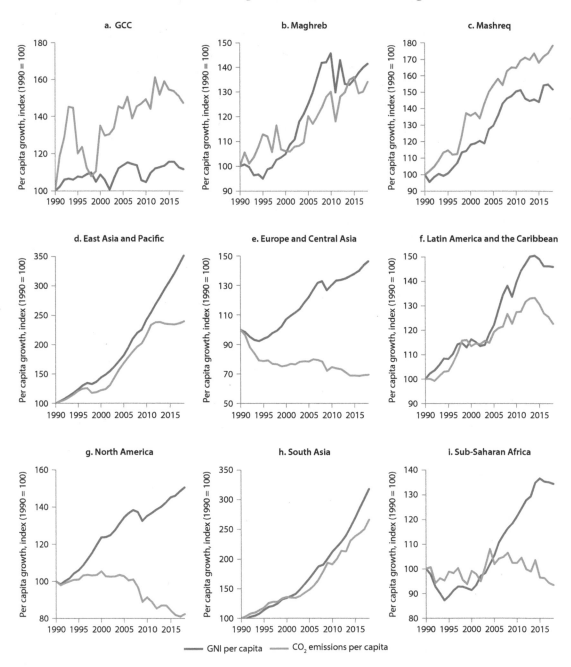

GNI per capita CO_2 emissions per capita

Sources: Based on Global Carbon Project (GCP) 2020 and the Human Development Data Center, United Nations Development Programme (http://hdr.undp.org/en/data).

Note: Gross national income (GNI) and carbon dioxide (CO_2) emission growth are indexed relative to 1990 (= 100). Panels a-c represent subregions of the Middle East and North Africa, as follows: The Gulf Cooperation Council (GCC) includes Bahrain, Kuwait, Oman, Qatar, Saudi Arabia, and the United Arab Emirates. The Maghreb subregion includes Algeria, Libya, Malta, Morocco, and Tunisia. The Mashreq subregion includes Djibouti, the Arab Republic of Egypt, Iraq, the Islamic Republic of Iran, Jordan, Lebanon, the Syrian Arab Republic, West Bank and Gaza, and the Republic of Yemen. "North America" (panel g) includes Canada and the United States.

FIGURE 2.13

Growth of GNI Per Capita in Relation to CO_2 Emissions Per Capita in Middle East and North Africa Economies, since 1990

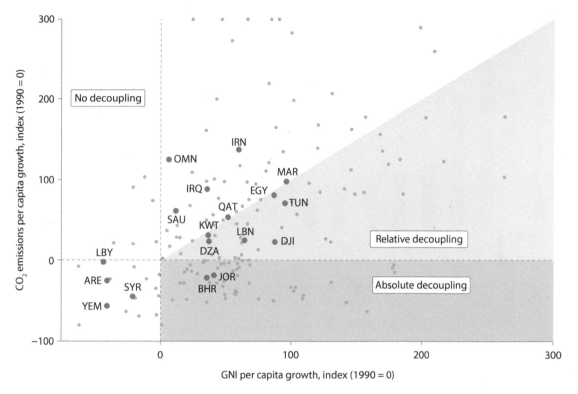

Sources: Based on Global Carbon Project (GCP) 2019 and the Human Development Data Center, United Nations Development Programme (http://hdr.undp.org/en/data).
Note: Gross national income (GNI) and carbon dioxide (CO_2) emission growth are indexed relative to 1990 (= 0) to assess the extent of each economy's decoupling of its economic growth from its carbon emissions. "Absolute decoupling" refers to emissions per capita that decrease while GNI per capita increases. "Relative decoupling" refers to emissions per capita that increase but more slowly than GNI per capita. Economies are labeled using ISO alpha-3 codes, as listed on the Abbreviations page in the front matter of this report. The light blue dots designate economies of other global regions. Data for Malta and West Bank and Gaza are unavailable.

Decoupling Air Pollutants from Growth

The Middle East and North Africa has also decoupled its growth the least of all world regions from air pollutant emissions. Unlike the situation with climate pollutants, the region has relatively decoupled its per capita GNI growth from its per capita air pollutant emissions of NO_X and SO_2—both of which are most directly related to industrial activity, power generation, and motorized transportation. However, air pollution decoupling has been slower in the Middle East and North Africa than in any other region (figure 2.14).

Although the region is decoupling per capita, high population growth means that overall air quality has still deteriorated (as shown earlier in figure 2.6). And, as with carbon emissions, the region's economies are

FIGURE 2.14

Progress in Decoupling Growth of GNI Per Capita from NO$_x$ and SO$_2$ Emissions Per Capita, by Global Region, since 1990

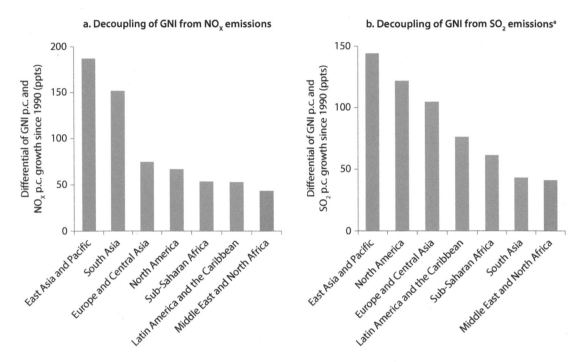

Sources: Based on Hoesly et al. 2018; data from the Human Development Data Center, United Nations Development Programme (http://hdr .undp.org/en/data); and the World Resources Institute's Climate Analysis Indicators Tool (CAIT) 2.0 Climate Data Explorer (http://cait.wri.org/). *Note:* The figures show the differential, in percentage points, between the per capita (p.c.) growth rates of gross national income (GNI) and the emissions of nitrogen oxide (NO$_x$) and sulfur dioxide (SO$_2$) in panels a and b, respectively. Growth rates are calculated in comparison to 1990 levels and the differences in growth rates computed. In panel a, for example, growth rates of both NO$_x$ emissions p.c. and GNI p.c. compared with their 1990 levels were computed. Averages of the respective growth rates for 2014–16 were computed to adjust for erratic changes, and the average growth rate of GNI p.c. was then subtracted from the average growth rate of NO$_x$ emissions p.c. The resulting statistic is the differential growth between the two growth rates, expressed in percentage points (ppts). For SO$_2$, the same procedure was applied, whereby the average of the growth rates for 2012–14 (due to data availability) was computed for both GNI and emissions p.c. "North America" includes Canada and the United States.
a. Data on SO$_2$ emissions in West Bank and Gaza are unavailable.

heterogeneous in their decoupling progress (figure 2.15). Similarly, oil importing economies have achieved better rates of decoupling than the oil exporters, even achieving absolute decoupling of NO$_x$ and SO$_2$ emissions from income growth. On the other hand, oil exporting countries achieved only relative decoupling in this respect and have been the slowest worldwide to do so.

Decoupling Black Carbon Emissions from Growth

The Middle East and North Africa has also not decoupled its income growth from black carbon emissions, among the most potent of

FIGURE 2.15

Extent of Decoupling Growth of GNI Per Capita from NO_x and SO_2 Emissions Per Capita, Middle East and North Africa Economies, since 1990

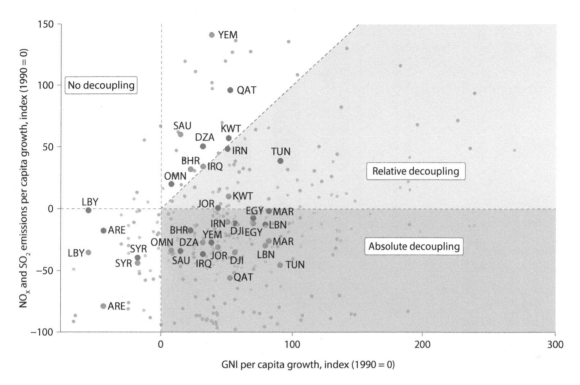

Sources: Based on Hoesly et al. 2018; data from the Human Development Data Center, United Nations Development Programme (http://hdr
.undp.org/en/data); and the World Resources Institute's Climate Analysis Indicators Tool (CAIT) 2.0 Climate Data Explorer (http://cait.wri.org/).
Note: Gross national income (GNI) and emissions growth are indexed relative to 1990 (= 0) to assess the extent of each economy's decoupling
of its economic growth from its nitrogen oxide (NO_x) and sulfur dioxide (SO_2) emissions. Dark blue and green dots denote NO_x emissions;
orange and light blue dots denote SO_2 emissions. "Absolute decoupling" refers to emissions per capita that decrease while GNI per capita
increases. "Relative decoupling" refers to emissions per capita that increase but more slowly than GNI per capita. Economies are labeled using
ISO alpha-3 codes, as listed on the Abbreviations page in the front matter of this report. Data for Malta and West Bank and Gaza are
unavailable. Light blue and green dots designate economies of other global regions.

climate and air pollutants. Black carbon emissions—linked mostly to
the burning of solid waste and agriculture but also (to a lesser extent
in the Middle East and North Africa) to road vehicle combustion and
power generation—have not been decoupled from economic growth
in either the Middle East and North Africa or Sub-Saharan Africa
(figure 2.16).

This particular type of particulate matter (PM) is disproportionately
harmful. Studies have shown that black carbon particles have more
harmful health effects than PM as a whole (the full range of microscopic
solid or liquid particles suspended in the air). Estimated increases in life
expectancy associated with abatement measures for black carbon are

FIGURE 2.16

Comparison of Global Regions in Decoupling Growth of GNI Per Capita from Black Carbon Emissions Per Capita since 1990

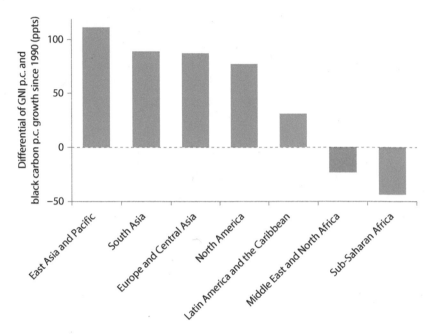

Sources: Based on Hoesly et al. 2018 and data from the Human Development Data Center, United Nations Development Programme (http://hdr.undp.org/en/data).

Note: The figure shows the differential, in percentage points (ppts), between the per capita (p.c.) growth rates of gross national income (GNI) and black carbon emissions. Growth rates are calculated in comparison to 1990 levels, and the differences in growth rates were computed. In particular, the growth rates of both black carbon emissions p.c. and GNI p.c. relative to their 1990 levels were computed. Averages of the respective growth rates for 2012–14 were computed to adjust for erratic changes, and the average growth rate of GNI p.c. was then subtracted from the average growth rate of black carbon emissions p.c. The resulting statistic is the differential growth between the two growth rates, expressed in percentage points. "North America" includes Canada and the United States.

about four to nine times higher than for an equivalent change in $PM_{2.5}$ mass (Janssen et al. 2011).[17] The World Health Organization also classifies black carbon as being especially harmful relative to other air pollutants (Janssen et al. 2012).

Globally, open burning of biomass (including agriculture-residue burning) accounts for nearly 37 percent of global black carbon emissions (Bond et al. 2013). However, the major source of black carbon in Cairo is traffic, even during large biomass-burning events (Mahmoud et al. 2008). As opposed to decoupling of CO_2, NO_x, and SO_2 emissions, neither oil exporting nor oil importing economies in the Middle East and North Africa have achieved any form of decoupling of black carbon emissions from income growth.

Intraregional Differences, Regional Implications

Oil exporters decoupled more slowly than oil importers in the Middle East and North Africa. The world's slowest decoupling rates, as presented above, are largely driven by the oil exporting countries, even though the non-oil exporting economies also show extremely low decoupling rates (as shown when splitting the sample accordingly). Although some of the region's oil producing countries have made efforts to diversify their economies, these often included investments in downstream industries of the oil and gas sectors, such as the petrochemical sector.

Moreover, the continued provision of subsidized fuels, particularly by the oil producing countries, disincentivizes the broader adoption of sustainable alternatives—such as cleaner mobility (public transportation and noncombustion vehicles), cleaner energy production, and less-resource-intensive production and consumption patterns in general. The reasons for the region's stubbornly high emission rates and lack of decoupling are reviewed in detail in chapter 3 but summarized here.

A failure to sufficiently decouple emissions growth from income growth has adverse implications for the climate and for residents' exposure to hazardous particles in the air. As noted earlier, the economies of the Middle East and North Africa have either failed or succeeded only partially in decoupling emissions growth from income growth, especially when compared with other regions. Reasons for this are manifold and include high emissions from outdated vehicle fleets, low incentives to switch to alternative modes of transport such as public transportation or nonmotorized options, weak regulatory frameworks for industrial emission control, and the region's unsustainable energy sources—about which chapter 3 (on air pollution) provides further insights. These deficiencies have also translated into only a limited decoupling between income growth and the PM levels that the region's residents are exposed to. Only Sub-Saharan Africa has a slower rate of decoupling of income growth from $PM_{2.5}$ exposure (figure 2.17).

Data limitations prevent investigation into possible decoupling processes regarding plastic consumption, marine-plastic pollution, and coastal erosion. Additional data on such possible decoupling processes would be highly desirable to help advance the "blueing" of the region's skies and seas.

FIGURE 2.17

Comparison of Global Regions in Decoupling Growth of GNI Per Capita from PM$_{2.5}$ Exposure since 1990

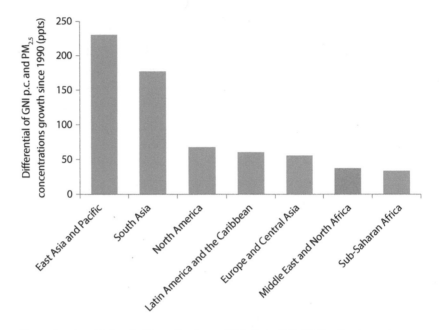

Sources: Based on data from the Human Development Data Center, United Nations Development Programme (http://hdr.undp.org/en/data) and the World Development Indicators database.
Note: The figure shows the differential, in percentage points (ppts), between the per capita (p.c.) growth rates of gross national income (GNI) and the mean annual exposure to PM$_{2.5}$ air pollution (fine particulate matter of 2.5 microns or less in diameter). The growth rates of both PM$_{2.5}$ p.c. and GNI p.c. relative to their 1990 levels were computed. Averages of the respective growth rates for 2015–17 were computed to adjust for erratic changes, and the average growth rate of GNI p.c. was then subtracted from the average growth rate of PM$_{2.5}$ exposure. The resulting statistic is the differential growth between the two growth rates, expressed in ppts. "North America" includes Canada and the United States.

SWITCHING TO A GREEN GROWTH PATH

The COVID-19 pandemic and its associated recovery programs provide a window of opportunity to kick-start or accelerate decoupling in the Middle East and North Africa and to shift toward a green growth path. After the initial relief phase, during which countries have focused on public health and social transfers, and as they move into the recovery phase, it is important to make this a GRID recovery, which will accelerate decoupling.

Economies in the Middle East and North Africa reacted rapidly to the onset of the COVID-19 crisis and have implemented large-scale programs in both the public health and social protection sectors to support households and firms, combined with strict containment measures (OECD 2020). However, the recovery from COVID-19 also offers opportunities for "building back better"—boosting climate-smart infrastructure and technologies, supporting adaptation measures, and avoiding carbon-intensive investments (Batini et al. 2021). Hence, recovery from COVID-19 provides an opportunity to reset policies toward green, renewable energy that will be more sustainable in the longer run.

The World Bank Group and other multilateral organizations have moved quickly to respond. Although most COVID-19 relief support measures have been "climate neutral," there has also been some support for greening recoveries (box 2.1).

BOX 2.1

Green Recovery Goals Guide COVID-19 Responses by Multilateral Organizations

World Bank Group
The World Bank Group supports governments around the world in their responses to the COVID-19 crisis, targeting the recovery of their health systems, populations, and economies. It is also supporting activities in some countries to facilitate a "green recovery." For example, it is making the following investments:

- *In Egypt,* a US$200 million operation aims to enhance air pollution management and upgrade waste management infrastructure in the Greater Cairo region (World Bank 2021c). Through measures to manage medical waste and foster public-private partnerships in solid waste management, the World Bank will support the response to COVID-19 while also advancing more sustainable patterns in the future.

- *In Morocco,* in cooperation with the French Development Agency (AFD), the World Bank is supporting a US$250 million "Green Generation Program-for-Results" project to facilitate a swift response to the COVID-19 pandemic while supporting activities in climate-smart agriculture that will help

prevent and mitigate the impacts of severe droughts (World Bank 2021e).

- *In Panama,* a US$300 million project supports the protection of human capital during the COVID-19 response and includes reforms to incorporate energy and environmental considerations into procurement processes, increase the competitiveness of renewable energies, and create an inventory and registry of carbon emissions to allow the future creation of a carbon market (World Bank 2021f).

- *In Mexico,* a US$750 million Development Policy Financing (DPF) program supports the government in various activities related to forest conservation, the implementation of a greenhouse gas (GHG) emission trading system, and the introduction of standardized air quality monitoring and a public communication campaign in large cities (World Bank 2021b). Additionally, this DPF program aims to expand access to resilient urban infrastructure and social housing for the poor while simultaneously creating jobs that contribute to climate change mitigation and adaptation.

(continued)

BOX 2.1

Green Recovery Goals Guide COVID-19 Responses by Multilateral Organizations (*Continued*)

Other COVID-19 response efforts supported by the World Bank Group include programs in Brazil, Colombia, Ecuador, Ghana, Guatemala, Haiti, Maldives, Pakistan, and Rwanda, among other countries.

Inter-American Development Bank
The Inter-American Development Bank (IDB) supports a range of projects in Latin America and the Caribbean to identify best practices to enhance the recovery process through resilient and sustainable green growth investments.

Examples, among others, include support for developing renewable energy projects in rural Bolivia that strengthen the use of alternative energy during the COVID-19 recovery process (IDB 2020b). In addition, a project in Barbados will set up an integrated coastal zone management (ICZM) scheme (IDB 2020a). This will support the country's post-COVID reopening and strengthen the transition to a blue economy while also addressing hazards related to coastal flooding and uncoordinated development.

Asian Development Bank
The Asian Development Bank (ADB) supports several projects to ensure a green recovery from the COVID-19 pandemic. In a regional project, it is supporting efforts to improve energy efficiency and ensure safe working conditions in public buildings by deploying efficient, clean, and smart centralized air-conditioning systems (ADB 2020c). In China, the ADB is providing US$150 million for a project in the greater Beijing-Tianjin-Hebei region in support of air quality improvement and GHG emission reduction through energy efficiency, cleaner transportation, renewable energy, and other measures (ADB 2020a). By employing "build back better" principles, this project aims to stimulate investment in post–COVID-19 recovery while scaling up clean energy technologies in the targeted region. Another project in China supports green development in rural areas through improvement of rural waste management services, promoting recovery from COVID-19 through inclusive and sustainable rural economic development.

In India, the ADB is providing liquidity support in the form of nonconvertible debentures amounting to US$22 million for Azure Power India Private Ltd., an independent solar power producer, to sustain renewable energy operations and simultaneously mitigate the impact of COVID-19 on economic activity (ADB 2020b).

Advantages in Common from Green Growth

Although the Middle East and North Africa is a highly heterogeneous region, reorienting its economies toward a greener economic model would benefit all of them, especially in the long term. This highly complex region comprises economies with (very) different income levels (from low to high); different economic structures (most notably, oil exporters versus oil importers); and different socioeconomic and demographic structures. Hence, it is important to acknowledge that there is no one-size-fits-all approach to addressing the region's varying developmental and environmental challenges. Nonetheless, initiating and promoting a green transition toward a more sustainable economic model holds promise for all of the region's economies despite their heterogeneity.

Green growth has at least three distinct advantages over business-as-usual brown growth: fewer costs from environmental degradation, more growth, and more jobs.

Cost savings. *Green growth reduces environmental degradation and the economic costs associated with it.* The cost of not decoupling growth from emissions and continuing to grow with large environmental externalities is high, as this report details in its chapters on air quality (chapter 3), marine-plastic pollution (chapter 4), and coastal erosion (chapter 5)—a cost upward of 3 percent of GDP in certain economies. The costs associated with the failure to decouple and its consequences show that the GRID approach would bring savings from avoided degradation costs.

Returns on investment. *Building back greener also brings higher returns on investments for the region's economies than would result from continuing traditional growth patterns.* Investing in a greener future both raises prospects for the restoration of the region's natural capital and presents opportunities for important economic benefits. Investments in clean energy and biodiversity conservation have a substantially higher output multiplier than investments that support brown growth (Batini et al. 2021). This implies that returns for each dollar spent in these sectors are considerably higher—confirming that reversing biodiversity loss and environmental degradation by changing development patterns is not at odds with economic advancement.

On the contrary, an increase in green spending will generate a disproportionately higher increase in economic activity, exceeding initial investments by 10–50 percent (Batini et al. 2021). Hence, building back greener has the potential to increase future growth, accompanied by a further increase in living standards. Thus, it is imperative to funnel policy actions away from options that are not eco-friendly and toward sustainable ones, to induce a switch away from industries that must be drastically reformed to meet climate goals (Hepburn et al. 2020).

Job creation. *Blue or green projects can fare better in creating jobs than traditional fiscal stimuli.* It is crucial for responses to the economic challenges caused by the COVID-19 pandemic to address the precarious labor market situation. To approach this issue, a stronger focus on green projects is warranted not only from an environmental perspective but also regarding green projects' potential to create more jobs than traditional fiscal support measures (box 2.2).

BOX 2.2

Job Creation from Green Growth Strategies

A recent survey of more than 200 officials from central banks, finance ministries, and other economic experts identified five policy objectives with high potential for both economic outcomes (such as job creation) and climate impacts: clean physical infrastructure, building-efficiency retrofits, investment in education and training, natural capital investment, and clean research and development (Hepburn et al. 2020). These are promising areas for projects seen as able to create more jobs, deliver higher short-term returns, and bring greater long-term savings than traditional fiscal stimuli can create.

The view that green policies have greater potential than traditional stimuli to create jobs is based on experience (as in the examples described below) as well as anticipation of a changing business environment due to the global commitments of major economic players to a greener orientation of global value chains (GVCs). Theoretical contributions also highlight the greater potential of green sectors for job creation—as, for example, in renewable energy projects that have delivered three times more jobs per dollar than comparable ones in the fossil fuel sector (Garrett-Peltier 2017).

Comprehensive studies of the clean-energy and other green sectors in the United States provide clear evidence that these sectors have an advantage in terms of not only the quantity of jobs created but also the quality of those jobs. The number of jobs in the clean-energy sector has been growing faster than in the fossil fuel industry while also paying more across all pay levels, with an average wage premium of around 25 percent (BW Research 2020). The 1990 Clean Air Act Amendments and their environmental policies led to strong employment growth in the United States (Walker 2011).

A more recent success story is that of the American Recovery and Reinvestment Act (ARRA), implemented in the United States in response to the 2008–09 Global Financial Crisis. It included large investments in green sectors such as waste treatment, public transportation, and energy, and it has led to substantial employment gains, at least in the longer term (Popp et al. 2020). Although the short-term employment gains were muted, long-run effects were positive and have been especially pronounced in areas with a higher endowment of preexisting green skills. The delayed materialization of employment gains implies that the full benefits of such programs may be visible with a lag—an important point for policy makers to consider.

Furthermore, green policies and investment in sustainable sectors have created jobs at an impressive rate around the world. In China, the introduction of an emission-trading scheme in seven pilot regions has been a driver for employment while also distinctly reducing carbon emissions, yielding a double employment-dividend effect (Yang, Jiang, and Pan 2020; Yu and Li 2021). In Navarre, Spain, early adoption of renewable energy technology

(continued)

BOX 2.2

Job Creation from Green Growth Strategies (*Continued*)

and support for the sector have transformed the region's energy profile and enormously increased sectoral employment rates (Faulin et al. 2006). And in India, access to frequent public transportation services has also increased nonagricultural employment substantially, with women especially profiting from such developments (Lei, Desai, and Vanneman 2019).

Moving toward a circular economy (an economic system restores its resources, as opposed to a linear "throwaway" system, as further discussed in chapter 4) also shows great potential for a double dividend in the sense of reducing the adverse effects of inadequate waste treatment while simultaneously creating jobs. It is estimated that repair—an integral part of a circular-economy approach—creates over 200 times as many jobs as traditional disposal methods, which lead to high plastic leakage to marine spaces and uncontrolled waste burning (Ribeiro-Broomhead and Tangri 2021). Although the options for repairing goods made out of plastic are often limited, it is estimated

that recycling creates 50 times more jobs than open dumps, landfills, and similar disposal methods, while remanufacturing goods creates almost 30 times as many jobs (Ribeiro-Broomhead and Tangri 2021).

In the European Union (EU), the number of jobs linked to the circular economy grew by 5 percent between 2012 and 2018, reaching around 4 million. Applying the principles of this concept to the whole EU could increase aggregate output by an estimated 0.5 percent by 2030, creating around 700,000 additional jobs (EC 2020). And in Indonesia, the government's move toward a circular economy as a green response to the COVID-19 pandemic is expected to have substantial positive effects. Compared with a business-as-usual scenario, it is estimated that the adoption of circular-economy opportunities could raise aggregate output by an additional US$42–US$45 billion (around 0.4 percent of Indonesian gross domestic product [GDP]) in 2030 while creating 4.4 million new jobs between 2021 and 2030 (Bappenas and UNDP 2021).

In sectors that negatively affect the environment—those following the traditional brown development path—green projects often bring higher economic returns than traditional projects. However, it is also important to acknowledge that such a transition would also lead to job reductions in displaced sectors. This includes the oil and gas sector (which traditionally has been a major source of value added and employment in oil-exporting countries of the Middle East and North Africa) as well as related downstream industries such as the petrochemical sector, which were used as a way of diversifying the economy in these countries (as briefly discussed in chapter 4, box 4.4).

Putting in place social compensation schemes for displaced workers during transition periods is hence important to cushion these adverse effects. Similarly, supporting the transition of workers out of these sectors into more sustainable ones through retraining programs, relocation assistance, and active promotion of new industries would facilitate a just transformation toward a greener growth model.

"Blueing" the Skies and Seas

The blue natural assets of the Middle East and North Africa are of particular importance for a GRID path. "Blueing" the skies would have the largest effect of any environmental factor on improving public health. Moreover, blueing the skies is synonymous with reducing GHG emissions and mitigating climate change because virtually everything that reduces the emission of air pollutants also reduces the emission of climate pollutants.

Blueing the seas is also of major importance because the blue economy is a significant sector in many of the region's economies. Tourism alone, which is crucially built on the quality of the marine natural capital of the Middle East and North Africa, accounts for upward of 15 percent of GDP in some of these economies. Moreover, tourism is the main focus for diversification away from the fossil-fuel revenue dependence of many countries in the region. Marine plastics and coastal erosion are critical threats to the tourism sector as well as to other sectors such as fisheries—a particularly critical sector in terms of inclusion given its importance to low-income households as a source of employment and income.

A failure to embark on a GRID path could leave the region's economies, especially the oil exporters, at risk of ending up with stranded assets. Carbon assets have been a source of prosperity in recent decades for about half the economies in the Middle East and North Africa. However, with a broadening of climate policies globally and the emergence of promising new technologies to make the decoupling of energy production from the use of fossil fuels competitive, the value of these assets is expected to diminish over time (Lange, Wodon, and Carey 2018).

More than 60 countries worldwide, including major players such as China, the European Union (EU), and the US have already pledged to achieve zero net emissions as well as decarbonization of value chains, with more to follow. Although global decarbonization efforts will not be in full force immediately, the global trend toward reduced fossil fuel use will happen gradually, and it is important for the region's economies to concomitantly start their energy transition so as not to end up with stranded assets and uncompetitive national economies.

Box 2.3 provides some details on the carbon wealth of the region's economies and how overreliance on their fossil fuel resources could soon become a pitfall. Hence, to reduce environmental pressures and

BOX 2.3

Carbon Wealth in the Middle East and North Africa and Its Potential Pitfalls

Carbon wealth (rents from underground oil-, natural gas-, and coal-based assets) make up a large share of the Middle East and North Africa's total wealth—almost 30 percent in 2014 (primarily in the form of oil and gas), surpassing other forms of capital such as produced or human capital (Lange, Wodon, and Carey 2018). For some of the region's major oil exporters (such as Iraq, Kuwait, or Saudi Arabia), yearly fossil fuel rents (the income derived from this wealth) are estimated to exceed 40 percent of gross domestic product (GDP). Because these figures do not consider the full value added of downstream and other connected industries, the total contribution of carbon-related sectors to GDP is in fact even larger.

The exploitation of these assets has been an engine for growth in recent decades, but this growth trajectory is not sustainable either environmentally or economically over the next couple of decades. For one thing, fossil fuels are a major contributor to greenhouse gas (GHG) emissions and hence to human-induced climate change. To meet the goal of keeping the rise in average global ground temperature below 2 degrees Celsius, an estimated 82 percent of known global coal deposits, 33 percent of known oil reserves, and 49 percent of known gas reserves must remain unburned (McGlade and Ekins 2015). To keep the temperate increase below 1.5 degrees Celsius, nearly 90 percent of the 2018 reserve base of coal deposits must remain unextracted, and around 60 percent of both oil and fossil

methane gas deposits must remain in the ground (Welsby et al. 2021).

Therefore, large parts of the fossil fuel reserves in Middle East and North Africa countries risk being stranded (Lange, Wodon, and Carey 2018), which puts future incomes derived from this wealth at risk. Downstream industries currently relying on cheap fossil fuel energy—that is, subsidized access to oil and gas or the energy derived from them—may come under increasing pressure in the coming decades, making an economic diversification critical. In a world that is taking serious steps toward net-zero emissions and decarbonization of value chains, this pressure increases the risk of being left with stranded assets.

One of the main developments driving the risk of stranded assets and nations is the push in the international agenda to move toward emission neutrality and decarbonization of global value chains (GVCs)—as more than 60 countries have already pledged to do, including major economies such as China, the European Union (EU), and the United States. The EU's European Green Deal is among the trailblazers in this respect. The Green Deal aims to make Europe climate neutral by 2050 and entails a fundamental overhaul of its energy system, which will affect its key energy suppliers (in the Middle East and North Africa, most notably Algeria and Libya).

The European Commission has also proposed establishing a border carbon adjustment mechanism in the Green Deal that introduces a tax or tariff on imports to

(continued)

BOX 2.3

Carbon Wealth in the Middle East and North Africa and Its Potential Pitfalls (*Continued*)

the EU based on the emissions embedded in those products (EC 2021). The rationale for this carbon tariff is twofold: First, it would prevent carbon leakage and disincentivize consumers from switching to foreign suppliers, hence also protecting domestic production. Second, it would incentivize other countries globally to also decarbonize their value chains. However, even for those oil exporting countries in the Middle East and North Africa for which Europe is *not* a main trading partner, the EU's transition away from fossil fuels and toward renewable energy sources will have an impact, if only a small one, because Europe accounts for around 20 percent of global crude oil imports.

The change in the European energy landscape will be incremental, with the European Commission projecting that about half the EU's energy in 2030 will still be provided from fossil fuels. The use of coal must be substantially reduced by

2030, while most of the change in oil and gas demand will happen between 2030 and 2050.

The decades after 2030 will be characterized by lower demand in the EU for oil (dropping by almost 80 percent) and for gas (dropping by around 60 percent) (Leonard et al. 2021). Oil- and gas-producing countries will feel the repercussions through both the direct channel of lower exports and the indirect channel of lower global prices for fossil fuels. It is hence important for the Middle East and North Africa's oil exporting countries—but also for its oil-dependent economies with carbon-intensive industries—to start changing their production base and their economic mix. One particularly promising way to do so is to intensify the expansion of renewable energy sources, for which the region holds great prospects (as further discussed within chapter 3's "Policy Review" section).

economic risks, concomitant actions to reduce the carbon intensity of current processes and production patterns and to prepare for economic diversification are crucial to make Middle East and North Africa economies in general—but the oil exporting ones in particular—ready for a decarbonized future.

CONCLUSION

Blue skies will improve residents' health and resilience in the Middle East and North Africa, reduce economic costs, and make the cities better places to live and work. Better air quality will reduce the impacts of respiratory diseases on lives and livelihoods. Those respiratory diseases include

COVID-19[18] (whose severity is exacerbated by air pollution) and other diseases to which air pollution has been linked, including cardiovascular diseases, ischemic heart diseases, diabetes, kidney diseases, and many others.[19]

With better health, people become more resilient. Improved air quality benefits everybody but especially the poor, who often work outdoors. Air pollution can also decrease earnings, both when workers are unwell and when they must take time off to care for family members, including children, who are exposed to air pollution. With a large share of the workforce in the informal sector in many Middle East and North Africa economies (Gatti et al. 2014), these sick days are equivalent to lost income. Another concern for gender inclusion is that women generally assume responsibility for the care of sick relatives. Thus, women's ability to work is particularly affected by family illness related to air pollution. Importantly, the benefits of better air quality are inclusive.

Reducing air pollution will also reduce the costs it incurs, which are upward of 3 percent of GDP in some of the region's economies (as further discussed in chapter 3). Moreover, enhancing air quality can make countries more attractive to tourists (Łapko et al. 2020), while the ability to manage pollution will influence their ability to grow and hence a city's competitiveness (Lozano-Gracia and Soppelsa 2019).

Key policies to improve air quality include a switch away from polluting fossil fuels (which also contribute to global warming); better public transportation and urban planning; and city greening—all elements of greener development and improved urban livability.

Blue seas, including well-managed coastlines and reduced plastic pollution, will increase the resilience and sustainability of coastal economies. Every Middle East and North Africa economy has a coastline; each has access to an ocean or a sea. Well-managed coastlines and marine ecosystems—free of plastic pollution—ensure the sustainability of coastal cities' key economic sectors such as beach tourism, ports, and fisheries, as well as the jobs dependent on those sectors. They also increase the resilience of coastal areas to the impacts of rising sea levels from climate change. Improved solid waste management, including management of plastics, reduces the incidence of flooding by unblocking drainage channels, further increasing resilience.

Blue skies and blue seas also demonstrate the extent to which management of all aspects of natural capital are interconnected. Poor land management, an energy mix that is overdependent on fossil fuels, and poor urban and transportation planning damage air quality, public health, and productivity. The result is that the Middle East and North Africa's skies are not blue. Poor management of water resources and water quality and inadequate management of solid waste and plastics reduce coastal resilience and the broader blue economy. The result is that the Middle East and North Africa's seas are not blue. Yet there are opportunities, as the region recovers from COVID-19, to make the policy changes necessary to "blue" its skies and seas.

Hence, it is important for the Middle East and North Africa region to conserve and revive its blue assets to improve livelihoods and reduce the burden of pollution on its residents and the environment. Given the opportunity for transformational change that recovery from the COVID-19 pandemic presents, it is crucial for the region's economies to induce a meaningful stimulus for reversing some of the damages to its blue assets caused by unsustainable past growth paths. These changes will not only benefit the region's environment and ecosystems but also help strengthen its economies to withstand the challenges ahead.

With more and more countries worldwide pledging to move toward greener economic development, the Middle East and North Africa would otherwise risk being left behind with stranded assets and outgrown business models. Getting ahead of this risk should be a priority for the region's policy makers. This report identifies some of the most severe deficiencies that also hinder the region's continued rise in prosperity and points to some of the most promising possibilities to overcome those deficiencies.

NOTES

1. The national capital accounting framework was developed by the Wealth Accounting and the Valuation of Ecosystem Services (WAVES) Secretariat. WAVES is a World Bank-led global partnership that aims to promote sustainable development by ensuring that natural resources are mainstreamed in development planning and national economic accounts. It is now part of the broader World Bank umbrella initiative, the Global Program for Sustainability. For more information, see "Wealth Accounting and WAVES" on the WAVES website: https://www.wavespartnership.org/en /wealth-accounting-and-WAVES.
2. Economic growth data are from the Human Development Data Center, United Nations Development Programme, http://hdr.undp.org/en/data.
3. The GCC comprises the countries of Bahrain, Kuwait, Oman, Qatar, Saudi Arabia, and the United Arab Emirates. The Maghreb economies include Algeria, Libya, Malta, Morocco, and Tunisia. The Mashreq economies include Djibouti, the Arab Republic of Egypt, Iraq, the Islamic Republic of Iran, Jordan, Lebanon, the Syrian Arab Republic, West Bank and Gaza, and the Republic of Yemen.
4. The information and communication technology (ICT) sector would benefit from increases in fixed and mobile broadband transmission capacity as well as greater competition, and small and medium enterprises still have only a limited online presence in most countries. The quality of trade and transportation infrastructure—as measured by the infrastructure performance subindex of the World Bank's Logistics Performance Index (https://lpi .worldbank.org/)—has improved since 2007. However, there are wide

differences between economies, and performance appears to have deterio-rated since 2015 in some of them. The poorest performing category in this regard was rail infrastructure, with only 12 percent of survey respondents rating its quality as "very high" or "high" (Arvis et al. 2018). Transportation presents opportunities for improvements in rail freight, urban metro trans-portation, and transitions to e-mobility that could also improve air quality. A 2019 OECD study, "Enhancing Connectivity through Infrastructure Investment," estimated total investment needs in the Middle East and North Africa region of US$100 billion over the next five years to upgrade and main-tain infrastructure, with transportation and electricity accounting for around 43 percent of total needs, followed by ICT (9 percent) and water and sanitation (5 percent) (OECD 2019). There are also constraints regarding "soft" infra-structure in the regulatory environment and in trade and customs facilita-tion. However, the region's economies have already started investing in their transportation infrastructure, particularly in extensions of the rail network.

5. Country-specific average years of schooling from United Nations Development Programme (UNDP) Human Development Center data (http://hdr.undp.org/en/data).

6. Literacy data are from the World Development Indicators database, http://wdi.worldbank.org.

7. Regional and global labor and business leadership data are from World Bank Enterprise Surveys: https://www.enterprisesurveys.org/en/enterprisesurveys.

8. Data on increased poverty rates and food prices are from, respectively, the World Development Indicators database (http://wdi.worldbank.org) and the Food and Agriculture Organization of the United Nations (FAO) Food Price Index (https://www.fao.org/worldfoodsituation/foodpricesindex/en/).

9. For comprehensive analyses of the Middle East and North Africa's economic development on a regional level as well as in individual economies, consult, for example, the World Bank's "MENA Quarterly Economic Brief" (https://www.worldbank.org/en/region/mena/publication/mena-quarterly-economic-brief) and semiannual "MENA Economic Update" series (https://www.worldbank.org/en/region/mena/publication/mena-economic-monitor).

10. Despite an overall increase, NO_X emissions in the Mashreq have decreased since 2010, driven mainly by decreases in the Islamic Republic of Iran and stagnant trends in Egypt. In the Islamic Republic of Iran, NO_X emissions have been reduced mainly in the agriculture sector, stemming from less-intensive use of fertilizers that account for a large amount of soil emissions. In Egypt, industrial NO_X emissions decreased beginning in 2010. This was probably because of advances such as the switch from burning heavy fuel oil (mazout) to using compressed natural gas in brick factories (see Higazy et al. 2019), while NO_X emissions stemming from agriculture remained stagnant. In addition, recent SO_2 emissions reductions in the Mashreq were driven primarily by the Islamic Republic of Iran and Syria (where the ongoing con-flict disrupted industrial activity, most probably causing the reduction in SO_2 emissions). In the Islamic Republic of Iran, the switch toward gas for energy production (away from heavy oils), desulfurization of flue gas, and increased use of cleaner fuels for energy production have helped reduce SO_2 emissions (Delfi et al. 2018). In the Maghreb, Morocco has contributed most to the subregion's stagnant or decreasing trends in SO_2 emissions. It has set some of the strictest sulfur limits in gasoline and diesel in recent years (see the

chapter 3 section on policies to reduce vehicle emissions), potentially contributing to these trends.

11. Land degradation is defined as the reduction or loss of the biological or economic productivity of lands, such as long-term loss of natural vegetation, arising from human activities. Although land management is discussed briefly here, it is not the focus of this report and has been addressed in two recent regional publications: "Sustainable Land Management and Restoration in the Middle East and North Africa Region: Issues, Challenges and Recommendations" (World Bank 2019c) and "Sand and Dust Storms in the Middle East and North Africa (MENA) Region: Sources, Costs and Solutions" (World Bank 2019b).

12. According to national reports to the Convention on Biological Diversity (https://www.cbd.int/countries/), the Middle East and North Africa countries with the greatest plant diversity, each with more than 3,000 species, include Algeria, Egypt, Lebanon, Morocco, Syria, and Tunisia. Animal diversity is highest, with more than 5,000 species each, in Algeria, Lebanon, Syria, and Tunisia; mammal diversity is particularly high in the Arabian peninsula.

13. Data on land under protection, by country, come from the World Database on Protected Areas, International Union for Conservation of Nature (IUCN): https://www.iucn.org/theme/protected-areas/our-work/world-database-protected-areas; and "Terrestrial Protected Areas (% of Total Land Area) – Country Ranking," Index Mundi: https://www.indexmundi.com/facts/indicators/ER.LND.PTLD.ZS/rankings.

14. Water stress occurs when water withdrawals for human, agricultural, and industrial uses are high relative to the renewable water resources—that is, when there's a high ratio of annual water withdrawals to average annual surface freshwater availability.

15. The region's five countries among the top 10 GHG emitters per capita are Bahrain, Kuwait, Qatar, Saudi Arabia, and the United Arab Emirates. Oman is ranked 12th among the world's top GHG-emitting countries. Except for the Islamic Republic of Iran, per capita emissions in the Mashreq and Maghreb subregions are less than one-fourth those in the GCC, as follows (by descending order of magnitude): Iraq, Lebanon, Algeria, Jordan, Tunisia, Egypt, Morocco, Syria, the Republic of Yemen, and Djibouti. GHG emission data are from the Carbon Dioxide Information Analysis Center (CDIAC) at the Oak Ridge (Tennessee) National Laboratory, Climate and Environmental Sciences Division, US Department of Energy: https://cdiac.ess-dive.lbl.gov/home.html.

16. The carbon emissions data presented in this section are based on territorial data by the Global Carbon Project (https://www.icos-cp.eu/science-and-impact/global-carbon-budget/2019) using a production-based approach. This means that the numbers do not adjust for international trade and emissions embodied in traded goods and services. However, when using data that incorporate such effects, leading to what are called consumption-based carbon emissions estimates, the section's findings are qualitatively unchanged. The Middle East and North Africa is the world's only region not decoupling at all when using these trade-adjusted estimates as well, while other regions have either absolutely or relatively decoupled. This is because—although there are differences in the levels of carbon emissions between these approaches for some countries (such as Bahrain or Qatar)—the differences in trends (changes over time) are not pronounced. The analysis presented

here, however, focuses on the evolution of carbon emissions over time, explaining the rather marginal differences between the two approaches regarding the decoupling process. One pragmatic reason for using the production-based approach is the much broader coverage of countries for this dataset, because the construction of consumption-based carbon budgets requires detailed information on trade between countries, which is often not available for individual countries. Such a broad coverage is preferable for the comparison of regions across the world.

17. $PM_{2.5}$ refers to fine particulate matter, having a diameter of 2.5 micrograms or less. (See also chapter 3, box 3.1 for a further description.)

18. Several studies focus on that relation and have found increases in mortality due to COVID-19 that are associated with higher air pollution in the United States (Wu et al. 2020); in several Asian cities (Gupta et al. 2020); and in Dutch (Cole, Ozgen, and Strobl 2020) and northern Italian municipalities (Coker et al. 2020).

19. For a review of the diseases linked to air pollution, see Chen, Goldberg, and Villeneuve (2008); Manisalidis et al. (2020); and Pope and Dockery (2006).

REFERENCES

ADB (Asian Development Bank). 2020a. "China, People's Republic of: Air Quality Improvement in the Greater Beijing-Tinjin-Heibei Region—Green Financing Scale Up Project." Project details, Project No. 51033-001, ADB, Manila.

ADB (Asian Development Bank). 2020b. "India: Azure Power COVID-19 Liquidity Support Project." Project details, Project No. 54241-001, ADB, Manila.

ADB (Asian Development Bank). 2020c. "Regional Support to Build Disease Resilient and Energy Efficient Centralized Air-Conditioning Systems." Project details, Project No. 54210-001, ADB, Manila.

Alemu, A. M. 2017. "To What Extent Does Access to Improved Sanitation Explain the Observed Differences in Infant Mortality in Africa?" *African Journal of Primary Health Care & Family Medicine* 9 (1): 1–9.

Al-Tokhais, A., and B. Thapa. 2019. "Stakeholder Perspectives towards National Parks and Protected Areas in Saudi Arabia." *Sustainability* 11 (8): 1–15.

Alwash, A., H. Istepanian, R. Tollast, and Z. Y. Al-Shibaany, eds. 2018. "Towards Sustainable Water Resources Management in Iraq." Publication No. IEI 300818, Iraq Energy Institute, London.

Argimon, M. 2019. "Climate Change Vulnerability on the Mediterranean's Southern Shore." Unpublished manuscript, World Bank, Washington, DC.

Arvis, J.-F., L. Ojala, C. Wiederer, B. Shepherd, A. Raj, K. Dairabayeva, and T. Kiiski. 2018. "Connecting to Compete 2018: Trade Logistics in the Global Economy." Sixth edition of the Logistics Performance Index, World Bank, Washington, DC.

Bappenas and UNDP (Ministry of National Planning and Development Indonesia and the United Nations Development Programme). 2021. "The Economic, Social, and Environmental Benefits of a Circular Economy in Indonesia." Report, Kementerian PPN/Bappenas and UNDP, Jakarta.

Batini, N., M. Di Serio, M. Fragetta, G. Melina, and A. Waldron. 2021. "Building Back Better: How Big Are Green Spending Multipliers?" Working Paper 2021/087, International Monetary Fund, Washington, DC.

Belhaj, F. 2018. "Fixing the Education Crisis in the Middle East and North Africa." Online news, World Bank, November 13. https://www.worldbank .org/en/news/opinion/2018/11/13/fixing-the-education-crisis-in-the-middle -east-and-north-africa.

Bjerde, Anna. 2020. "Fulfilling the Aspirations of MENA's Youth." *Arab Voices* (blog), January 13. https://blogs.worldbank.org/arabvoices/fulfilling -aspirations-menas-youth.

Bond, T. C., S. J. Doherty, D. W. Fahey, P. M. Forster, T. Berntsen, B. J. DeAngelo, M. G. Flanner, et al. 2013. "Bounding the Role of Black Carbon in the Climate System: A Scientific Assessment." *Journal of Geophysical Research: Atmospheres* 118 (11): 5380–52.

BW Research. 2020. "Clean Jobs, Better Jobs: An Examination of Clean Energy Job Wages and Benefits." Report for E2, ACORE, and CELI (Environmental Entrepreneurs, American Council on Renewable Energy, and Clean Energy Leadership Institute), BW Research Partnership, Carlsbad, CA.

Chen, H., M. S. Goldberg, and P. J. Villeneuve. 2008. "A Systematic Review of Relation between Long-Term Exposure to Ambient Air Pollution and Chronic Disease." *Review of Environmental Health* 23: 243–97.

CMEMS (Copernicus Marine Environment Monitoring Service). 2021. "Mediterranean Sea Chlorophyll-a Trends (1997–2019)." From the Ocean Monitoring Indicators, Copernicus Programme, European Union, Luxembourg. https://marine.copernicus.eu/access-data/ocean-monitoring -indicators/mediterranean-sea-chlorophyll-trend.

Coker, E. S., L. Cavalli, E. Fabrizi, G. Guastella, E. Lippo, M. L. Parisi, N. Pontarollo, M. Rizzati, A. Varacca, and S. Vergalli. 2020. "The Effects of Air Pollution on COVID-19 Related Mortality in Northern Italy." *Environmental and Resource Economics* 76 (4): 611–34.

Cole, M. A., C. Ozgen, and E. Strobl. 2020. "Air Pollution Exposure and COVID-19 in Dutch Municipalities." *Environmental and Resource Economics* 76 (4): 581–610.

Damania, R., S. Desbureaux, A.-S. Rodella, J. Russ, and E. Zaveri. 2019. *Quality Unknown: The Invisible Water Crisis.* Washington, DC: World Bank.

Delfi, S., M. Mosaferi, A. Khalafi, and K. Zoroufchi Benis. 2018. "Sulfur Dioxide Emissions in Iran and Environmental Impacts of Sulfur Recovery Plant in Tabriz Oil Refinery." *Environmental Health Engineering and Management Journal* 5 (3): 159–66.

EC (European Commission). 2020. *Circular Economy Action Plan: For a Cleaner and More Competitive Europe.* Luxembourg: Publications Office of the European Union.

EC (European Commission). 2021. "Proposal for a Regulation of the European Parliament and of the Council Establishing a Carbon Border Adjustment Mechanism." Proposal COM (2021) 564 final, July 14, EC, Brussels.

Egypt Today. 2021. "Egypt Has 146 Wastewater Treatment Plants, 2 to Be Added." *Egypt Today,* April 8. https://www.egypttoday.com/Article/1/100667 /Egypt-has-146-wastewater-treatment-plants-2-to-be-added.

El-Kogali, S. E. T., and C. Krafft, eds. 2019. *Expectations and Aspirations: A New Framework for Education in the Middle East and North Africa.* Washington, DC: World Bank.

Elsharkay, H., H. Rashed, and I. Rached. 2009. "Climate Change: The Impacts of Sea Level Rise on Egypt." Paper presented at the 45th International Society of City and Regional Planners (ISOCARP) World Planning Congress, Porto, Portugal, October 18–22.

FAO (Food and Agriculture Organization of the United Nations). 2014. "Food Losses and Waste in the MENA Countries: Status, Prospects and Strategies for Reduction." PowerPoint presentation, Workshop on Agricultural Trade and Food Security in the Euro-Med Area, Akdeniz University, Turkey, September 25–26, FAO, Rome.

FAO (Food and Agriculture Organization of the United Nations). 2020. *The State of World Fisheries and Aquaculture 2020: Sustainability in Action.* Rome: FAO.

Faulin, J., F. Lera, J. M. Pintor, and J. García. 2006. "The Outlook for Renewable Energy in Navarre: An Economic Profile." *Energy Policy* 34 (15): 2201–16.

García, N., I. Harrison, N. Cox, and M. F. Tognelli. 2015. *The Status and Distribution of Freshwater Biodiversity in the Arabian Peninsula.* Gland, Switzerland: International Union for Conservation of Nature.

Garrett-Peltier, H. 2017. "Green versus Brown: Comparing the Employment Impacts of Energy Efficiency, Renewable Energy, and Fossil Fuels Using an Input-Output Model." *Economic Modelling* 61: 439–47.

Gatti, R., D. F. Angel-Urdinola, J. Silva, and A. Bodor. 2014. "Striving for Better Jobs: The Challenge of Informality in the Middle East and North Africa." MENA Knowledge and Learning Quick Notes Series, No. 49, World Bank, Washington, DC.

GCP (Global Carbon Project). 2019. "Supplemental Data of Global Carbon Budget 2019" (Version 1.0) [Data set]. GCP, a project of Future Earth. https://www.icos-cp.eu/science-and-impact/global-carbon-budget/2019.

GCP (Global Carbon Project). 2020. "Data Supplement to the Global Carbon Budget 2020" (Version 1.0) [Data set]. GCP, a project of Future Earth. doi:10.18160/gcp-2020. https://www.icos-cp.eu/science-and-impact/global-carbon-budget/2020.

Gupta, A., H. Bherwani, S. Gautam, S. Anjum, K. Musugu, N. Kumar, A. Anshul, and R. Kumar. 2020. "Air Pollution Aggravating COVID-19 Lethality? Exploration in Asian Cities Using Statistical Models." *Environment, Development and Sustainability* 23: 6408–17.

GWI (Global Water Intelligence). 2016. "Global Water Market 2017: Meeting the World's Water and Wastewater Needs until 2020." Report, Global Water Intelligence, Oxford, UK.

Hepburn, C., B. O'Callaghan, N. Stern, J. Stiglitz, and D. Zenghelis. 2020. "Will COVID-19 Fiscal Recovery Packages Accelerate or Retard Progress on Climate Change?" *Oxford Review of Economic Policy* 36 (Suppl 1): S359–S381.

Higazy, M., K. S. M. Essa, F. Mubarak, M. El-Sayed, A. M. Sallam, and M. S. Talaat. 2019. "Analytical Study of Fuel Switching from Heavy Fuel Oil to Natural Gas in Clay Brick Factories at Arab Abu Saed, Greater Cairo." *Scientific Reports* 9 (1): 1–10.

Hoesly, R. M., S. J. Smith, L. Feng, Z. Klimont, G. Janssens-Maenhout, T. Pitkanen, J. J. Seibert, et al. 2018. "Historical (1750–2014) Anthropogenic Emissions of Reactive Gases and Aerosols from the Community Emission Data System (CEDS)." *Geoscientific Model Development Discussions* 11: 369–408.

IBP (International Business Publications USA). 2016. *Saudi Arabia Ecology, Nature Protection Laws and Regulation Handbook. Vol. 1: Strategic Information and Laws.* Washington, DC: IBP USA.

IDB (Inter-American Development Bank). 2020a. "BA-T1068: Improving: Institutional Frameworks for Integrated Coastal Zone Management, National Risk Information Planning Systems and Sustainable Climate-Resilient Coastal Infrastructure." Project details, Project No. BA-T1068, IDB, Washington, DC.

IDB (Inter-American Development Bank). 2020b. "BO-T1356: Support to Change the Energy Matrix in Bolivia." Project details, Project No. BO-T1356, IDB, Washington, DC.

Jambeck, J. R., R. Geyer, C. Wilcox, T. R. Siegler, M. Perryman, A. Andrady, R. Narayan, and K. Lavender Law. 2015. "Plastic Waste Inputs from Land into the Ocean." *Science* 347 (6223): 768–71.

Janssen, N. A. H., G. Hoek, M. Simic-Lawson, P. Fischer, L. Van Bree, H. Ten Brink, M. Keuken, et al. 2011. "Black Carbon as an Additional Indicator of the Adverse Health Effects of Airborne Particles Compared with PM10 and PM2.5." *Environmental Health Perspectives* 119 (12): 1691–99.

Janssen, N. A. H., M. E. Gerlofs-Nijland, T. Lanki, R. O. Salonen, F. Cassee, G. Hoek, P. Fischer, B. Brunekreef, and M. Krzyzanowski. 2012. *Health Effects of Black Carbon.* Geneva: World Health Organization.

Lange, G.-M., Q. Wodon, and K. Carey, eds. 2018. *The Changing Wealth of Nations 2018: Building a Sustainable Future.* Washington, DC: World Bank.

Łapko, A., A. Panasiuk, R. Strulak-Wójcikiewicz, and M. Landowski. 2020. "The State of Air Pollution as a Factor Determining the Assessment of a City's Tourist Attractiveness—Based on the Opinions of Polish Respondents." *Sustainability* 12 (4): 1466.

Larsen, B. 2011. "Cost Assessment of Environmental Degradation in the Middle East and North Africa Region – Selected Issues." Working Paper No. 583, Economic Research Forum, Giza, Egypt.

Lei, L., S. Desai, and R. Vanneman. 2019. "The Impact of Transportation Infrastructure on Women's Employment in India." *Feminist Economics* 25 (4): 94–125.

Leonard, M., J. Pisani-Ferry, J. Shapiro, S. Tagliapietra, and G. B. Wolff. 2021. "The Geopolitics of the European Green Deal." Policy Contribution 04/2021, *Bruegel*, February 2.

Lozano-Gracia, N., and M. E. Soppelsa. 2019. "Pollution and City Competitiveness: A Descriptive Analysis." Policy Research Working Paper 8740, World Bank, Washington, DC.

Luijendijk, A., G. Hagenaars, R. Ranasinghe, F. Baart, G. Donchyts, and S. Aarninkhof. 2018. "The State of the World's Beaches." *Scientific Reports* 8 (1): 1–11.

Mahmoud, K. F., S. C. Alfaro, O. Favez, M. M. Abdel Wahab, and J. Sciare. 2008. "Origin of Black Carbon Concentration Peaks in Cairo (Egypt)." *Atmospheric Research* 89 (1–2): 161–69.

Manisalidis, I., E. Stavropoulou, A. Stavropoulos, and E. Bezirtzoglou. 2020. "Environmental and Health Impacts of Air Pollution: A Review." *Frontiers in Public Health* 8 (14): 1–13.

McGlade, C., and P. Ekins. 2015. "The Geographical Distribution of Fossil Fuels Unused when Limiting Global Warming to 2°C." *Nature* 517 (7533): 187–90.

MedPAN, UNEP/MAP, and RAC/SPA (Mediterranean Protected Areas Network; United Nations Environment Programme, Mediterranean Action Plan; and Regional Activity Centre for Specially Protected Areas). 2016. "The 2016 Status of Marine Protected Areas in the Mediterranean: Main Findings." Brochure, MedPAN, Marseille, France; and RAC/SPA, Tunis, Tunisia.

Naifar, I., F. Pereira, R. Zmemla, M. Bouaziz, B. Elleuch, and D. Garcia. 2018. "Spatial Distribution and Contamination Assessment of Heavy Metals in Marine Sediments of the Southern Coast of Sfax, Gabes Gulf, Tunisia." *Marine Pollution Bulletin* 131 (Pt A): 53–62.

Nonay, C. 2020. "Working Together in the Midst of an Active Conflict." *The Water Blog*, December 10. https://blogs.worldbank.org/water /working-together-midst-active-conflict.

OECD (Organisation for Economic Co-operation and Development). 2017. *Women's Economic Empowerment in Selected MENA Countries: The Impact of Legal Frameworks in Algeria, Egypt, Jordan, Libya, Morocco and Tunisia.* Paris: OECD Publishing.

OECD (Organisation for Economic Co-operation and Development). 2019. "Enhancing Connectivity through Infrastructure Investment." In *Middle East and North Africa Investment Policy Perspectives*, 169–88. Paris: OECD Publishing.

OECD (Organisation for Economic Co-operation and Development). 2020. "COVID-10 Crisis Response in MENA Countries." Policy update paper, OECD Policy Responses to Coronavirus (COVID-19), updated November 6, 2020, OECD, Paris.

Pope, C. A. III, and D. W. Dockery. 2006. "Health Effects of Fine Particulate Air Pollution: Lines that Connect." *Journal of the Air & Waste Management Association* 56 (6): 709–42.

Popp, D., F. Vona, G. Marin, and Z. Chen. 2020. "The Employment Impact of Green Fiscal Push: Evidence from the American Recovery Act." Working Paper 27321, National Bureau of Economic Research, Cambridge, MA.

Ribeiro-Broomhead, J., and N. Tangri. 2021. "Zero Waste and Economic Recovery: The Job Creation Potential of Zero Waste Solutions." Report, Global Alliance for Incinerator Alternatives (GAIA), Berkeley, CA.

Roudies, N. 2013. "Vision 2020 for Tourism in Morocco: Focus on Sustainability and Ecotourism." Presentation, Expert Group Meeting on Sustainable

Tourism: Ecotourism, Poverty Reduction and Environmental Protection, New York, October 29.

Saab, N., A. Badran, and A.-K. Sadik, eds. 2019. "Environmental Education for Sustainable Development in Arab Countries." Annual report of the Arab Forum for Environment and Development (AFED), Beirut, Lebanon.

Sagynbekov, K. 2018. "Childhood and Maternal Health in the Middle East and North Africa." Report, Milken Institute, Santa Monica, CA.

Scott, D. A. 1989. "Birds in Iran." In *Encyclopaedia Iranica*, edited by E. Yarshater, Vol. 4: fascicle 3, 265–72. New York: Routledge and Kegan Paul. https://iranicaonline.org/articles/birds-in-iran.

SPNL (Society for the Protection of Nature in Lebanon). 2018. "Fighting Land and Ecosystem Degradation in Lebanon." Online report, July 9, SPNL, Beirut. https://www.spnl.org/fighting-land-and-ecosystem-degradation-in-lebanon/.

Tolba, M. K., and N. W. Saab, eds. 2009. *Arab Environment Climate Change: Impact of Climate Change on Arab Countries*. Beirut, Lebanon: Arab Forum for Environment and Development (AFED).

Wada, Y., and M. F. P. Bierkens. 2014. "Sustainability of Global Water Use: Past Reconstruction and Future Projections." *Environmental Research Letters* 9 (10): 104003.

Walker, W. R. 2011. "Environmental Regulation and Labor Reallocation: Evidence from the Clean Air Act." *American Economic Review* 101 (3): 442–47.

WAVES (Wealth Accounting and the Valuation of Ecosystem Services). n.d. "Wealth Accounting and WAVES." Web page, WAVES Secretariat, World Bank, Washington, DC. https://www.wavespartnership.org/en/wealth-accounting-and-WAVES.

Welsby, D., J. Price, S. Pye, and P. Ekins. 2021. "Unextractable Fossil Fuels in a 1.5°C World." *Nature* 597 (7875): 230–34.

Wendling, Z. A., J. W. Emerson, A. de Sherbinin, D. C. Esty, et al. 2020. "Environmental Performance Index 2020." Report, Yale Center for Environmental Law & Policy, New Haven, CT.

World Bank. 2010. "World Bank Supports Irrigation Modernization for Egyptian Farms." Press release, December 14.

World Bank. 2014. *Turn Down the Heat: Confronting the New Climate Normal*. Washington, DC: World Bank.

World Bank. 2017a. *The Toll of War: The Economic and Social Consequences of the Conflict in Syria*. Washington, DC: World Bank.

World Bank. 2017b. "Tunisia – Integrated Landscapes Management in Lagging Regions Project." Project Appraisal Document No. PAD1520, World Bank, Washington, DC.

World Bank. 2018. *Beyond Scarcity: Water Security in the Middle East and North Africa*. Washington, DC: World Bank.

World Bank. 2019a. "New Program to Support Youth Employment in Morocco Focusing on Building Skills, Promoting Entrepreneurship and Expanding the Private Sector." Press release, May 10.

World Bank. 2019b. "Sand and Dust Storms in the Middle East and North Africa (MENA) Region: Sources, Costs and Solutions." Environmental study, Report No. 33036, World Bank, Washington, DC.

World Bank. 2019c. "Sustainable Land Management and Restoration in the Middle East and North Africa Region: Issues, Challenges, and Recommendations." Report, World Bank, Washington, DC.

World Bank. 2020a. *Convergence: Five Critical Steps Toward Integrating Lagging and Leading Areas in the Middle East and North Africa.* Washington, DC: World Bank.

World Bank. 2020b. *Trading Together: Reviving Middle East and North Africa Regional Integration in the Post-Covid Era.* MENA Economic Update, October 2020. Washington, DC: World Bank.

World Bank. 2021a. "Assistance Strategy for the West Bank and Gaza for the Period FY22–25." Report No. 156451-GZ, World Bank, Washington, DC.

World Bank. 2021b. "Environmental Sustainability and Urban Resilience DPF." Details page, Project No. P174000, World Bank, Washington, DC.

World Bank. 2021c. "Greater Cairo Air Pollution Management and Climate Change Project." Details page, Project No. P172548, World Bank, Washington, DC.

World Bank. 2021d. "MENA Crisis Tracker – 11/8/2021." Newsletter, Office of the Chief Economist, Middle East and North Africa Region, World Bank, Washington, DC.

World Bank. 2021e. "Morocco Green Generation Program-for-Results." Details page, Project No. P170419, World Bank, Washington, DC.

World Bank. 2021f. "Panama Pandemic Response and Growth Recovery Development Policy Operation." Details page, Project No. P174107, World Bank, Washington, DC.

World Bank and IMF (International Monetary Fund). 2021. "From COVID-19 Crisis Response to Resilient Recovery: Saving Lives and Livelihoods while Supporting Green, Resilient and Inclusive Development (GRID)." Document No. DC2021-0004 for the April 9, 2021, Meeting of the Development Committee (Joint Ministerial Committee of the Boards of Governors of the Bank and the Fund on the Transfer of Real Resources to Developing Countries), Washington, DC. https://www.devcommittee.org/sites/dc/files /download/Documents/2021-03/DC2021-0004%20Green%20Resilient%20 final.pdf.

Wu, X., R. C. Nethery, B. M. Sabath, D. Braun, and F. Dominici. 2020. "Exposure to Air Pollution and COVID-19 Mortality in the United States." *Science Advances* 6 (45): 1–6.

Yang, X., P. Jiang, and Y. Pan. 2020. "Does China's Carbon Emission Trading Policy Have an Employment Double Dividend and a Porter Effect?" *Energy Policy* 142: 111492.

Yu, D.-J., and J. Li. 2021. "Evaluating the Employment Effect of China's Carbon Emission Trading Policy: Based on the Perspective of Spatial Spillover." *Journal of Cleaner Production* 292: 126052.

Blue Skies for Healthy and Prosperous Cities

OVERVIEW

Ambient air pollution (AAP)—or outdoor air pollution—causes severe adverse health effects including premature deaths in the Middle East and North Africa. The region's urban air is among the world's most polluted, exposing its residents to air pollution levels 10 times higher than considered safe. No capital or major city in the region for which data are available meets guidelines on maximum safe particulate matter (PM) concentrations. Furthermore, these capitals and major cities are often considerably more polluted than their income levels would suggest.

This chapter documents the effects of AAP in terms of lives lost and reduced health as well as reduced labor supply and productivity. AAP is estimated to induce several hundred thousand deaths in the Middle East and North Africa every year. It also significantly affects morbidity, increasing the risk for several potentially deadly diseases and leading to large numbers of hospitalizations, sick days, and other adverse health impacts. As a result, AAP is one of the largest threats to the livelihoods of large portions of the region's population. It also disproportionately affects the poor, who often cannot afford protective measures and often perform manual labor outdoors, hence increasing their intake of air pollutants through heavy breathing. Gender inequality is yet another concern because it is usually women who take care of family members who become sick (predominantly the youngest and the oldest), further leading to lost productivity and income.

As such, the effects of elevated AAP are in stark contrast to those of a green, resilient, and inclusive development (GRID) framework[1] along many of its dimensions. They contribute to a degradation of livability in the Middle East and North Africa.

Given the severe negative impacts of AAP, it is alarming that not much is known about its exact sources in most of the region's cities. This paucity of information severely hampers policy makers' ability to formulate responses to effectively and efficiently lower concentrations of air pollutants. Proper and thorough research on the various sources of air pollution should include source apportionment studies, emissions inventories, and dispersion modeling—all of which should be supported and expanded to uncover the main contributors to air pollution in the region's cities. Despite some advances in this respect, overall knowledge remains sparse, and this calls for broad research efforts.

Despite the limited information about the sources of air pollution in the Middle East and North Africa, this chapter identifies several important priority policies that the region's decision makers should approach in a timely manner:

- First, raising public awareness about air pollution and its source—and how individual actions can contribute toward solving it—is important to both inform and mobilize the population, whose support is critical to the success of most policies.

- Second, the pervasiveness of fossil fuel subsidies and low environmental taxes are natural points of departure for reforms that simultaneously lower the tax burden in other areas, in the sense of "eco-social" tax reforms. Reforms of such subsidies and taxes have the dual advantage of decreasing air pollution and easing strained public budgets. They also have ramifications for the issue of marine plastics by raising the prices of feedstock and energy input for the production of plastics. It is vital to plan such subsidy reforms with a view toward the effects on especially low-income households, and to make proactive provisions, such as reducing income taxes, or offering compensation, which is especially important for low-income households.

- Third, strengthening public transportation systems is critical to induce a modal shift away from personal, motorized transportation and toward more sustainable transportation patterns.

- Fourth, more stringent industrial emissions standards should be implemented and the existing ones enforced properly.

- Fifth, strengthening solid waste management (SWM) in the region's economies is imperative to reduce uncontrolled burning of both municipal and agricultural waste and has cross-benefits by reducing the amount of plastics flowing into the region's seas (as chapter 4 will discuss).

Once the sources of air pollution are identified through the increased deployment of suitable studies, more-targeted policies can be implemented to tackle polluted air on a more sector-specific basis. The concluding section of this chapter discusses several such policies deemed suitable in the Middle East and North Africa context, separated along the lines of the sectors whose emissions are most likely to contribute to dirty skies: vehicles, industrial processes (including energy production), and uncontrolled burning of waste. Policy makers in the region should give these possibilities the very serious consideration they deserve, support research, and act to mitigate the AAP obstacle to a GRID path. To get blue skies and healthy, prosperous cities in the region tomorrow, the time to act on the threats posed by air pollution is today.

HOW POLLUTED ARE THE CITIES' SKIES?

Visible or not, dissatisfaction with bad air is real. Air quality is key to human well-being, but most Middle East and North Africa residents, especially in densely populated urban areas, live with elevated levels of air pollution that damage productivity, health, and quality of life. And they are increasingly unhappy with the situation: The 2018 Gallup World Poll showed that people in some of the region's economies (among those from which Gallup collected data) are particularly unhappy with air quality.[2] They include 6 in 10 *Kuwaitis* and *Lebanese*; 5 in 10 *Algerians* and *West Bank and Gaza residents*; 4 in 10 *Jordanians* and *Saudi Arabians*; and 3 in 10 *Iranians, Moroccans,* and *Tunisians*.

Air pollution is typically higher in urban agglomerations because of the greater concentration of economic and human activities. Generally, the adverse health implications of AAP are especially critical in population centers because they are hot spots for emissions, particularly from traffic and industry (see, for example, von Schneidemesser et al. 2019). Urbanization is quickly moving forward in the Middle East and North Africa, and hence the population's exposure to higher pollution levels exacerbates the need for effective abatement policies.

However, air pollution is also substantial in the region's rural areas and is reinforced by human sources and natural causes like sandstorms. The human sources include burning of agricultural waste (mostly in the region's low- and middle-income economies). Notably, however, these rural sources often influence air pollution in cities like those in the Nile Delta, where crop residue burnings substantially deteriorate the air quality in Cairo (box 3.14).

Ambient (outdoor) air pollution is the main concern for most Middle East and North Africa economies, with household air pollution being of major concern in only a few. Most of them have quasi-universal access to nonsolid fuels for domestic energy needs (UNEP 2017). In Djibouti, the Republic of Yemen, and some parts of rural Morocco, however, burning of solid fuels for cooking, heating, or both is still common, leading to indoor air pollution that constitutes a health threat in these countries in addition to outdoor air pollution. The focus of this report will be on AAP, highlighting the importance of identifying its sources and summarizing policy options for combating it in the Middle East and North Africa.

Global Comparisons

The Middle East and North Africa exhibits the world's second highest levels of air pollution. Only South Asia has higher concentrations of fine particulate matter of 2.5 microns or less in diameter ($PM_{2.5}$), which is considered to have the largest health effects globally. Nearly the entire population of the Middle East and North Africa is exposed to levels of air pollution deemed unsafe (World Bank data based on GBD 2018).

The region's air quality is affected by naturally caused dust storms linked to windblown geological dust and salt from the arid and semiarid landscapes (World Bank 2019b). Yet human activities that affect air quality range from industries to road-transport emissions and operation of power plants, among other sources, especially in urbanized areas (Saab and Habib 2020). The newly revised guidelines from the World Health Organization (WHO) stipulate that the mean annual exposure to $PM_{2.5}$ should not exceed 5 micrograms per cubic meter of air ($\mu g/m^3$), a measurement further discussed in box 3.1.[3] Yet the average resident of the Middle East and North Africa is exposed to air with concentrations more than 10 times this threshold (figure 3.1). In fact, no region meets these guidelines; even North America and Western Europe exceed them slightly.

Comparing the Region's Major Cities

Generally, air pollution concentrations are high throughout the Middle East and North Africa, but there is also substantial variance between the region's economies. Comparing the capital cities (or other major cities when data for capitals were not available), reveals that Cairo exhibits the region's highest concentrations of $PM_{2.5}$, with Baghdad and Riyadh following close behind (figure 3.2). At the other end of the spectrum, the air in Tehran, Amman, and Marrakech is substantially less polluted. However, PM concentrations in the sample's cleaner cities are still about six times the WHO recommended limit, carrying severe health risks for inhabitants and affecting tourism in areas with those conditions (Sajjad, Noreen, and Zaman 2014).

FIGURE 3.1

Ambient Air Pollution (PM$_{2.5}$) in Urban Areas, by World Region, 2016

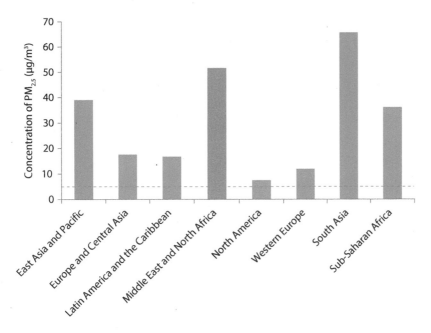

Source: Based on 2016 data from the Global Health Observatory database, World Health Organization (WHO), https://apps.who.int/gho/data/view.main.
Note: Particulate matter (PM) is made up of solid or liquid matter associated with Earth's atmosphere and suspended as atmospheric aerosol (the particulate/air mixture). PM$_{2.5}$ is a fine particle of 2.5 microns or less in diameter. The orange line denotes the 2021 WHO annual mean threshold of 5 μg/m³ (micrograms per cubic meter of air). Because PM$_{2.5}$ has health impacts even at very low concentrations, no threshold has been identified below which no damage to health is observed. The WHO guidelines aim to achieve the lowest PM concentrations possible (WHO 2021). "North America" includes Canada and the United States.

Cairo residents suffer from health impacts on their respiratory and cardiovascular systems due to AAP (Aboel Fetouh et al. 2013). Around 11 percent of premature mortalities there can be attributed to PM$_{2.5}$ pollution in the Greater Cairo region (Wheida et al. 2018). Elevated air pollution and related health impacts are also observed in other major cities and towns in the region for which data on PM$_{2.5}$ concentration are available (figure 3.3). None of these cities meets the WHO guidelines. Only Al-Jahra (Kuwait) and Salé (Morocco) exhibit air pollution levels somewhat close to the WHO threshold.

Comparisons with Cities and Countries of Comparable Income

Most of the region's cities exhibit disproportionately high air pollution concentrations relative to other cities with similar incomes. Figure 3.4

FIGURE 3.2

Ambient Air Pollution (PM$_{2.5}$) in Capital Cities of Selected Middle East and North Africa Countries, 2018

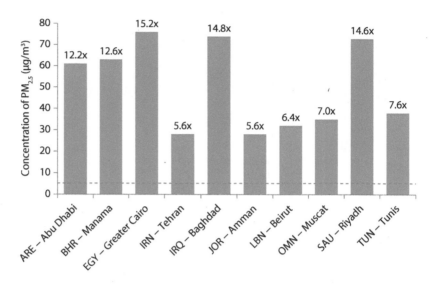

Source: Based on 2018 data from the Ambient Air Quality Database of the World Health Organization's (WHO) Global Health Observatory, https://www.who.int/data/gho/data/themes/air-pollution/who-air-quality-database.
Note: The orange line denotes the 2021 WHO annual mean threshold for PM$_{2.5}$ (particulate matter of 2.5 microns or less in diameter) of 5 µg/m³ (micrograms per cubic meter of air). Values above bars indicate multiples of that threshold found in each respective city. The selected countries are those for which ground monitoring data for capital cities were available. Countries are labeled using ISO alpha-3 codes, as listed on the Abbreviations page in the front matter of this report.

shows the PM$_{2.5}$ concentration of capitals and other major cities around the world plotted against the income level of the countries where they are located. The blue line indicates the average or expected degree of air pollution in a city given its level of income (measured as gross domestic product [GDP] per capita).

Within the Middle East and North Africa, some economies in the Maghreb and Mashreq subregions exhibit lower AAP levels than others with similar income levels, with the Arab Republic of Egypt and Iraq being notable exceptions. However, the Gulf Cooperation Council (GCC) countries (Bahrain, Oman, and Saudi Arabia) exhibit much higher pollution levels than their income levels would suggest.

In general, air quality has deteriorated and emission levels have risen in the Middle East and North Africa over the past two decades (see chapter 2, figure 2.6). However, there have also been some exceptions. The region's economies as well as areas within them have shown spatial heterogeneity, according to recent trend analysis of different proxy measures for PM air pollution derived from different satellite products.

FIGURE 3.3

Ambient Air Pollution (PM$_{2.5}$) in Non-Capital Major Cities of Selected Middle East and North Africa Countries, 2018

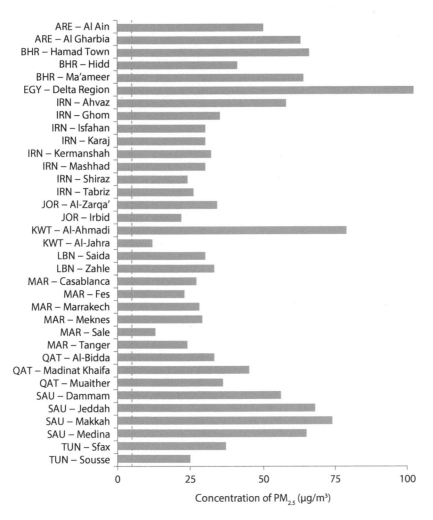

Concentration of PM$_{2.5}$ (µg/m³)

Source: Based on 2018 data from the Ambient Air Quality Database of the World Health Organization's (WHO) Global Health Observatory, https://www.who.int/data/gho/data/themes/air-pollution/who-air -quality-database.
Note: The orange line denotes 2021 WHO annual mean threshold for PM$_{2.5}$ (particulate matter of 2.5 microns or less in diameter) of 5 µg/m³ (micrograms per cubic meter of air). The selected countries are those for which ground monitoring data for capital cities were available. Countries are labeled using ISO alpha-3 codes, as listed on the Abbreviations page in the front matter of this report.

Air quality has generally improved in Egypt (despite some pockets where it appears to have worsened), while it has deteriorated in pockets of Iraq, Saudi Arabia, and the Syrian Arab Republic—in some periods more than others. Jordan's air quality, on the other hand, has appeared steady during the studied period, 2001–18 (Shaheen, Wu, and Aldabash 2020).

FIGURE 3.4

Global Comparison of Ambient Air Pollution (PM$_{2.5}$) in Capital or Other Major Cities in Relation to Countries' Income Levels

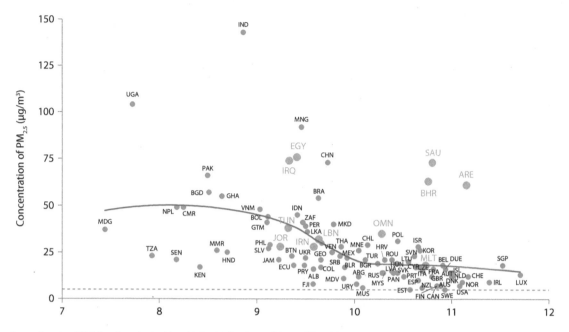

Source: Based on 2018 data from the Ambient Air Quality Database of the World Health Organization's (WHO) Global Health Observatory, https://www.who.int/data/gho/data/themes/air-pollution/who-air-quality-database; and the World Bank's 2021 World Development Indicators database, https://databank.worldbank.org/source/world-development-indicators.
Note: The orange line denotes the 2021 WHO annual mean threshold for PM$_{2.5}$ (particulate matter of 2.5 microns or less in diameter) of 5 µg/m³ (micrograms per cubic meter of air). The blue line is fitted using locally weighted regressions. The selected countries are those for which ground monitoring data were available. Countries are labeled using ISO alpha-3 codes. Those in the Middle East and North Africa are listed on the Abbreviations page in the front matter of this report.

Clearer Skies from the COVID-19 Lockdowns

During and in the wake of the COVID-19 pandemic, many of the region's economies administered strict lockdowns and other containment measures that brought significant emissions reductions and improved air quality (at least in terms of some air pollutant concentrations) in cities across the region. Directly after the onset of the COVID-19 in March 2020, observed nitrogen dioxide (NO$_2$) levels were substantially lower than in December of the previous year and have increased little since (World Bank 2020a).

In Cairo and Alexandria, Egypt, NO$_2$ emissions decreased by 15 percent and 33 percent, respectively. In addition, carbon dioxide (CO$_2$) and other greenhouse gas (GHG) emissions decreased by approximately 5 percent and 4 percent, respectively, in each governorate (Mostafa, Gamal, and Wafiq 2020). And environmental noise and pollution of beaches, surface water, and groundwater also decreased.

In Morocco, studies reported stark decreases of air pollutants in several cities, including Marrakech, Casablanca, and Salè (Khomsi et al. 2020; Otmani et al. 2020)—a finding confirmed on a national level (Sekmoudi et al. 2020).

In Tunisian cities, the imposed measures, including a general lockdown, caused marked decreases, lowering NO_2 and sulfur dioxide (SO_2) levels by around 50 percent in Sfax, 40 percent in Tunis, and 20 percent in Sousse. Furthermore, $PM_{2.5}$ concentrations dropped by around 20 percent, 7 percent, and 23 percent, respectively, in those cities while the concentrations of particles with a diameter of 10 microns or less (PM_{10}) did not significantly change (Chekir and Salem 2020). As those authors put it, "In other words, one side effect of this horrific virus is cleaner Tunisian air" (Chekir and Salem 2020, 7).

In the Islamic Republic of Iran, Tehran exhibited significant decreases in SO_2 (varying by a range of 5–28 percent in different parts of the city); NO_2 (by 1–33 percent); PM_{10} (by 1.4–30 percent); and carbon monoxide (CO) (by 5–41 percent), while ozone and $PM_{2.5}$ levels increased by a range of 0.5–103 percent and by 2–50 percent, respectively (Broomandi et al. 2020). Those authors attribute the latter two findings to unfavorable meteorological conditions and conclude that air quality improved overall, noting that these experiences "clearly showed that it is possible to have significant air pollution reduction in megacities by effective traffic control programs along with the promotions of green commuting and the technologies to expand remote working" (Broomandi et al. 2020, 1800).

As the measures to curtail COVID-19 are lifted, air pollution levels are expected to bounce back. The question is, to what extent? These significant reductions demonstrate that it is possible to rapidly reduce air pollution. Furthermore, the hope is that emissions will rebound to lower than prepandemic levels in the medium term as governments focus on greening their recoveries from COVID-19.

THE HEALTH AND ECONOMIC IMPACTS OF DIRTY SKIES

AAP has emerged as a major cause of illness and death globally (Boogard, Walker, and Cohen 2019) and in the Middle East and North Africa specifically. It has also been found to disproportionately affect lower-income groups (Hajat, Hsia, and O'Neill 2015; Miranda et al. 2011). Various air pollutants affect human health, but PM—especially the very small $PM_{2.5}$—has been found to be especially hazardous, having the largest health effects globally (GBD 2018). Box 3.1 compares $PM_{2.5}$ with PM_{10}.

BOX 3.1

Different Sources and Health Effects of Different PM Diameters

Particulate matter (PM) comes in different sizes, often stemming from different sources. The two most often used distinctions are PM_{10} (particles with a diameter equal to or less than 10 microns) and $PM_{2.5}$ (fine inhalable particles with a diameter equal to or less than 2.5 microns). Figure B3.1.1 shows comparisons to the size of a human hair and grains of sand.

These differences in size translate into different health effects. Generally, $PM_{2.5}$ is the outdoor air pollutant globally associated with the largest health effects (GBD 2018). Although PM_{10} also poses a threat to human health, its bigger size typically leads to it being deposited on the surfaces of the upper region of the lung, having less severe repercussions than $PM_{2.5}$, which is deposited in the lower parts of the lung. Short- or long-term exposure to PM_{10} or $PM_{2.5}$ can have substantial health effects, with the effects of long-term exposure generally being more severe.

FIGURE B3.1.1

Size Comparisons for PM_{10} and $PM_{2.5}$ Particles

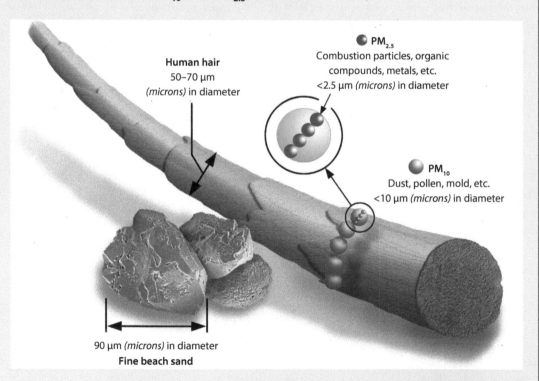

Human hair
50–70 μm
(microns) in diameter

$PM_{2.5}$
Combustion particles, organic
compounds, metals, etc.
<2.5 μm *(microns)* in diameter

PM_{10}
Dust, pollen, mold, etc.
<10 μm *(microns)* in diameter

90 μm *(microns)* in diameter
Fine beach sand

Source: "Particulate Matter (PM) Basics," Particulate Matter (PM) Pollution, US Environmental Protection Agency (EPA) website: https://www.epa.gov/pm-pollution/particulate-matter-pm-basics.

(continued)

BOX 3.1

Different Sources and Health Effects of Different PM Diameters (*Continued*)

World Health Organization (WHO) guidelines most commonly refer to annual and 24-hour mean concentrations of these pollutants. For $PM_{2.5}$, it stipulates a threshold of 5 micrograms per cubic meters of air ($\mu g/m^3$) and 15 $\mu g/m^3$, respectively, whereas PM_{10} concentrations should not exceed 15 $\mu g/m^3$ and 45 $\mu g/m^3$, respectively (WHO 2021).[a]

The differences in size also mean that the sources are usually different. PM_{10} particles often originate from airborne soil, dust, or sea salt, whereas $PM_{2.5}$ typically results from anthropogenic sources such as emissions from combustion vehicles, industrial processes, or cooking stoves. Thus, human activity influences $PM_{2.5}$ concentrations more than PM_{10}. Furthermore, because of their size, $PM_{2.5}$ particles stay in the air longer than the coarser PM_{10} particles.

a. Following updated evidence on the harmful effects of AAP, especially PM, the WHO in September 2021 issued revised guidelines regarding the maximum concentrations of both $PM_{2.5}$ and PM_{10}—the first revision since 2005 (WHO 2021). The previous guidelines for maximum exposure to annual mean and 24-hour mean concentrations of $PM_{2.5}$ had been 10 $\mu g/m^3$ and 25 $\mu g/m^3$, respectively, whereas for PM_{10}, they had been 20 $\mu g/m^3$ and 50 $\mu g/m^3$, respectively. See also "Ambient (Outdoor) Air Pollution," an online WHO fact sheet (updated September 22, 2021): https://www.who.int/news-room/fact-sheets/detail/ambient-(outdoor)-air-quality-and-health.

Risk Exposure in the Middle East and North Africa

In the Middle East and North Africa, AAP in 2019 constituted the fourth most significant risk factor for premature mortality. According to 2020 data and research from the Institute for Health Metrics and Evaluation (IHME) at the University of Washington,[4] the region's risk for premature mortality stemming from AAP[5] is surpassed only by metabolic risk—a broad category including risks from high cholesterol or kidney dysfunctions, among others; dietary risks; and health risks stemming from tobacco consumption in 2019 (figure 3.5). Notably, these three factors are within individuals' control, whereas air pollution is a risk that requires collective action such as, for example, government regulations and safeguards.[6]

Risk exposure to AAP has been increasing in the region, while risk exposure to other factors has been decreasing. Globally, AAP was among the group of factors whose risk exposure increased the most in the past decade (2010–19). The risk exposure to AAP has grown mainly for disadvantaged, lower-income groups (GBD 2020). The risk exposures from AAP and a high body mass index (which often affect health outcomes in similar ways) have both been increasing in recent decades (figure 3.6). On the other hand, great strides have been made to reduce several other risks to which the average Middle East and North Africa resident is exposed: household air pollution; unsafe water sanitation and handwashing services; and (to a lesser degree) smoking.

FIGURE 3.5

Share of Total Mortality Risk from Most Prevalent Causes in the Middle East and North Africa, 2019

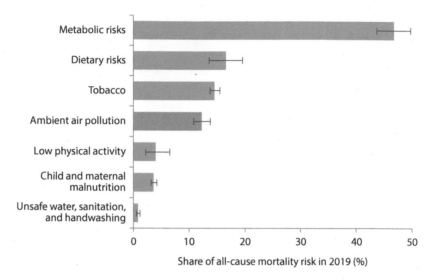

Share of all-cause mortality risk in 2019 (%)

Source: Data from the Global Burden of Disease (GBD) tool of the Institute for Health Metrics and Evaluation (IHME) Global Health Data Exchange (http://ghdx.healthdata.org/gbd-results-tool).
Note: Error bars indicate the confidence intervals.

FIGURE 3.6

Trends in Risk Exposure, by Cause, in the Middle East and North Africa, 1990–2019

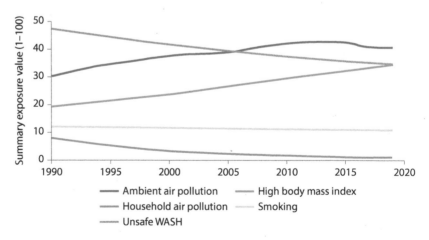

Source: Data from the Global Burden of Disease (GBD) tool of the Institute for Health Metrics and Evaluation (IHME) Global Health Data Exchange (http://ghdx.healthdata.org/gbd-results-tool).
Note: The summary exposure value (SEV) measures a population's exposure to a risk factor that considers the extent of exposure by risk level and the severity of that risk's contribution to disease burden. SEV takes the value of zero when no excess risk exists and the value one when the population is at the highest level of risk. The IHME reports SEV on a scale of 0–100 percent to emphasize that it is a risk-weighted prevalence ("Terms Defined," IHME website: http://www.healthdata.org/terms-defined/).
Unsafe WASH = unsafe water, sanitation, and handwashing.

The Burden of Disease: Mortality and Morbidity Effects of AAP

AAP can induce a range of diseases and disorders that may eventually lead to death (WHO 2019). In the Middle East and North Africa, the major cause of premature deaths associated with polluted air are ischemic heart diseases, followed by strokes and diabetes (figure 3.7).[7] Other AAP-related diseases that significantly contribute to deaths in the region are lower respiratory infections, chronic obstructive pulmonary disease, and different types of cancer. Furthermore, polluted air significantly contributes to the deaths of newborns and their mothers in the Middle East and North Africa.

Exposure to excessively polluted air has been linked to higher mortality rates in many Middle East and North Africa economies, including Egypt, Kuwait, Lebanon, and the United Arab Emirates (see, for example, Al-Hemoud et al. 2018; Amini et al. 2019). Death rates attributable to AAP are high throughout the region, but there are variations (figure 3.8). Egypt leads this trend, with more than 150 per 100,000 people estimated to prematurely die from AAP-induced causes in 2019. Following are two GCC countries, Oman and Qatar (slightly more than 125 such deaths per 100,000 inhabitants), closely trailed by Iraq (122 deaths per 100,000). The lowest death rates—still troublingly

FIGURE 3.7

AAP-Induced Causes of Death in the Middle East and North Africa, 2019

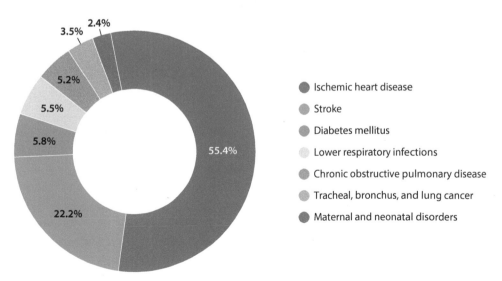

Source: Data from the Global Burden of Disease (GBD) tool of the Institute for Health Metrics and Evaluation (IHME) Global Health Data Exchange (http://ghdx.healthdata.org/gbd-results-tool).
Note: AAP = ambient air pollution.

FIGURE 3.8

Death Rates Attributable to AAP in the Middle East and North Africa, by Economy, 2019

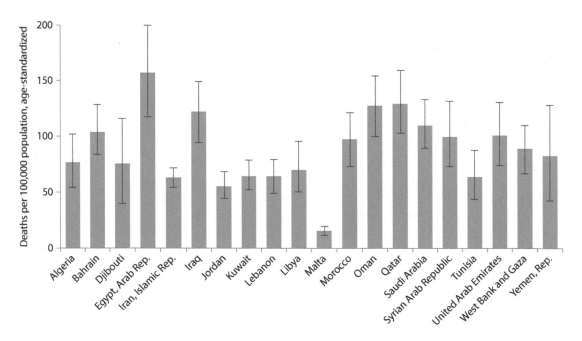

Source: Data from the Global Burden of Disease (GBD) tool of the Institute for Health Metrics and Evaluation (IHME) Global Health Data Exchange (http://ghdx.healthdata.org/gbd-results-tool).
Note: Age-standardization accounts for demographic (age structure) differences between populations to make disease-related data comparable across those populations. Error bars indicate confidence intervals. AAP = ambient air pollution.

high—are recorded in Jordan and Tunisia.[8] In addition—and in contrast to the other economies—Djibouti and the Republic of Yemen exhibit high death rates attributable to household air pollution (not shown).

The estimated numbers of deaths directly related to AAP are substantial. Estimates by the IHME link more than 270,000 deaths in the Middle East and North Africa in 2019 to this type of risk.

Air pollution also leads to increased hospitalizations in the region. In Egypt, increases of PM_{10} at concentrations of 10 ug/m³ were associated with increased hospitalization rates of 1–2 percent for chronic obstructive pulmonary disease (COPD) and bronchitis (Heger, Zens, and Meisner 2019). Similarly, in the Islamic Republic of Iran, days with higher air pollution were also associated with more hospitalizations in Tehran (Khalilzadeh et al. 2009), and days with high $PM_{2.5}$ concentrations led to increased admissions to hospital emergency departments (Heger and Sarraf 2018).

More specifically, air pollution is considered a major environmental risk factor associated with a range of diseases including asthma, lung cancer, and ventricular hypertrophy, among others (Ghorani-Azam, Riahi-Zanjani, and Balali-Mood 2016). Air pollution is also associated with acute strokes, and long-term increases in $PM_{2.5}$ concentrations are related to ischemic strokes in the Islamic Republic of Iran (Alimohammadi et al. 2016). Similar to the higher risk of morbidity for infants, elevated air pollution led to increased hospitalizations of children due to respiratory diseases in Isfahan, Islamic Republic of Iran (Mansourian et al. 2010).

Subregional Comparisons

Subregional comparisons show that the Mashreq economies are the worst affected, with around 188,000 fatalities due to excessively polluted air in 2019. Estimations by the IHME suggest that, in Egypt alone, more than 90,000 such deaths in 2019 were related to AAP—almost 250 per day (figure 3.9). Maghreb countries are also affected severely, with around

FIGURE 3.9

Total AAP-Related Deaths in the Middle East and North Africa, by Subregion and Economy, 2019

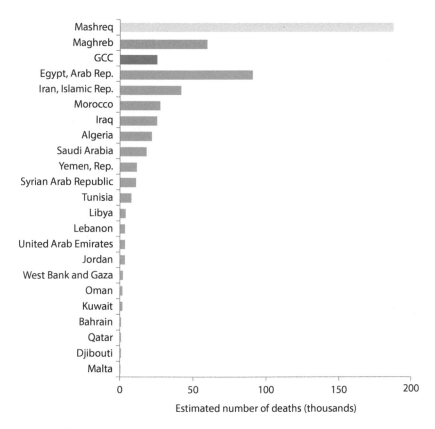

Source: Data from the Global Burden of Disease (GBD) tool of the Institute for Health Metrics and Evaluation (IHME) Global Health Data Exchange (http://ghdx.healthdata.org/gbd-results-tool).
Note: The subregions (three top bars) are as follows: The Gulf Cooperation Council (GCC) includes Bahrain, Kuwait, Oman, Qatar, Saudi Arabia, and the United Arab Emirates. The Maghreb subregion includes Algeria, Libya, Malta, Morocco, and Tunisia. The Mashreq subregion includes Djibouti, the Arab Republic of Egypt, the Islamic Republic of Iran, Iraq, Jordan, Lebanon, the Syrian Arab Republic, West Bank and Gaza, and the Republic of Yemen. AAP = ambient air pollution.

60,000 AAP-related deaths in 2019, Morocco accounting for almost half of them. And among the GCC countries, an estimated 25,000 persons died prematurely in 2019 because of excessive PM concentrations, most of them in Saudi Arabia.

Morbidity due to AAP has been rising throughout the Middle East and North Africa. When looking at years lived with disability (YLDs) as a measure of morbidity over time by subregion, figure 3.10 shows a continuous upward trend in morbidity in all subregions. Consistent with the fact that increases in GHG emissions and air pollution were highest in GCC countries, morbidity has been rising in this subregion the fastest, while in the Maghreb and the Mashreq, morbidity was similarly trending upward.

Comparisons by Age Group

Air pollution especially harms the old and the young (including infants). A comparison of deaths by age group in each economy (figure 3.11) shows that older cohorts are affected most severely, largely because of exposure over a longer period. However, a large number of deaths occur in newborns and children under the age of five, especially in countries like Algeria, Egypt, Iraq, and the Republic of Yemen.

FIGURE 3.10

Trends in Morbidity Due to AAP in the Middle East and North Africa, by Subregion, 1990–2019

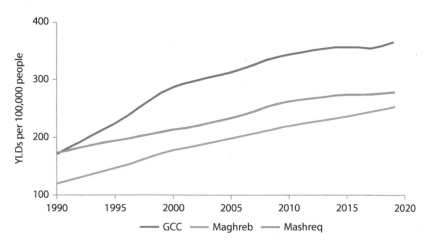

Source: Data from the Global Burden of Disease (GBD) tool of the Institute for Health Metrics and Evaluation (IHME) Global Health Data Exchange (http://ghdx.healthdata.org/gbd-results-tool).
Note: The Gulf Cooperation Council (GCC) includes Bahrain, Kuwait, Oman, Qatar, Saudi Arabia, and the United Arab Emirates. The Maghreb subregion includes Algeria, Libya, Malta, Morocco, and Tunisia. The Mashreq subregion includes Djibouti, the Arab Republic of Egypt, the Islamic Republic of Iran, Iraq, Jordan, Lebanon, the Syrian Arab Republic, West Bank and Gaza, and the Republic of Yemen.
AAP = ambient air pollution; YLDs = years lived with disability.

FIGURE 3.11

AAP-Related Deaths in the Middle East and North Africa, by Economy and Age Group, 2019

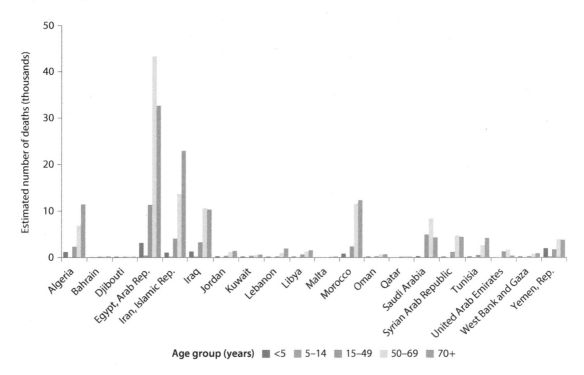

Source: Data from the Global Burden of Disease (GBD) tool of the Institute for Health Metrics and Evaluation (IHME) Global Health Data Exchange (http://ghdx.healthdata.org/gbd-results-tool).
Note: AAP = ambient air pollution

Children are more susceptible than adults to air pollution because they inhale more air per pound of body weight and have narrower airways than adults. Children's lungs are still developing, and children are also active outdoors; consequently, they have higher exposure to AAP. They also often breathe through their mouths, especially when exercising, which exposes the more-sensitive areas of their lungs.

Children's exposure to air pollutants (including PM, SO_2, NO_2, nitrogen oxide [NO_x], and CO) has been consistently linked to a higher incidence of various respiratory diseases, such as childhood asthma, and related symptoms such as wheezing (Gasana et al. 2012). Hence, living or attending schools near pollution sources such as industrial sites or roads with high traffic density potentially entails substantial health risks for children.

In addition, stunting and wasting in children is significantly linked to high-AAP events in the Middle East and North Africa. Children of mothers who were exposed to more days of particularly high AAP during their pregnancies are more likely to have worse health outcomes when growing up (Heft-Neal et al., forthcoming). In particular, the probability of children under the age of five being stunted (having a low height for a certain age) or wasted (having low weight for a certain height) has been investigated in Egypt, Jordan, and Morocco. The main results indicate that there are indeed negative effects of exposure to elevated AAP concentrations during pregnancy on subsequent children's health outcomes. Every day that a pregnant women is exposed to a particularly highly polluted day increases the risk of her child being stunted or wasted by 0.89 percent and 0.61 percent, respectively (Heft-Neal et al., forthcoming), as shown in figure 3.12.[9] Similarly, AAP increases the risk of pregnancy loss, with maternal exposure to elevated $PM_{2.5}$ levels accounting for 7.1 percent of the total annual pregnancy loss in South Asia (Xue et al. 2021).

As for risk exposure across age groups, the elderly and young children are especially susceptible to air pollution and vulnerable to various diseases resulting from exposure to it (Pope and Dockery 2006). Exposure to high PM levels, besides increasing mortality, has been found to

FIGURE 3.12

Estimated Effect of Increased AAP Exposure during Pregnancy on Probability of Stunting or Wasting of Children Born 2002–14, Selected Middle East and North Africa Countries

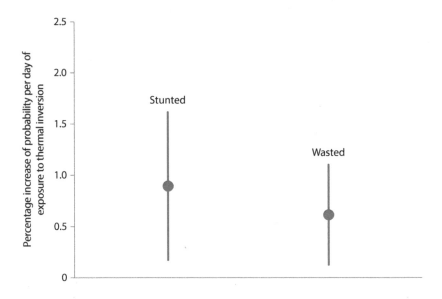

Source: Adapted from Heft-Neal et al., forthcoming.
Note: Data were gathered from the Arab Republic of Egypt, Jordan, and Morocco. Dots denote the estimated average effect of each day of exposure to thermal inversions during pregnancy on the probability of born children (under 5 years of age) being stunted (having a low height for a certain age) or wasted (having low weight for a certain height). Lines denote the 95 percent confidence interval surrounding these point estimates. Thermal inversions are natural phenomena in which atmospheric temperature deviates from the norm (that is, air temperature normally decreases with altitude). During inversion periods, the atmospheric temperature profile is inverted, meaning that air becomes warmer at higher altitudes. Inversion distorts the dispersion of air pollutants, trapping them close to the ground and hence temporarily increasing pollution concentrations. AAP = ambient air pollution.

increase substantially the number of hospitalizations (as a measure of morbidity). AAP is a major source of many potentially fatal diseases (as shown in figure 3.7) and increases the frequency of hospital stays or days staying home from work. Common pollutants such as CO, NO_2, and PM have been found to increase hospital admissions for cardiovascular and all cardiac diseases, cardiac failure, ischemic heart disease, and myocardial infarction.[10] Short-term exposure to higher $PM_{2.5}$ levels in particular increases the risk of hospital admissions for cardiovascular and respiratory diseases (Dominici et al. 2006).

Global Comparisons

The Middle East and North Africa has the world's highest rates of morbidity and mortality due to AAP. Measured in terms of disability days in a lifetime (figure 3.13, panel a), it surpasses even South Asia, which has the highest $PM_{2.5}$ concentration (as shown earlier in figure 3.1). Similarly, the Middle East and North Africa emerges as the region with the highest AAP-related mortality rates worldwide, trailed closely by South Asia (figure 3.13, panel b).

FIGURE 3.13

Global Morbidity and Mortality Rates Related to Air Pollution, by Region, 2019

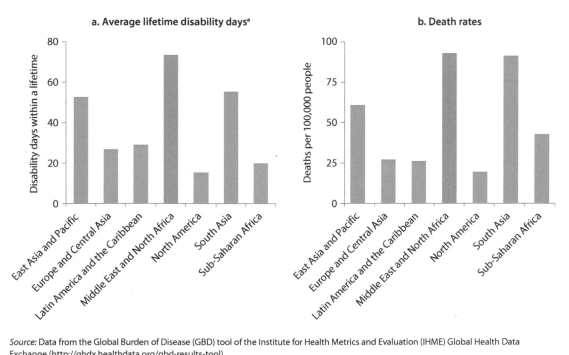

Source: Data from the Global Burden of Disease (GBD) tool of the Institute for Health Metrics and Evaluation (IHME) Global Health Data Exchange (http://ghdx.healthdata.org/gbd-results-tool).
Note: "North America" includes Canada and the United States.
a. The data are calculated from information on years lived with disability (YLD)—years lived with any short-term or long-term health loss—taking into account the different life expectancies at birth in different regions.

That both death rates and YLD—years lived with any short-term or long-term health loss—are higher in the Middle East and North Africa even though South Asia has higher $PM_{2.5}$ concentrations has much to do with the fact that many air pollution-induced diseases (those shown in figure 3.7) are affected by comorbidities that also raise the risk of suffering from such diseases. For example, the relation between obesity and the effects of air pollution has been heavily studied, and it has been found that AAP may aggravate obesity-induced diseases (see, for example, Sun et al. 2009; Yang et al. 2018). The Middle East and North Africa has much higher obesity rates than South Asia and substantially higher rates of physical inactivity, according to WHO Global Health Observatory data. These differences contribute to the higher AAP-related YLD and death rates in the Middle East and North Africa even though AAP is higher in South Asia.[11]

Impacts of Air Pollution on COVID-19 Risks

Increased AAP has been linked to a significantly higher probability of severe illness or death from COVID-19. Exposure to polluted air has decisive effects on the risk of suffering from a serious course of, or dying from, COVID-19 in the United States (Wu et al. 2020); several Asian cities (Gupta et al. 2020); the Netherlands (Cole, Ozgen, and Strobl 2020); and northern Italian municipalities (Coker et al. 2020). Higher AAP (a one-unit increase in $PM_{2.5}$ concentrations) is associated with an almost 10 percent increase in deaths from COVID-19 (Coker et al. 2020; Wu et al. 2020).

The *prevalence* of COVID-19 cases is also positively related to AAP, as is the *number of hospitalizations* (Cole, Ozgen, and Strobl 2020). In a panel of 120 Chinese cities, an increase of 10 µg/m³ in $PM_{2.5}$ and PM_{10} was associated with an increase of daily confirmed cases of 2.2 percent and 1.8 percent, respectively (Zhu et al. 2020).

Conversely, decreased AAP during lockdowns may have saved lives. Decreased air pollution, mainly because of reduced traffic mobility and hence reduced emissions from the transportation industry, may have had a mortality benefit considering deaths linked to polluted air.[12] It is estimated that during the lockdown period, almost 300 deaths were avoided because of air pollution reductions in Casablanca and Marrakech, with more than 60 percent of these avoidable deaths being related to cardiovascular diseases (Khomsi et al. 2020).

Competitive Disadvantage: Air Pollution's Economic Costs

The many health impacts of air pollution also have profound implications for the economy throughout the Middle East and North Africa. Among the many adverse impacts on human capital, air pollution has been linked to lower cognitive performance. Long-term exposure to air pollution has negative effects on individuals' performance on verbal and math tests. These effects become more pronounced as people age and are especially prevalent for males and those who are less educated. Considering the impact of decreased cognitive performance on the decision making of elderly people, the brain-aging process induced by air pollution may entail substantial health and economic costs (Zhang, Chen, and Zhang 2018).

Labor Supply and Productivity

Various channels can drive the response of labor supply to increases in air pollution. As noted earlier, exposure to polluted air is associated with adverse health outcomes that could reduce labor supply when workers are either too sick to work or must take care of sick relatives. Indeed, neighborhoods near pollution sources (for example, industrial sites such as refineries) show an increase in labor supply once pollution levels drop, because of a closure of pollutant sources (Hanna and Oliva 2015).

For people who are informally employed, these health impacts have severe consequences, because absence from work corresponds to a loss of income. In the Middle East and North Africa (excluding the GCC countries), almost two-thirds of employment is in the informal sector (Gatti et al. 2014), leading to loss of pay for affected people who cannot work, not only for themselves but for the families they support.

Beyond the lost workdays due to illness, paralleling air pollution's effect of decreasing cognitive abilities, it also decreases labor productivity (Graff Zivin and Neidell 2012; He, Liu, and Salvo 2019), which depresses economic output. This finding implies that even people who were not hospitalized or unable to show up for work were not able to unlock their potential in terms of contributing to value-adding activities.

Targeted policies to limit or reduce air pollution not only would improve air quality and decrease adverse health effects on the affected population but also could generate considerable environmental and economic side benefits such as increasing labor productivity (OECD 2016). Hence, environmental protection policies to increase air quality can also be viewed as investments in human capital (Graff Zivin and Neidell 2012).

GDP Losses

Globally, AAP causes staggering welfare losses. Economic losses stemming from premature mortality alone (that is, without considering the morbidity impacts) exceeded an estimated US$3.5 trillion in 2013 (World Bank and IHME 2016). In the Middle East and North Africa, AAP incurs large economic costs through both premature mortality and forgone labor output. Between 1990 and 2013, the region's annual welfare losses from AAP more than doubled, totaling more than US$141 billion in 2013 (World Bank and IHME 2016). These losses are equivalent to about 2 percent of regional GDP, with forgone labor output adding another 0.1 percent of GDP to these costs.

Focusing on the impacts of air pollution from the use of fossil fuels only (that is, excluding waste burning and other sources), $PM_{2.5}$ pollution was responsible for 1.8 billion days of work absence worldwide and millions of new cases of child asthma and preterm births, causing huge welfare costs on a global scale. The economic costs of these impacts of air pollution from fossil fuels approximated US$2.9 trillion in 2018—around 3.3 percent of global GDP (Myllyvirta 2020).

In the Middle East and North Africa, the annual costs related to air pollution are substantial. Estimated welfare losses in 2013 ranged from around 0.4 percent in Qatar to more than 3 percent in Egypt, Lebanon, and the Republic of Yemen (table 3.1). Note that although table 3.1

TABLE 3.1

Estimated Annual Costs of Air Pollution in Middle East and North Africa Economies, 2013

Share of GDP (%)

Economy	Total welfare loss	Total forgone labor output
Algeria	1.74	0.07
Bahrain	1.41	0.07
Egypt, Arab Rep.	3.58	0.27
Iran, Islamic Rep.	2.48	0.12
Iraq	2.67	0.22
Jordan	1.34	0.13
Kuwait	1.38	0.06
Lebanon	3.58	0.20
Libya	2.86	0.26
Morocco	1.55	0.18
Oman	1.73	0.07
Qatar	0.42	0.02
Saudi Arabia	2.05	0.12
Tunisia	2.83	0.18
United Arab Emirates	0.93	0.09
West Bank and Gaza	1.65	—
Yemen, Rep.	3.45	—

Source: World Bank and IHME 2016.
Note: The percentages include the costs of both ambient and household air pollution. Forgone labor output includes costs from premature deaths and increased morbidity (in the form of sick days). Data for Djibouti and the Syrian Arab Republic are unavailable. — = not available. GDP = gross domestic product.

includes costs stemming from both ambient and household air pollution, the costs due to household air pollution are negligible in the region except in the Republic of Yemen and potentially in Djibouti (where estimates of such costs are missing but exposure to household air pollution is still an issue). For the region in total, costs due to household air pollution amounted to US$5 billion compared with US$141 billion caused by AAP.

In absolute values, Egypt, the Islamic Republic of Iran, and Saudi Arabia bear the region's highest costs from air pollution, estimated to total more than US$30 billion in each of these countries in 2013. Total forgone labor output has been generally lower but amounted to more than 0.2 percent of GDP in Egypt, Iraq, Lebanon, and Libya. Recent reports on the health and economic costs of $PM_{2.5}$ pollution related to fossil fuel showed that the costs from premature deaths and increased morbidity (in the form of sick days) and hence forgone labor output are substantial in all Middle East and North Africa economies, ranging from 0.4 percent to 2.8 percent of GDP (Farrow, Miller, and Myllyvirta 2020).

In addition, polluted air detracts from cities' attractiveness to tourists and affects their competitiveness. City tourism is one of the most dynamic forms of tourism and has grown considerably in recent decades globally. The Middle East and North Africa is no exception. Degraded skies can damage tourists' image of a certain city and hence reduce their willingness to visit it (Łapko et al. 2020). Similarly, when cities grow and generate more output, they often face growth-related challenges such as congestion and the proper management of pollution in general and of air pollution more significantly. How cities approach these challenges directly influences their ability to grow further and to generate additional jobs and economic output. Hence, proper air pollution management is an important factor in a city's competitiveness (Lozano-Gracia and Soppelsa 2019).

The Cost-Benefit Equation

Inaction has costs—in lives lost, decreased health, and the related economic losses—that far outweigh the costs of action. The positive net benefits (benefits minus costs) of tackling polluted air have consistently been shown in a host of scientific studies around the world and for many different interventions. For example, instruments like the US Clean Air Act that proposed control requirements on major air pollutants have produced US$4 in benefits for every US$1 of cost (EPA 2001). The introduction of Canada-wide standards for PM_{10}, $PM_{2.5}$, and ozone resulted in net benefits of US$3.6 billion per year (Pandey and Nathwani 2003).

Myriad studies also quantify the net benefit of more-specific measures such as the introduction of tighter emission and fuel standards,[13] place-based restrictions on vehicles like low emission zones (LEZs) or congestion charges,[14] or the removal of fossil fuel subsidies,[15] among many others. Of course, the cost-effectiveness of individual measures is case-dependent and can vary, but *not* tackling the issues presented by air pollution usually costs more. Recent studies compared the costs of air pollution abatement policies with the resulting health benefits from cleaner air, revealing the cost-effectiveness of different measures for more than 60 countries (Wagner et al. 2020). Such exercises can be useful guides to select the most cost-effective policies and to help inform policy makers' decisions.

Inclusion of the health benefits of reduced air pollution in cost-benefit analyses of climate change policies flips the trade-off between climate damages and the costs of their mitigation. GHGs such as CO_2 and air pollutants often share the same sources. Considering the health benefits of air quality improvement measures in the assessment of carbon-emission strategies often reveals larger benefits of action relative to the costs of inaction (Scovronick et al. 2019). Air pollution policies can also contribute to meeting the Paris Agreement climate targets (Markandya et al. 2018). Factoring climate benefits into cost-benefit analyses of the air pollution abatement options would further increase their already pronounced benefits.

Given the toll of air pollution every year, action to tackle it is imperative. AAP causes so much human suffering every year and carries such severe adverse economic effects—amounting to a substantial portion of annual GDP—that these human and economic costs are undermining the livability of countries and cities while also reducing their economic efficiency. These impacts will only be exacerbated by the looming threat of climate change, to which many sources of air pollutants contribute in the form of climate pollution. Hence, a swift reaction by policy makers is needed to avoid these recurring costs of polluted air.

The next section shows that policy makers in the Middle East and North Africa can pursue many options to address AAP. However, the current paucity of information on the sources of air pollution hampers the identification of the most cost-effective options. Hence, one of the most important ways forward is to increase efforts to identify these sources. Despite this lack of knowledge, the section presents some priority recommendations that policy makers can pursue even without exact source information.

POLICY REVIEW: HOW TO GET CLEAR BLUE SKIES

This section discusses the steps toward "blueing" the region's skies, focusing on source identification and policy design. Although evidence of exceedingly high air pollution is available, information about the sources of this pollution is sparse in the Middle East and North Africa. Consequently, it is imperative to increase efforts to obtain source information to effectively design pollution-abatement measures, following the credo "you can only manage what you can measure." However, the region's policy makers can pursue certain measures in the meantime, given that some pollution sources are at work in every city and town. Hence, such measures can be considered priority recommendations to curb air pollution that governments can act on right away.

The next subsection concisely describes these priority measures and some of the most important ramifications for policy makers to consider. After that, a subsection on the sources of air pollution presents the very limited available evidence on sources in specific countries and cities, which highlights the need for more studies to identify air pollution sources. Finally, the "Detailed Sectoral Measures for Reducing Air Pollution" subsection presents a full range of options for decreasing AAP that are deemed most suitable in the Middle East and North Africa context. It also presents more detailed information and context about the policy options discussed in the earlier "Priority Recommendations" subsection, including regional and international examples.

Priority Recommendations: Which Actions Can Tackle Air Pollution without Knowing More about the Sources?

Given air pollution's massive toll every year in lost lives and reduced quality of life and productivity, it is imperative that governments in the Middle East and North Africa respond to this crisis swiftly. Although a broad range of policy measures can be taken (as further detailed in the "Sectoral Measures" subsection below), they are best taken once more specific knowledge is available about the *sources* of air pollution—information that most of the region's economies and cities currently lack. However, (local) governments can pursue some pathways in the absence of such detailed information, making those pathways priority recommendations for addressing air pollution. The following priority measures are pertinent to the entire region:

- Providing regular ground monitoring data on air pollution (and climate pollution) to raise awareness about the health consequences, accompanied by guidance for effective changes in individual behavior

- Removing distortive emissions subsidies and increasing the use of an environmental fiscal policy (while unburdening other factors, such as labor incomes, and providing compensation, especially to low-income households)

- Bringing about a modal shift from personal motorized transportation to personal nonmotorized transportation while also improving the quality of public transit

- Controlling emissions from industries

- Enhancing the treatment of solid waste, both agricultural and municipal.

This section identifies some of the most promising approaches, which later sections will discuss in more detail within a broader presentation of possible measures to fight air pollution.

No. 1: Close the Knowledge and Information Gaps

It is critical to frequently acquire and disseminate broad knowledge about the levels of air pollution in the region's cities. This requires the installation of ground monitoring stations across cities to continuously measure concentrations of criteria air pollutants.[16] The lack of appropriate equipment, especially for $PM_{2.5}$ monitoring stations (shown in map 3.1), impedes effective dissemination of air quality information to the public. Furthermore, gaps in the knowledge about air pollution levels obstruct the detailed investigation of air pollution's effects on morbidity and mortality.

City-level assessments of the human and economic costs of elevated pollution in the Middle East and North Africa remain sparse, and regional assessments such as those presented in the previous two sections often rely on estimates of PM concentrations derived from satellite images and advanced machine learning techniques. Although these methods have proven reliable, their accuracy can be strongly increased by in situ verification of concentration levels. Information on the local impacts of air pollution can be disseminated accurately, frequently, and in an easily accessible way. Local information facilitates the formulation of the most appropriate policy responses for the specific local pollution profile.

Public information plays a key role in building consensus for policy change and in mitigating pollution's impacts when air quality reaches critical levels. Information must be provided in an efficient, timely, and easily accessible manner—that is, the residents should view it as a service provided as part of the social contract between them and the (local) government. Regional examples are Abu Dhabi's traffic light system or Tehran's use of billboards and newspapers to keep residents informed.

Letting people know that air pollution in their cities regularly reaches critical limits, affecting them and their loved ones, will nurture demand for change and build support for people to adapt their behavior to contribute to bluer skies. Campaigns should include information about specific actions individuals can take to lower air pollution, such as changing their mode of transportation, conserving energy where possible, or treating their waste appropriately (that is, not burning it). Clear communications about the dangers of air pollution and ways to change it can help generate broad political and public support and prevent misinformation about new price increases or regulations (for example, reduced fuel subsidies, tighter vehicle emission standards, or traffic controls).

No. 2: Address Fossil Fuel Pricing

Fossil fuels must be priced at a level that provides incentives to reduce consumption and emissions from the transportation, energy, and industry sectors while concurrently contributing revenue to public budgets. Many Middle East and North Africa economies continue to be among the world's heaviest subsidizers of all sorts of fossil fuel products (Coady et al. 2019; IMF 2017)—whether those products are gasoline or diesel at the pump station (Ross, Hazlett, and Mahdavi 2017), input factors for energy production (Poudineh, Sen, and Fattouh 2018), or cheap feedstock for petrochemical and plastic producers. This provision of fossil fuel products at artificially low prices is a disincentive for consumers and reliant industries to increase their conservation efforts or to switch to more sustainable alternatives such as noncombustion transportation or renewable energies.

Apart from reductions in air pollution, efficiently pricing fossil fuels has clear benefits for climate pollution—decreasing carbon as well as other GHG emissions. Efficient pricing also has ramifications for the production and use of plastic products, which have distinct price advantages over greener alternatives thanks partly to the heavy subsidization of the petrochemical sector through cheap inputs (that is, fossil fuels).

Subsidy reform. Reducing or abolishing fossil fuel subsidies is an important way forward on the path to fight air and climate pollution, with co-benefits for other environmental issues such as marine-plastic pollution. Furthermore, the looming prospects of rising global oil prices from historical lows can substantially drive up the already high public costs of these subsidy programs. Slashing subsidies can provide important relief for these budgets but could also be accompanied by reductions on the burden for other factors such as income taxes, in the sense of an "eco-social" tax reform.

Subsidy reforms are often unpopular, but clear communications can soften the pushback. Governments need to analyze the impact of subsidy reform on households of different income levels. Often, fuel subsidies do not benefit the low-income households for which they are ostensibly intended. Subsidies in the Middle East and North Africa have been found to be regressive (Fattouh and El-Katiri 2013), disproportionately benefiting better-off households. Subsidized fuels are also frequently smuggled outside the country.[17]

Nonetheless, reforms to fossil fuel subsidies may have side effects on lower-income households that must be anticipated. For example, reduction or removal of fuel subsidies may increase food prices as well. Effective support programs for low-income households may therefore be necessary to soften the blow of subsidy reforms on them. Specific mitigation measures may include the following:

- Retaining the savings of removed subsidy programs as an important source of income to implement support programs for low-income households or using savings to reduce the tax burden on labor incomes.

- Introducing reforms gradually, with appropriate phase-in periods to help spread out their impacts

- Reducing the burdens of phasing out fuel subsidies and depoliticizing them by implementing an automatic pricing formula that links domestic fuel prices to global oil prices (Coady, Flamini, and Sears 2015)

- Executing an effective communications strategy—as part of a comprehensive reform plan with clear objectives, timelines, and sequencing measures—to mitigate the immediate impacts of such reforms and foster broad political and public support, with transparency being key (Coady, Flamini, and Sears 2015).

Carbon taxation or pricing. Taxing emissions or creating a market for them can also be an important source for government income and has proven effective in reducing emissions' negative externalities while resulting in cleaner air. As of 2021, none of the region's economies has planned or implemented either an emissions trading scheme or a carbon tax (World Bank 2021b), lagging every other region of the world in this respect. Pursuing such programs could contribute to more-balanced public budgets, and such programs are an important impetus for further environmental taxation.

Decarbonization of value chains. In addition, with much of the world striving to decarbonize their value chains in the coming decades, overt reliance on fossil fuels could leave the region's economies with stranded assets in the future. Large sectors of high-income countries are moving, or at least striving toward, decarbonization of not only their production processes but also their value chains. In contrast, 10 out of 20 Middle East and North Africa economies strongly rely on their oil and gas reserves for external trade of fossil fuels directly.[18] Some of them (most notably the GCC countries but also Iraq and the Islamic Republic of Iran) are expanding their downstream industries such as petrochemicals.

Although these sectors are presently contributing heavily to government revenues, the upcoming momentum toward cleaner supply chains will eventually affect these income sources. This leaves countries at risk to end up with stranded assets in these industries, burdening future development. Some countries, especially in the GCC, have recognized this potential threat to stable income and have intensified their efforts to initiate a shift toward a more pluralistic model of income generation. However, some GCC countries and other Middle East and North Africa economies, such as Algeria and Libya, have been reluctant to initiate fuel subsidy reforms, making their climb to appropriate pricing steeper than for others that have already made some progress in this respect, such as Egypt and the Islamic Republic of Iran.

No. 3: Shift Transportation System Priorities

Substantial efforts are necessary to induce a modal shift from individual transportation to expanded provision and use of public transportation. Personal motorized transportation and the emissions stemming from it contribute substantially to polluted air in virtually every city in the Middle East and North Africa because of high traffic intensities and often-outdated vehicle fleets (Abbass, Kumar, and El-Gendy 2018). Hence, measures to increase the share of people relying on public or preferably nonmotorized transit options are essential to curb transportation-related emissions. Compounding the negative externalities, in many of the region's cities, public transportation systems are weak at best, with large shares of the population relying on personal combustion vehicles, leading to traffic congestion (Waked and Afif 2012).

Strengthening public transportation systems and incentivizing residents to use them requires large investments in public vehicle fleets and infrastructure. Upgrading the existing public transportation fleet to more efficient and environmentally less harmful options such as electric or hybrid vehicles comes as a natural extension to promoting use of public transportation to lower the impacts on air quality. Furthermore, to induce a modal switch toward nonmotorized options such as bicycles or walking requires safety measures for nonmotorized road users and pedestrians.

Making public transportation affordable for low-income households is also one way to soften the blow that might be induced by subsidy reforms. Moreover, providing the poor with a means to commute cheaply increases their prospects to engage in jobs in more-productive sectors that might otherwise be unreachable for them. Efforts to expand the transportation system and make travel affordable require careful planning and substantial investments—further reasons for (local) governments to initiate these processes in a timely manner.

No. 4: Implement Emissions Standards and Enforce Compliance

In the Middle East and North Africa, industrial emissions, including from energy production, are a major source of air pollutants—especially given the generally lax regulations and oversight in the region. Most of the region's economies lack incentives for clean production (only 4 in 20 have them; see UNEP 2017), and most firms are lagging in obtaining internationally recognized, voluntary environmental standards (except in some GCC countries such as the United Arab Emirates).

International emissions standards (such as the mandatory use of fume scrubbers or exhaust filters) must be introduced into legal and regulatory frameworks, and the capacity of regulatory and inspection agencies must be enhanced to ensure compliance. Similar to reducing emissions from industries directly, mandating minimum energy-efficiency standards could curb the overuse of energy-intensive processes. Given that the region's energy mix is heavily based on fossil fuels, conserving energy in industrial processes is an important step to get clean air in the region.

Specifically, governments can start supporting companies in several ways:

- *Providing support for retrofitting facilities* to upgraded standards as well as implementing measures to support the switch to renewable energy sources, which in turn may incentivize transitions to cleaner technologies

- *Offering subsidies but also special credit lines* for small and medium-sized enterprises (SMEs)

- *Providing technical assistance* for companies to transition to less-polluting facilities and to identify the regulations with which they must comply

- *Providing certifications* to complying facilities, which they can also use to signal their willingness to adopt emission-reducing technology upgrades and to present themselves as "clean-producing" companies.

Finally, an important step toward clean value chains is the production and provision of energy derived from renewable energy sources instead of fossil fuels. To achieve this, it is vital to increase the production capacity of solar or wind plants as well as to invest in infrastructure for efficient storage and distribution of the produced energy. The extension of alternative energy sources is important to reduce emissions stemming not only from the industrial sector but also from the residential building sector.

No. 5: Strengthen Solid Waste Management Systems

Strengthening the management of municipal and agricultural waste is an urgent need. As the subsection on detailed measures will show, the (illegal) burning of waste substantially contributes to polluted air in the region. SWM is inadequate, especially in the Mashreq and Maghreb subregions, and a substantial proportion of waste is either disposed of in uncontrolled dump sites or burned.

Regarding agricultural waste, a combination of properly enforced penalties for waste burning in conjunction with incentives for the productive use of agricultural waste have worked well in some countries of the region, such as Egypt. In a similar vein, the introduction of technologies that turn otherwise discarded agricultural waste into fertilizer is a viable way to help curb air pollution stemming from this sector. Some of these technologies have the important co-benefit of substantially improving agricultural yields of plots where they are applied, which can again act as an incentive for farmers to use such technologies more intensively. Accompanying these efforts, information campaigns about the dangers of burning crop residues, and the benefits of alternatives to it, are important in informing farmers how to make such changes.

The poor conditions of SWM in large parts of the Middle East and North Africa are also a major contributor to the continuing flow of plastics into the regions' seas. Plastic pollution has major implications for the environment and human health—and for the economy. Chapter 4 investigates this issue in more detail. Strengthening the collection and treatment of municipal waste is identified as a priority recommendation to stem the marine-plastic tide. Hence, there are important cross-sector benefits in this area, and chapter 4 presents detailed information about the state of SWM in the region and possible ways forward.

One Must Measure What One Would Manage: The Sources of Air Pollution

The Middle East and North Africa lags other regions of the world in monitoring air pollutants—especially $PM_{2.5}$, the most significant pollutant for health. Regular monitoring of air quality is crucial for understanding the gravity of the issues in different locations and for identifying hot spots. This is especially the case for ground monitors that measure $PM_{2.5}$, the pollutant that is the most consistent and robust determinant of mortality in large range studies of long-term exposure to air pollutants (Cohen et al. 2017).

Regular monitoring enables understanding of pollution trends over time, which is crucial for evaluating the effectiveness of interventions. Regular, day-by-day monitoring of pollution levels also provides the information that empowers individuals to adopt behaviors that avert problems (such as avoiding exposure on particularly bad days, using air purifiers, wearing masks, and so forth). Exposure avoidance is particularly important for sensitive groups (such as asthmatics or people with other respiratory diseases such as COPD)—who, by reducing their exposure, can also reduce the negative health effects significantly.

Despite the importance of monitoring, many Middle East and North Africa economies lag other regions of the world in this regard because the region has a low number of ground monitors to measure PM compared with the monitors in East Asia, Europe, Latin America, North America, and South Asia (Shaddick et al. 2018), as shown in map 3.1.

Know the Sources

To prepare effective pollution abatement programs, the first and foremost question is, *"What are the sources?"* Especially in this section about policy measures for "blueing" the Middle East and North Africa's skies, the importance of knowing the sources of air pollution in a given country, city, or other area becomes apparent. Without such knowledge, a worst-case situation arises in which expensive pollution abatement measures are

MAP 3.1

Global Distribution of Ground Monitors for Measuring PM$_{10}$ and PM$_{2.5}$

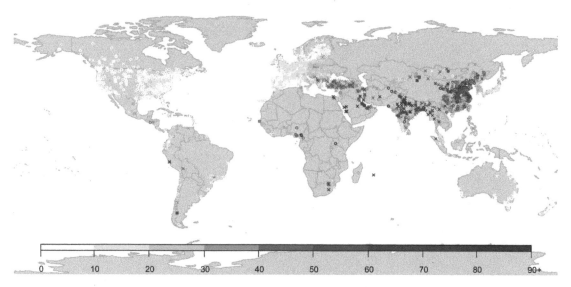

Source: Shaddick et al. 2018. License: Creative Commons Attribution CC BY-NC 4.0.

Note: Circles denote ground monitors for PM$_{2.5}$, crosses for PM$_{10}$. Colors denote the annual average PM$_{2.5}$ concentrations in μg/m³ (micrograms per cubic meter of air), converted from PM$_{10}$ where PM$_{2.5}$ data were not available. Data are from 2014 (46 percent); 2013 (36 percent); 2012 (9 percent); and 2006–11 and 2015 (9 percent total for these last two time periods). PM$_{2.5}$ is particulate matter (PM) of 2.5 microns or less in diameter, whereas PM$_{10}$ is 10 microns or less in diameter.

taken in areas that do not contribute much to the overall problem. More often than not, the findings from source apportionment analysis, emissions inventory, and dispersion modeling challenge the experts' prior assumptions about the relative contributions of various sources.[19] The uncertainty surrounding the sources makes it so important to carry out source analyses before embarking on costly pollution abatement pathways.

Evidence from global modeling of the Middle East and North Africa indicates that road vehicles, waste burning, and industries are the three most important sources of $PM_{2.5}$ in the region (figure 3.14). In North Africa, the fourth most important source is agriculture, particularly agricultural waste burning. In the Middle East, it is power plants. However, the anthropogenic sources of AAP vary widely from country to country and from city to city and also temporally. The proportions shown in figure 3.14 are based on a global and regional modeling effort, and they by no means replace detailed source apportionment campaigns based on collecting ground-monitored data in the main cities of the region's economies.

Knowing the local sources of air pollution is necessary to formulate effective policy responses. These sources vary both geographically and temporally. For example, the sources of $PM_{2.5}$ in the Greater Cairo area

FIGURE 3.14

Decomposition of National Sources of $PM_{2.5}$ Concentrations in the Middle East and North Africa, by Subregion, 2018

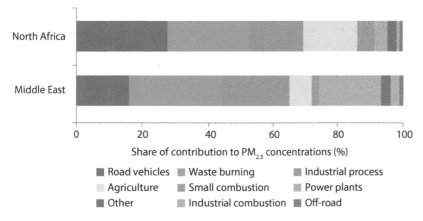

Source: Adapted from Wagner et al. 2020, using the GAINS model of the International Institute for Applied Systems Analysis (IIASA) for assessing emission reduction strategies.
Note: The bars show percentages, by national source, of population-weighted exposure to concentrations of $PM_{2.5}$, which is particulate matter (PM) 2.5 microns or less in diameter. Transboundary sources are not considered. North Africa includes four Maghreb countries: Algeria, Libya, Morocco, and Tunisia. Middle East includes the Mashreq and Gulf Cooperation Council (GCC) economies. The Mashreq subregion includes Djibouti, the Arab Republic of Egypt, the Islamic Republic of Iran, Iraq, Jordan, Lebanon, the Syrian Arab Republic, West Bank and Gaza, and the Republic of Yemen. The GCC includes Bahrain, Kuwait, Oman, Qatar, Saudi Arabia, and the United Arab Emirates.

vary depending on the time of year (figure 3.15). Open burning, mostly from crop residues from the rice harvest, is much more prevalent in the fall after the harvest season. Thus, the share of PM$_{2.5}$ concentration in fall 2010 was much higher than in summer 2010.

Knowing the sources has allowed the government to launch a program to tackle air pollution stemming from agricultural waste burning, which included establishment of a market for crop residues and awareness programs about the dangers of burning crop residues and alternative options. For a more detailed description of the "black cloud" phenomenon and policies to tackle it, see box 3.14.

Despite the importance of source identification analyses—such as source apportionment analyses and emissions inventory work—only a few cities in the Middle East and North Africa have performed such studies, and openly available information on the sources of pollutants is sparse for the region's cities and economies. In a global database containing source apportionment studies (Karagulian et al. 2015), only seven studies for four cities in the Middle East and North Africa were available, some dating back to the early 2000s.[20] In the past few years, not many more have been added to that list. By comparison, in the European Union (EU), North America, and some higher-income Asian countries, such source studies are frequently carried out, and time series information is often available. This lack of sampling, testing, and results reporting in the Middle East and North Africa leads to a lack of

FIGURE 3.15

Decomposition of PM$_{2.5}$ Sources in Greater Cairo, Summer and Fall 2010

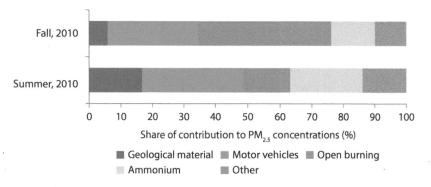

Source: Heger et al. 2019, based on World Bank 2013.
Note: Geological material includes sand and soil dust either occurring naturally (for example, sand from the desert) or stemming from human activity (for example, road dust, sand from construction activity). Particulate matter (PM) pollution from ammonium can occur through industrial but also agricultural activities (for example, fertilizer). PM$_{2.5}$ is PM that measures 2.5 microns or less in diameter.

understanding of the source profile of PM air pollution for most of the region's economies and cities.

The source information is likely the single most significant piece of missing evidence needed for policy making on air pollution management in the Middle East and North Africa. To help policy makers obtain accurate source attribution, further studies and more detailed data are required (Karagulian et al. 2015). To effectively and efficiently tackle the air pollution challenges that so many of the region's cities face, knowledge about where pollution comes from is of paramount importance and cannot be based on globally modeled information. It is strongly recommended that the region's economies and cities invest in carrying out source identification studies to understand their individual source profiles.

Models for Emulation

Regular measurement and reporting of air pollution sources has been adopted in the United States and Western Europe to support policy makers in their decision-making processes. Such measurement and reporting are highly encouraged for Middle East and North Africa economies. Source apportionment can be accomplished through three main approaches: emissions inventories, source-oriented models, and receptor-oriented models.

The US Environmental Protection Agency (EPA) provides a comprehensive, detailed estimate of air pollutant emissions—the National Emissions Inventory (NEI)—every three years. It collects data from state, local, and tribal air agencies and blends that information with data from other sources. The European Environment Agency (EEA) disseminates annual emissions inventories with air emissions accounts provided by each member state. This agency reports the amount of air pollutant and GHGs emissions broken down by 64 industries plus households. It also releases data on intensity ratios—emissions per unit of value added or production output.

For the EU and associated states, more detailed analyses of pollution sources are easily accessible and updated annually. For example, figure 3.16 shows sector-contribution shares of anthropogenic activities for various air pollutant emissions in the European Economic Area. For $PM_{2.5}$ emissions (sixth bar), the largest sector contribution came from the "commercial, institutional, and households" sector, mainly consisting of household emissions (95 percent) from sources such as wood combustion for heating. Industrial processes, when including energy production and its use in industries, contributed around 25 percent of the $PM_{2.5}$ pollution. Close behind as a source of $PM_{2.5}$ air pollution is the road transportation sector. In the Middle East and

FIGURE 3.16

Decomposition of Air Pollutant Sources, by Sector, in the EEA-33 Countries, 2017

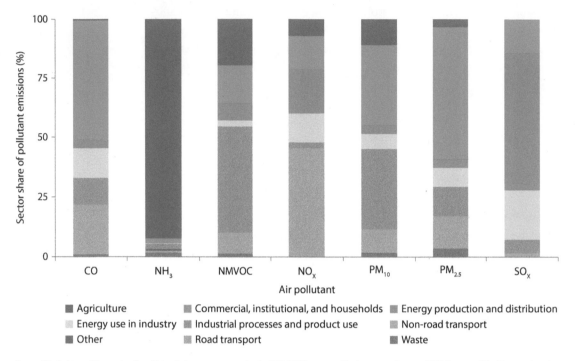

Source: "Emissions of the main air pollutants by sector group in the EEA-33," European Environment Agency (EEA) (last modified September 4, 2019): https://www.eea.europa.eu/data-and-maps/daviz/share-of-eea-33-emissions-5.
Note: The EEA-33 country group comprises the 27 EU member states and 6 other countries within the European Economic Area: Iceland, Liechtenstein, Norway, Switzerland, Turkey, and the United Kingdom. CO = carbon monoxide; NH_3 = ammonia; NMVOC = non-methane volatile organic compounds (such as benzene, ethanol, and so on); NO_x = nitrogen oxides; PM_{10} = particulate matter of 10 microns or less in diameter; $PM_{2.5}$ = particulate matter of 2.5 microns or less in diameter; SO_x = sulfur oxides.

North Africa, road transportation generally plays a bigger role than in the EU, given the age of much of the region's vehicle fleet and low use of public transportation in the region (Waked and Afif 2012).

Progress within the Region

There has been some movement in the right direction with initiatives for detailed emissions inventories. Some Middle East and North Africa countries are implementing national emissions inventory projects to improve knowledge about air pollution sources. For example, the United Arab Emirates recently released its first-ever emissions inventory, and Abu Dhabi has implemented an air quality index to communicate air pollution levels easily and in a timely manner to its inhabitants (box 3.2). Egypt is developing an air emissions inventory, has completed a point-source and area module, and is currently developing mobile-source and biogenic and geogenic modules.[21] It is highly recommended that other economies in the region follow suit and adopt a similar framework for emission reporting to tackle the issue of air pollution efficiently on a national scale.

BOX 3.2

Air Quality Monitoring in Abu Dhabi and the United Arab Emirates

Good air quality is recognized as essential for the health and well-being of the population in the Emirate of Abu Dhabi and as an intrinsic part of what makes it an attractive place to live and work. The emirate has established 20 stationary and 2 mobile stations for monitoring air quality. These stations monitor 17 pollutants and selected meteorological indicators on an hourly basis. There are 41 stations within the United Arab Emirates as a whole. Air quality standards are based on US Environmental Protection Agency (US EPA) guidelines, including particulate matter (PM), with hourly, daily, and annual standards.

In February 2020, the United Arab Emirates disseminated its first-ever National Air Emissions Inventory Report, which provides comprehensive data on sources of air pollutants such as SO_2, NO_X, CO, and PM (MOCCAE 2019). This report revealed that, among the anthropogenic sources of air pollution in the United Arab Emirates, industry is the major contributor to PM_{10} (76 percent) and $PM_{2.5}$ (68 percent), followed by road transportation (13 percent and 19 percent, respectively). CO stems mostly from road transportation (78 percent) and industry (17 percent). SO_2 is emitted mainly from operations in the oil and gas sector (79 percent) and NO_X from road transportation (38 percent) as well as from power generation and desalination processes and oil and gas operations (both 15 percent).

The detailed analysis carried by the Environment Agency – Abu Dhabi (EAD) in the course of its Air Emissions Inventory for 2018 enables an assessment of the varying nature of dominant air pollution sources within a city. Depending on the location of industries, energy production sites, or major transportation hubs such as ports, the dominant sources in certain localities of Abu Dhabi differ. For example, the Musafah district is primarily affected by industrial sources, while in Shahama on the outskirts of Abu Dhabi, agricultural sources of air pollution dominate. Vehicular emissions occur wherever there is traffic (that is, near roads). However, cities (or some parts of them) are more affected by both higher traffic and impaired air exchange in street canyons—that is, narrow roads with high surrounding buildings, making populations living there more exposed to these sources. The results in the EAD report showcase the high spatial variation of air pollution sources within a city and highlight the need for detailed studies to better understand these variations (Sanderson 2018).

Ways Forward

Enhancing the monitoring, reporting, and public dissemination of information about air pollution sources should be a key priority for Middle East and North Africa economies. A first step toward combating air pollution in an evidence-based manner is to install more air pollution tracking systems, together with supporting the responsible agencies (the respective environmental protection agencies). Furthermore, easier access and heightened public dissemination of such analyses will be important to raise awareness and spread knowledge about air pollution and its sources. This allows for more effective investigations by both academics and policy makers while also integrating the broader population into the discussion about air pollution and suitable counteractive measures as well as evaluation of their effectiveness. Importantly, this evidence base empowers the making of effective pollution abatement policy on a local scale while also increasing the acceptance of these policies.

Regular source identification and apportionment are key to tackling air pollution and to judging the effectiveness of any mitigation measures. Identifying the sources of air pollution is a precursor to formulating suitable, targeted measures. However, it does not stop there. If the goal is to evaluate measures taken in the past, determine their effectiveness, and revise them if necessary, then *regular* measurements of air pollutants and identification of their sources are required. This involves sophisticated analyses carried out as frequently as possible and accounting for seasonal variations. Regular apportionment analyses as well as emissions inventories would allow for timely, effective identification of policies that tackle AAP most efficiently and would allow for flexible reformulations of policy strategies over time if necessary.

Picking the most economical solutions across sectors requires source information but also detailed modeling exercises. Aside from knowing the sources, it is important to compare pollution abatement costs across the sectors and within sectors. Given limited fiscal space, it is also important to invest in the most cost-effective options. Marginal abatement cost curves—displaying the relative costs of particular interventions together with the benefits they bring—can be a useful tool for prioritizing solutions across a suite of options (much like the ones presented in the next chapter).

Earlier, the chapter presented policies that can be implemented even without a clear picture regarding the exact sources of air pollution. The next section presents the full range of promising policies most suitable for the Middle East and North Africa that could target different sources of air pollution. Which policies to pick depends crucially on information about the relative importance of the sources along with national and regional factors such as institutional and political economy considerations.

Detailed Sectoral Measures for Reducing Air Pollution

Because most of the region's economies and cities lack the source information described above, this section's objective is to describe (rather than prescribe) the "universe" of policies that countries and cities may take, depending on what the source information reveals. The reviewed policies include market-based incentives, regulations, technologies, and engineering solutions. This section also provides examples of effective interventions in the Middle East and North Africa and reviews international best practices. It then provides a descriptive review of a broad range of measures for reducing air pollution that the region's cities and economies may want to ponder.

Which combination of measures is optimal for which economy and city is a highly contextual question (depending, among other factors, on its institutions and political economy). The policy selection therefore requires in-depth analysis of local conditions and opportunities, as shown by recent detailed, city-level analyses in Egypt and the Islamic Republic of Iran.[22] Hence, the need for more detailed source identification studies and their dissemination should be reiterated. This would allow for the employment of specific measures in each location in an evidence-based manner to tackle the main sources of pollution and contribute to effective, efficient air pollution management in the Middle East and North Africa.

Before getting into specific options, it should be noted that for air quality management (AQM) to be effective, a governance and institutional framework is important to bring together the various stakeholders affected and establish clear guidelines and rules to follow. The specific options for tackling air pollution described below target the most common sources of air pollution. Some of these options can be differentiated by source and describe measures to (a) reduce vehicle emissions; (b) lower industrial emissions, including those from energy production and use; and (c) mitigate or eliminate air pollution resulting from uncontrolled burning of agricultural or solid waste. Furthermore, some crucial measures are independent of the exact source of air pollution, including (d) raising public awareness about air pollution; (e) increasing energy efficiency in the residential sector; and (f) greening cities and their infrastructure.

The policy options presented in the following subsections are an extensive but not exhaustive list deemed to be most appropriate within the Middle East and North Africa context, both to limit further degradation of the region's air and to provide authorities with tools that are efficiently enforceable. Several of the subsections—on vehicle emissions, industrial emissions, and greening cities—include tables that summarize the proposed policy options, describe their expected effectiveness and costs for implementation, and assess the time horizon in which they could be implemented based on regional and international experiences. However, the Middle East and North Africa's very heterogeneity implies that these aspects may vary across countries, necessitating additional economy-specific assessments.

Prerequisites for Implementing Effective Policies

Legal and regulatory frameworks. This process entails the enshrinement of a series of laws, acts, and regulations within a transparent *legal and regulatory framework*. This framework would define institutional roles and responsibilities of government officials and agencies for AQM and establish compliance, reporting, and enforcement mechanisms as well as legal instruments that adopt previously agreed-upon national air quality standards.

Governmental coordination. Given the various governmental stakeholders involved in an effective AQM framework, *horizontal and vertical coordination* among them becomes even more important. Establishing functional arrangements to coordinate these stakeholders across sectors (horizontal) and between different levels of government (vertical) with clear descriptions of functions, responsibilities, and management rights is necessary to develop plans to reduce air pollution and effectively execute those plans.

Nested planning processes. Air pollution is a transnational issue, with the actions taken by one country possibly affecting—or even likely to affect—others. It is also an intranational issue involving various administrative units within a country. Having modalities and planning procedures in place to allow for *nested planning* across national and subnational boundaries is crucial to avoid frictions across those boundaries. Ideally, coordination between countries would lead to unified approaches to tackle air pollution, and synergies could arise based on sound evidence developed through consultation with stakeholders, government agencies, and technical experts.

Committed executive structures. Many policy options described below involve some sort of regulation, ban, or technology requirements. To ensure compliance, institutions in charge of implementing these AQM policies in the form of *committed executive branches* must be clearly defined. This process includes the establishment of penalty schemes and prosecution procedures for noncompliant companies and residents as well as staff trained to properly administer such procedures. It also requires clear structures for financing those activities.

Public accountability and transparency. Related to the current lack of comprehensive source information and insufficient dissemination of information to the public, upholding *accountability and transparency* standards by disclosing information, tracking and evaluating progress, promoting public participation, and holding responsible institutions accountable is key to ensure broad public support. Furthermore, having reliable data on air quality and its sources of degradation is essential for assessments of health impacts and economic analyses, which in turn are key for evidence-based policy making and decision support.

Policies to Reduce Vehicle Emissions

Motorized vehicles are a main source of air pollution in the Middle East and North Africa. This has been and remains true because of the region's relatively outdated vehicle fleet, low-quality fuel, and the lack of comprehensive regulations to control exhaust emissions (Waked and Afif 2012). Low, often subsidized fuel prices, an underdeveloped public transportation system, and increasing demand for motor vehicles due to population and economic growth have exacerbated this trend even though the average emissions per kilometer driven have been decreasing (Abdallah et al. 2020). Hence, the region's policy makers should focus on reducing motor vehicle emissions to effectively combat the adverse health, environmental, and economic effects of these emissions.

This subsection reviews the main options for reducing emissions from urban mobility, including

- Properly pricing fuel by removing distortive fossil fuel subsidies

- Introducing environmental fiscal reforms

- Improving vehicle technology and strengthening maintenance and inspection

- Improving fuel quality and supporting vehicle fuel switching

- Strengthening public transportation

- Implementing place-based policies (such as LEZs).

A broad range of measures are available to combat air pollution caused by vehicles (see tables 3.2 and 3.3). These tables also include additional measures not discussed in the text, because their purpose is to present a broad set of available policy options. Successful examples within the

Middle East and North Africa are discussed in addition to evidence from international best practices.

Removing Fossil Fuel Subsidies and Pricing Fuel Appropriately

Current subsidies and pricing. Most economies in the Middle East and North Africa heavily subsidize fossil-based fuels. Reducing or removing such subsidies would result in substantial fiscal savings. Although the region's economies have started to reform these subsidies, the International Energy Agency (IEA) estimated that transport fuel subsidies amounted to 4 percent of 2019 GDP in Algeria, 2.85 percent in the Islamic Republic of Iran, 1.85 percent in Saudi Arabia, 1.59 percent in Egypt, 1.32 percent in Iraq, and 0.12–0.66 percent in Bahrain, Kuwait, and Qatar.[23]

The Middle East and North Africa has the world's lowest fuel prices. As figure 3.17 shows, the 2018 regional average (US$0.74 per liter for gasoline and US$0.69 for diesel) were less than half the 2018 average prices in Europe (US$1.48 and US$1.46 for gasoline and diesel, respectively). Between 2016 and 2018, fuel prices also increased the least in the Middle East and North Africa—by only US$0.07 and US$0.13 for gasoline and diesel, respectively.

In recent years, these prices have increased in most Middle East and North Africa economies but only slightly. In most of them, fuel prices increased between 2016 and 2018 (figures 3.18 and 3.19), owing largely

FIGURE 3.17

Average Diesel and Gasoline Pump Prices Per Liter, by Global Region, 2016 and 2018

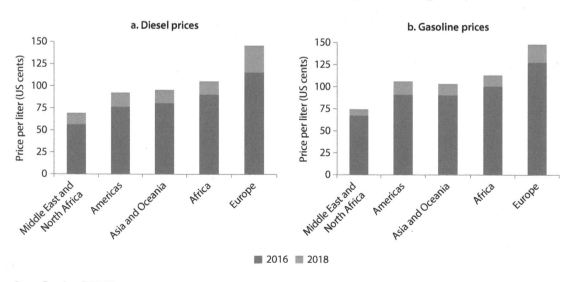

Source: Based on GIZ 2019.
Note: Regions differ from World Bank convention because GIZ definitions were used in the original research.

to fuel subsidy reform programs. The exceptions were the Islamic Republic of Iran, Iraq, Libya, and Tunisia (regarding diesel prices) as well as the Islamic Republic of Iran, Tunisia, and the Republic of Yemen (regarding gasoline prices).

Most of the region's economies had substantially lower fuel prices than the world average in 2018 (indicated by the dotted orange line in figures 3.18 and 3.19). Notably, GCC prices were among the lowest recorded despite those countries' much higher incomes than elsewhere in the region. These low fuel prices do not encourage fuel conservation via a price-based signal.

Benefits of fuel taxes and subsidy reforms. Cheap fuel leads to overuse and disincentivizes the use of cleaner, green-growth alternatives such as public transportation, nonmotorized transportation, and other options based on noncombustion engines (Asare and Reguant 2020). Higher fuel prices, on the other hand, drive up the relative costs of private motorized trips. Increasing fuel prices, including fuel taxes, could reduce consumption, lower the contribution of transportation to air pollution, and free up fiscal space for spending on important public goods—including support for those affected by the COVID-19 pandemic.

Eliminating fuel subsidies also helps to level the playing field for development of renewable energy sources. A combined framework

FIGURE 3.18

Average Diesel Pump Prices Per Liter in the Middle East and North Africa, by Economy, 2016 and 2018

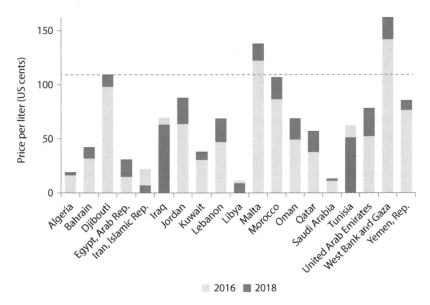

Source: Based on GIZ 2019.

Note: The orange line denotes the 2018 world average price. Data for the Syrian Arab Republic are unavailable.

FIGURE 3.19

Average Gasoline Pump Prices Per Liter in the Middle East and North Africa, by Economy, 2016 and 2018

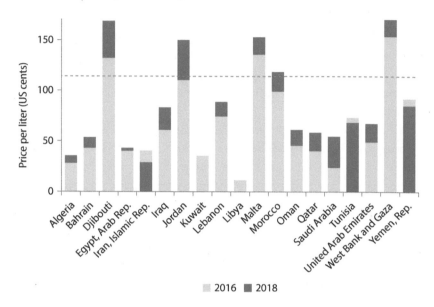

Source: Based on GIZ 2019.
Note: The orange line denotes the 2018 world average price. Data for the Syrian Arab Republic are unavailable.

that gradually phases out fossil fuel subsidies while incentivizing investments in renewables could induce a switch in the energy mix away from fossil-fuel-based energy and toward renewable sources, including the use of electric cars. It would also help foster a switch to public transportation.

Moreover, increasing fuel prices improves air quality. For example, fuel subsidy reforms in Egypt led to price increases that in turn reduced PM_{10} concentrations in the Greater Cairo area by almost 4 percent (Heger, Zens, and Meisner 2019).[24] Similarly, in the Islamic Republic of Iran, removing fuel subsidies led to improvements in Tehran's air quality (Kheiravar 2019). Both maximum and average daily concentrations of major pollutants were reduced significantly, and the policies were found to be generally effective. A scientific study also investigated the effects of an increase in energy prices, including gasoline, in the Islamic Republic of Iran. It showed that the average Iranian household would reduce its energy consumption by 2 percent, 16 percent, 29 percent, 38 percent, and 45 percent if energy prices were hiked by 10 percent, 50 percent, 100 percent, 150 percent, and 200 percent, respectively (Khatibi et al. 2020). These consumption reductions would lead to decreasing emissions of various pollutants like CO_2, NO_X, SO_X, and PM.

These results are promising examples of how the reduction or removal of distortive fuel price subsidies and resulting price increases can contribute to cleaner skies in the Middle East and North Africa. However, it is also important to note that the effectiveness of such measures could vary across the region and should be assessed by regular, objective evaluations to guide policy makers in an evidence-based manner.

In addition to reducing carbon emissions, phasing out fossil fuel subsidies frees up government funds for other purposes. Globally, removing fossil fuel subsidies and efficiently pricing fuel could have decreased carbon emissions by 21 percent and reduced the number of deaths related to fossil-fuel-induced air pollution by a staggering 55 percent in 2013, while raising government revenue and social welfare (Coady et al. 2017). An updated study found similar results for 2015, associating the hypothetical removal of fossil fuel subsidies with an estimated reduction of 28 percent in global carbon emissions and 46 percent fewer deaths related to fossil-fuel-related air pollution (Coady et al. 2019).

The COVID-19 pandemic has shown that constrained public budgets can be an impediment to meeting the needs of the public at times of unexpected shocks to the health system, the economic system, or both. Hence, restructuring the social transfer system away from unsustainable fossil fuel subsidies and toward sectors and programs that especially benefit the poor and vulnerable in society would be desirable. In a similar vein, a reduction in fuel subsidies could also be accompanied by income tax reductions that would help relieve the tax burden on workers, in the sense of an "eco-social" tax reform. However, given the high share of informality in some Middle East and North Africa economies, such a plan could benefit only a certain portion of the public.

Implementing subsidy reform. To cushion some of the side effects of fuel subsidy reforms, especially on the poor, such reforms should be accompanied by measures to support households during the transition away from subsidies. Although fuel and energy subsidy reform is desirable from the perspective of lowering the fiscal burden and enabling green growth, increased fuel prices can also cause hardship for beneficiaries, especially low-income households.

Fuel and energy subsidies protect domestic households from volatile prices for these goods in international markets (IMF 2017). However, the bulk of these subsidies are often captured primarily by the richest households in the Middle East and North Africa because of their higher consumption (Fattouh and El-Katiri 2015). Therefore, if the intent is to protect low-income households from volatile prices or to reduce the cost of their basket of elementary goods, this can be achieved much more cost-effectively than through fuel subsidies, which predominantly benefit wealthier households.

Targeted social protection measures such as cash transfers or food subsidies are good tools for cushioning the effect of higher fuel prices (Breisinger et al. 2019). An excerpt from Coady, Flamini, and Sears (2015) provides some lessons from international experience about starting to reform fuel subsidy programs (box 3.3). Learning from these international experiences—and also from more regional ones—is important for Middle East and North Africa economies to avoid frictions and make reforms as effective but also as acceptable as possible.

BOX 3.3

Reforming Fuel Subsidies: Lessons from International Experience

Many countries that have successfully reformed energy subsidies have incorporated specific measures into their subsidy reform strategies to overcome these barriers. While there is no single recipe for success, analysis of international reform experiences suggests the following six reform ingredients can help address reform barriers and increase the likelihood that reforms will achieve their objectives, thus helping to avoid policy reversals:

- *Develop a comprehensive reform plan.* The reform plan should have clear objectives. It should identify specific measures that will achieve these objectives and include a timeline for implementing and assessing these measures. A comprehensive plan will incorporate many of the measures discussed below. Designing and executing such a reform plan therefore needs careful advance planning.

- *Develop an effective communication strategy.* An extensive public communication campaign can help generate broad political and public support, help prevent

misinformation, and should be undertaken throughout the reform process. Transparency is a key component of a successful communication strategy.

- *Appropriately phase and sequence price increases.* Phasing in price increases and sequencing them differently across energy products may be desirable. The appropriate phasing and sequencing of price increases will depend on a range of factors, including the magnitude of the price increases required to eliminate subsidies, the economy's fiscal position, the political and social context in which reforms are being undertaken, and the time needed to develop an effective social safety net and communication strategy. However, gradual reform can create additional reform challenges, including lower budgetary savings in the short term, distortion in consumption patterns due to sequencing of reform by energy product, and the risk that opposition may build up over time.

- *Improve the efficiency of energy state-owned enterprises (SOEs).* Improving the efficiency

(continued)

BOX 3.3

Reforming Fuel Subsidies: Lessons from International Experience (*Continued*)

of SOEs (refineries, distribution companies, and so on) can reduce the fiscal burden of the energy sector. Energy producers often receive substantial budgetary resources—consisting of both current and capital transfers—to compensate for inefficiencies in production, distribution, and revenue collection. Improvements in efficiency can strengthen the financial position of these enterprises and reduce the need for such transfers. It will also help assure consumers that price increases are not simply being used to protect inefficient and poorly governed producers.

- *Implement targeted mitigating measures.* Well-targeted measures to mitigate the impact of energy price increases on the poor are critical for building public support for subsidy reforms. The degree to which compensation should be targeted is a strategic decision that involves trade-offs between fiscal savings, capacity to

target, and the need to achieve broad acceptance of the reform. Subsidy reform involving SOE restructuring may require temporary, sector-specific social measures to support employees and enterprises.

- *Depoliticize energy pricing.* Successful and durable reforms require a depoliticized mechanism for setting energy prices. Establishing an automatic pricing formula for fuel products that links domestic energy prices to international energy prices can help distance the government from the pricing of energy and make it clearer that domestic price changes reflect changes in international prices that are outside the government's control. Price-smoothing rules can help prevent large price increases. How much smoothing the government chooses to implement will depend on its preference between lower price volatility and higher fiscal volatility.

Source: Coady, Flamini, and Sears 2015.

Current subsidy reform efforts. Liberalization of fuel prices has progressed in certain Middle East and North Africa economies, while for others, there is still a steep climb ahead. Although the region is the world's heaviest subsidizer of fossil fuels, some of its countries have made big strides in slashing the subsidies. For example, Egypt, Jordan, and Morocco have all implemented or started to implement ambitious plans to reduce fuel subsidies and liberalize their fuel prices. On the other hand, fuel price liberalization in a host of countries such as Algeria, the Islamic Republic of Iran, Libya, and some of the GCC countries still faces headwinds and has not progressed much in this respect. In this latter category, the path toward less subsidization of fossil fuels is conceivably longer. Furthermore, for countries such as these that are dependent on oil exports, a move toward less-carbon-intensive value chains requires comprehensive commitments to stronger diversification of their economies so as not to be left with stranded assets.

Some countries have taken advantage of historically low global oil prices to initiate reforms to their fuel subsidy programs (box 3.4). The 2020 slump in oil prices, resulting partly from the COVID-19 crisis, opened up a window of opportunity for countries to reform their subsidy programs and minimize public backlash. Although many of the Middle East and North Africa economies have unfortunately not used this opportunity, others have implemented more or less far-reaching reforms.

BOX 3.4

Slashing Fuel Subsidies during Periods of Low Global Oil Prices Reduces Public Discontent

Low oil prices reduce the knock-on impacts on the population from removing fuel price subsidies (Benes et al. 2015), because low global prices are passed through to consumers (Coady et al. 2019).

To further reduce knock-on impacts and public discontent, compensation for fuel subsidy removals (especially to the poor) and clear communication are important (see also box 3.3). Plans to protect the poorest and most vulnerable parts of the population from the effects of increased fuel prices are necessary if the subsidy reforms are to have a chance at success. For example, in the Islamic Republic of Iran, the major subsidy reform plan in 2010 was supplemented by cash transfers, with almost 90 percent of the general population receiving about US$40 a month to reduce the economic pressure caused by the price increases (Fassihi 2010). Clear, well-targeted communication

campaigns that explain the reasoning behind the reforms are a key element in subsidy reduction programs.

In 2020, several Middle East and North Africa countries took advantage of low oil prices and began price reforms.

Tunisia introduced an "automatic monthly price adjustment mechanism" that liberalizes domestic prices of gasoline and diesel and lets them fluctuate with international prices (Cockayne and Calik 2020). Undertaking this measure when oil prices were low allowed the government to cut retail prices, avoiding the political pressure often accompanying such reforms.

Algeria raised the prices of gasoline and diesel because of fiscal pressure induced by the slump in global oil prices. The price of gasoline was increased by around 7.5 percent and diesel by more than 20 percent in June 2020 (*Dzair Daily* 2020), although domestic

(continued)

BOX 3.4

Slashing Fuel Subsidies during Periods of Low Global Oil Prices Reduces Public Discontent (*Continued*)

fuel prices continue to be subsidized (Ahmed 2020).

In **Libya**, the Ministry of Economy proposed reforms of fuel subsidies in March 2020, citing their distortive nature; furthermore, around 40 percent of subsidized fuel is smuggled outside the country.[a] At US$0.11 per liter, the price of gasoline is extremely cheap, even in the regional context. The proposal was to replace the fuel subsidies with direct cash subsidies, raise fuel prices, and potentially reduce fuel consumption by 30–40 percent (Zaptia 2020a). Similar proposals have been made by the Libyan policy reform think tank, the Economic Salon (Zaptia 2020b). This volume's drafting team could not verify the implementation of these reforms. However, conflicts between tribal leaders and the Tripoli government in 2020 led to blockages of oil facilities, leading to soaring prices and increasing black-market sales in some parts of the country. Even though the official price of US$0.11 per liter of fuel was maintained, prices charged by gas stations varied depending on the location and were much higher during these periods (Westcott 2020).

Lebanon may have to stop subsidizing fuel, food, and medicine as its central bank reserves diminish (Reuters 2020). The country's subsidy system is inefficient, with only 20 percent of subsidies reaching Lebanese citizens in need and the rest going to better-off residents or leaving the country through smuggling to Syria. However, removing fuel subsidies would be perceived as hitting the country's small industrial sector (Rayess 2020).

Syrian officials announced in May 2020 that automobile fuel subsidies will be reduced to tackle the country's deepening economic crises. Cars with engine displacements of 2,000 cubic centimeters or more as well as owners of more than one car were excluded from receiving subsidies (AP 2020). In October 2020, the government increased prices by more than 100 percent for diesel and more than 50 percent for gasoline amid a fuel shortage that also led to tighter rationing of both subsidized and unsubsidized fuel (*Atalayar* 2020).

In contrast, Saudi Aramco, the **Saudi Arabian** major oil company, slashed gasoline prices by almost 50 percent in May 2020 (Khalid 2020) after the kingdom introduced sharp increases on almost all fossil fuels in the course of a reform program in 2018 and smaller price increases in 2019.

a. On August 6, 2020, the US Treasury Department imposed sanctions on three Libyans and a Malta-based company, accusing them of acting as a network to smuggle drugs and Libyan fuel into Malta and thereby contributing to instability in Libya (Psaledakis 2020).

COVID-19 brought about some backtracking on subsidies. In response to the COVID-19 pandemic, some countries *increased* fossil fuel subsidies, contrary to green growth objectives. For instance, in Egypt, the government increased subsidies for the aviation and industry sectors (Moisio et al. 2020). Similarly, Saudi Arabia and the United Arab Emirates (temporarily) increased electricity subsidies for households and the industrial sector. These subsidies amounted to US$240 million in Saudi Arabia and around US$5.7 million in the United Arab Emirates (including water subsidies). In both countries, as in the rest of the Middle East and North Africa, electricity is still mainly produced from fossil fuels; hence, these measures can be considered fossil fuel subsidies (Moisio et al. 2020). In addition, Malta announced a €900 million (7 percent of GDP) package to stimulate economic recovery that includes a reduction in fuel prices. The hope is that these were just temporary measures and that the countries will return to green growth trajectories.

Introducing Environmental Fiscal Reforms

Environmental fiscal reforms in the form of green taxes in the Middle East and North Africa can be a suitable tool for decreasing pressure on fiscal budgets and promoting more sustainable patterns of consumption. Currently, environmental taxes play a negligible role in the tax-policy mix of most of the region's economies. Although such taxes in 2018 accounted for around 10 percent of total tax revenue in several EU member states (such as Bulgaria, Estonia, Greece, and Latvia), in the Middle East and North Africa countries for which data are available, these taxes contributed around 5 percent and 6 percent of tax revenues in Tunisia and Egypt, respectively, and 1.7 percent in Morocco. In the GCC countries, pricing CO_2 emissions could generate significant tax revenues (Saidi 2019) that they could use to combat the adverse fiscal effects of the COVID-19 crisis. Box 3.5 discusses international experiences with implementation of environmental fiscal reforms.

Several Middle East and North Africa countries introduced environmental taxes in recent years. In its budget law for the 2019–20 fiscal year, the Islamic Republic of Iran introduced a tax on goods that cause environmental damage in their manufacture or use. A 2 percent tax will be imposed on domestically produced paint, coating, primer, tire, tubes, plastic and electronic toys, plastic containers, polyethylene terephthalate, and melamine, with a 3 percent tax on imports of the same products. Furthermore, locally produced light bulbs, computers, linoleum, cellophane, and nylon will be taxed at 3 percent and imports thereof at 4 percent (Eghtesad Online 2019).

Algeria introduced an environmental (pollution) tax on all motor insurance policies of around US$12.50 for passenger vehicles and

BOX 3.5

Environmental Fiscal Reform: International Experiences

Countries around the world have recently implemented environmentally related tax measures (Enache 2020). Among those implementing or increasing environmental taxes in 2020, Ireland and Sweden raised their carbon taxes, and Iceland introduced a tax on fluorinated greenhouse gas (GHG) emissions. Ireland has been active in other areas related to environmental taxation—increasing electricity taxes on businesses, replacing the previously applicable surcharge on diesel vehicle registration of 1 percent with a nitrogen dioxide (NO_2)-based surcharge, and prolonging registration tax relief for hybrid vehicles to the end of 2020.

Among other recent international examples, Lithuania implemented a new pollution tax on cars in July 2020 to encourage purchases of new and less-polluting vehicles. The Netherlands and Poland extended their special tax treatment of hybrid cars. Denmark, Italy, and Sweden approved new or increased taxes on plastic packaging and plastic bags, with Italy delaying the implementation to January 2021. The United States granted tax credits for bio and alternative fuels—hence not raising taxes but trying to shift consumer demand. Latvia abolished exemptions regarding coal, coke, and brown coal used for electricity generation from its Natural Resources Tax and increased other tax rates, such as for the mining of sand.

In the Middle East and North Africa, environmental taxes could be more effective than income taxes in raising tax revenues, given the high degree of informal employment

in many of the region's economies. Almost two-thirds of the region's workers (excluding those in the Gulf Cooperation Council [GCC] countries) are employed in the informal sector (Gatti et al. 2014), implying that the tax base of income taxes is actually rather small in these countries.

Environmental taxes have the advantage of not being linked to the employment status of the polluting persons or entities. Hence, they can effectively increase tax revenues while simultaneously putting a price on externalities that otherwise would have to be borne by people who haven't caused them. The experiences of several countries are further discussed below.

Sweden

Sweden is an international role model when it comes to greening the tax system. With the introduction of its carbon tax—pricing carbon at SKr 250 (around US$30)[a] per ton in 1990–91 (the first such tax in the world)—Sweden simultaneously reduced other taxes such as its income tax. Through its *grön skatteväxling* (which can loosely be translated as "green tax-switch") the marginal tax rate on top incomes was reduced from 80 percent to 50 percent and the corporate tax rate from 57 percent to 30 percent (Jonsson, Ydstedt, and Asen 2020). The carbon tax has steadily increased since its inception, and at the beginning of 2021, a ton of carbon dioxide (CO_2) was priced at SKr 1,200 (around US$144).[b]

In 2019, the Swedish government again proposed a similar scheme in which

(continued)

BOX 3.5

Environmental Fiscal Reform: International Experiences (*Continued*)

environmental taxes were increased while taxes on jobs and entrepreneurship were simultaneously reduced. Although changes to the carbon tax were not part of this new package, it included a new excise tax on waste incineration and a tax on plastic bags. An additional surtax was levied on high-income earners, and tax deductions were introduced for employers of new labor-market entrants between the ages of 15 and 18 (Deloitte 2019).

Belgium

Other countries have used similar approaches to shift the tax burden from income to environmentally harmful activities. Belgium introduced a tax plan that reduced employers' social security contributions gradually, from 32.4 percent in 2016 to 25 percent in 2018. It also increased the tax-free amount and tax-deductible business expenses to alleviate the tax burden on labor. To finance these tax cuts, the value added tax on electricity and the excise duty on diesel were increased. Further measures gradually increased excise duties on alcoholic drinks and tobacco and taxed capital more strictly (EC 2017).

Colombia

The Colombian government introduced tax incentives in 2014 to spur investment in renewable energy sources. These include a special deduction equal to 50 percent for investments in renewable energy or energy efficiency, an accelerated depreciation rule,

and exemptions from value added taxes on goods and services and customs duty on imported goods.

In 2020, these incentives were expanded to include broader activities related to plant expansion or process improvements. Importantly, these changes include a special deduction scheme, incentivizing the switch to greener alternatives by applying the deduction to nonelectrical uses of renewable energy sources (such as switching away from fossil fuels in the transportation sector or replacing them with biofuels in the industrial sector). When the investment is made to comply with environmental standards, companies do not benefit from the deduction; hence, they are encouraged to go beyond the environmental regulations required by law (EY 2020; OECD 2020).

Mexico

The government of Mexico reformed its tax system and introduced several measures targeting a greener fiscal system. The General Law on Climate Change in April 2012 decreased fossil fuel subsidies and introduced additional taxes. Since August 2014, the Hydrocarbons Revenue Law has imposed a special tax regime on companies that engage in oil exploration and production. It includes exploration-phase fees and royalties, with monthly taxes to be paid by companies to municipal and state governments.

With the country's 2016 tax reform, taxpayers investing in energy efficiency and renewable energy equipment are eligible to

(continued)

BOX 3.5

Environmental Fiscal Reform: International Experiences (*Continued*)

obtain a 100 percent up-front deduction for the costs of these investments. However, companies are eligible to deduct costs of investments only if they comply with regulations. This can be seen as a way to incentivize companies to go beyond minimum requirements and enhance their energy mix.

The tax reform of 2016 also abolished fuel subsidies and increased taxes on transport fuel to reflect the external cost of fuel more closely. Furthermore, a new carbon tax was introduced that covers a larger share of emissions with a price, as was the case before

but now including coal albeit at a rather low level (Arlinghaus and van Dender 2017).

The tax reform also included a special mining right royalty of 7.5 percent of net profits derived from the sale or transfer of extraction activities. An additional 0.5 percent tax is levied on gross income from the sale of gold, silver, and platinum (Deloitte 2016). These efforts have helped the government raise substantial additional revenues, slowly shifting its income toward more environmentally friendly sources and shifting it away from reliance on oil exports (Arlinghaus and van Dender 2017).

a. Based on the SKr/US$ exchange rate as of January 13, 2021.
b. For more information about Sweden's carbon tax, see the dedicated page "Sweden's carbon tax" by the Government Offices of Sweden: https://www.government.se/government-policy/swedens-carbon-tax/.

US$25 for other vehicles and rolling machines per year in its 2020 draft budget law. Of these tax proceeds, 70 percent were supposed to benefit the state, while the remaining 30 percent were supposed to be allocated to the Solidarity and Guarantee Fund for Local Communities (*Atlas Magazine* 2020a). Unfortunately, because of pressure by insurers—whose collection of motor premiums was reduced by 10 percent in the first half of 2020 (also strongly affected by lockdown measures)—the 2021 draft finance bill abolished the tax (*Atlas Magazine* 2020b).

Putting taxes in place that target fossil fuels and the emissions caused by them can be successful in lowering air pollution. For example, the Swedish carbon tax, which primarily targets the consumption of gasoline and motor diesel, has reduced CO_2 emissions from the transportation sector (Andersson 2019). The carbon tax was implemented together with a value added tax on fuel. Emissions subsequently declined by almost 11 percent, of which 6 percentage points were attributable to the carbon tax. Consumers appeared to respond more strongly to changes in the tax rate than to equivalent gasoline price changes stemming from market price fluctuations.

In another case, Germany's 1 percent fuel-price hike decreased the number of kilometers driven by 0.4 percent, thus contributing to lower vehicle emissions and higher air quality (Frondel and Vance 2018).

Improving Vehicle Technology, Emissions Control, and Inspection

Increasing vehicle fuel efficiency requirements and setting vehicle emission limits can decrease air pollution. Continuous renewal of the vehicle fleet with vehicles adhering to emission standards, such as those set by the EU (see box 3.6 on the Euro emissions standards and their adoption in the Middle East and North Africa), have led to a decrease in average emissions per driven kilometer in Middle East and North Africa cities (Abdallah et al. 2020). The effects of the penetration of Euro standards in global markets—following from the fact that vehicles exported from Japan, the EU, and the US must fulfill these standards regardless of the domestic regulation in the destination country—are referred to as the "Euro effect" in some scientific studies. $PM_{2.5}$ emissions were more than

---

BOX 3.6

Vehicle Technology and Related Regulations in the Middle East and North Africa

Mandating increased efficiency for newly produced cars and trucks and setting standards for their maximum emissions are ways to reduce their environmental impact and decrease the air pollution they cause. The European Union (EU) puts forward directives for various types of vehicles—the so-called Euro standards—that regulate maximum emissions and introduce a set of approval tests that these vehicles must pass before being eligible for sale.

New-Vehicle Technology Regulations

Starting in 1992 with the Euro 1 standard, these regulations became mandatory in all EU member states, and new vehicles had to be equipped with a closed-loop, three-way catalyst. Subsequent adaptions of the regulations took place in the following decades and required reductions of pollutant emissions from newly introduced cars, with Euro 6 being in effect since September 2014. Exported vehicles from the EU also must meet these standards.

Some countries outside the EU followed suit, adopting the same or similar regulations. Not all economies in the Middle East and North Africa have adopted measures to restrict imported cars to fulfill the newest Euro emissions standards, but in

(continued)

BOX 3.6

Vehicle Technology and Related Regulations in the Middle East and North Africa (*Continued*)

most of them, certain minimum standards that correspond to earlier stages of the regulation must be met.

Many of the region's economies have lower standards or do not set any vehicle emissions at all. Morocco is one of the few that has adopted the Euro 4 standard for light-duty vehicles, while Algeria and Jordan have adopted the Euro 3 standard.[a] In other countries, either lower standards for light-duty vehicles apply (for example, in the Islamic Republic of Iran and Saudi Arabia) or no policy has been set (for example, in Lebanon, Oman, and the Republic of Yemen).

Some countries have introduced vehicle emissions standards for heavy-duty vehicles, such as the Islamic Republic of Iran (Euro IV), but the fleet of such vehicles in Tehran largely meets Euro III or lower standards (Heger and Sarraf 2018). However, more and more cities in the region are mandating minimum limits for buses in their public transportation fleets. For example, the Euro III emission standard is obligatory for buses in Ajman, Cairo, Jeddah, Madinah, Makkah, Riyadh, and Tunis, while Abu Dhabi and Dubai mandate at least Euro IV standards for buses. Amman in Jordan is prescribing the Euro V standard and Bahrain the Euro VI standard.

Fuel Content Regulations

For emissions technology to work efficiently, the fuel used must fulfill certain criteria, such as specific sulfur limits. As the EPA notes, high sulfur content in gasolines significantly impairs the effectiveness of emission control systems because of increased sulfur compounds in the exhaust.[b] Taking diesel particulate filters (DPFs) as an example, higher sulfur content in the fuel leads to higher production of sulfate particulate in the combustion process, in turn leading to quicker saturation of the DPFs. This increases the back pressure in the exhaust system, which impairs the life cycles of DPFs, implying that they have to be regenerated or replaced more often. Studies have shown that when fuel has a sulfur content of 50 parts per million (ppm) instead of 10 ppm, the regeneration temperature needed rises by around 50 degrees Celsius, implying that in the case of vehicles in city traffic, reliable regeneration can no longer be ensured. Furthermore, the increased back pressure increases the combustion system's fuel consumption, undermining efforts to lower emissions caused by DPFs. For the emission-decreasing technology prescribed in emissions standards to work efficiently, it is necessary to regulate the quality of the fuel used to run the vehicles, as the subsection on "Upgrading Fuel Quality" discusses.

a. Euro standards for light-duty vehicles are denoted with Arabic numerals (for example, Euro 6), while standards for heavy-duty vehicles are denoted with Roman numerals (for example, Euro VI).
b. "Gasoline Sulfur," Gasoline Standards, EPA: https://www.epa.gov/gasoline-standards/gasoline-sulfur.

60 percent lower in 2010 than they would have been in a world where no such regulations had been put in place (Crippa et al. 2016).

New-vehicle technology. Most economies in the Middle East and North Africa already have restrictions on the age of imported used cars. Mandating higher technology levels for new cars would decrease emissions from the region's car fleet and could incentivize car-producing markets to adopt these regulations or stricter ones.

Several international empirical studies have examined the effect of exhaust emissions standards on air pollution. For example, the introduction of tougher emissions standards (from pre-Euro regulations to Euro 4) significantly reduced exhaust emissions from gasoline-powered cars in London's vehicle fleet. For diesel cars, the evidence was less clear; in general, emissions from diesel-fueled cars were found to be substantially higher than those from gasoline-fueled cars (Rhys-Tyler, Legassick, and Bell 2011). Air pollution regulations in India requiring catalytic converters for new vehicles have had considerable positive effects on air quality. PM concentrations were reduced by 19 percent five years after the implementation of the regulation. Its adoption is also associated with a decline in infant mortality—though only modestly, with the statistical relationship not being significant (Greenstone and Hanna 2014).

Emission controls. Tackling air pollution from vehicles by raising emissions standards carries economic benefits for vehicle owners. Improved vehicle emissions standards and the accompanying technology upgrades increase cars' fuel efficiency and are highly viable from an economic perspective. Cost-benefit assessments for several already implemented or planned emissions standard improvements in various countries have shown that the annual benefit-cost ratio of such measures ranges from a minimum of 1.4:1 up to 16:1 when considering fuel costs and decreased health costs stemming from reduced air pollution.

Relating this to costs per vehicle, the additional costs for equipping vehicles with technologies to fulfill higher emissions standards are amortized within an estimated time period of 1.5 to 5 years, depending on the emissions standard considered (Kodjak 2015). This also applies to measures such as the introduction of diesel particulate filters that can lead to lower fuel consumption, carrying economic benefits for vehicle owners.

Inspections. Regular checks are necessary to ensure vehicles' compliance with emissions standards. Highlighting the importance of such checks, 25 percent of cars in Tehran failed to comply with maximum emission levels or had other deficiencies in their mandatory yearly checks (*Tehran Times* 2019). The introduction of mandatory vehicle inspections has led to a 44 percent decrease in black carbon emissions and mitigated air pollution significantly, according to Tehrani officials.

From a regulatory viewpoint, mandating regular checks (using fines or banning use of the vehicle as the consequence of noncompliance) is a simple but effective policy instrument, but it has to be enforced effectively. For example, practices such as removing catalytic converters or adjusting engine parameters are still widespread in Lebanon, highlighting the need for stricter regulations on regular inspections (Abdallah et al. 2020). In addition, achieving comprehensive, regular vehicle checks involves setting up vehicle inspection garages and training staff to perform the checks.

To incentivize regular inspections, monetary as well as nonmonetary benefits can be employed. Monetary incentives include tax breaks for upgrades and repairs, while nonmonetary benefits can be granted in the form of allowing only inspected vehicles to enter LEZs, as is the case in Tehran. To lower the perceived costs to vehicle owners, the city of Tehran provides mobile vehicle inspection units to reduce waiting times and ease access to official inspections (*Financial Tribune* 2018).

Introducing mandatory vehicle inspections has the dual advantage of job creation in this sector while alleviating air pollution from outdated vehicles. From an economic point of view, the jobs created from setting up the necessary garages would add employment opportunities for the population and an accompanying rise in living standard. Furthermore, the government tax revenues stemming from these additional jobs and services could be considerable. The additional costs of inspections for vehicle owners could be subsidized to reduce the refusal rates, perhaps by lowering the tax rates applicable to these inspections and any repair services needed to fix car deficiencies.

Upgrading Fuel Quality and Supporting Vehicle Fuel Switching

Fuel quality standards. The quality and content of fuel for motorized vehicles play significant roles in the emission of hazardous pollutants. Upgrading fuel quality by establishing maximum limits for lead and sulfur contents in fuels—two ingredients that emit as particularly harmful pollutants in the combustion process—should therefore be the primary concern of regulators worldwide. With the Euro directives on fuel quality, the EU has set standards for automotive fuels and lately also marine ones. The Middle East and North Africa generally lags other regions in setting comprehensive and appropriate fuel quality standards (box 3.7).

The adoption of Euro IV standards in Tehran's transportation sector, with an emphasis on taxis, has considerably reduced air pollutants and raised air quality (Ghadiri, Rashidi, and Broomandi 2017). The effectiveness of such policies has also been shown in a host of studies of Chinese provinces, which between 2013 and 2017 adopted regulations that mimic those in Europe, Japan, and the United States. The adoption of these standards significantly improved air quality, particularly in reductions of PM and ozone (Li, Lu, and Wang 2020; Wei 2019; Yang, Jiang, and Pan 2020).

BOX 3.7

Fuel Quality Standards in the Middle East and North Africa

Similar to its emissions standards, the European Union (EU) mandates fuel quality standards, particularly regarding the maximum limits of certain contents such as sulfur and lead. The first set of standards were voluntary (but observed by all fuel suppliers in Europe), and the first mandatory regulations were introduced in 1998, which, among other indicators, limited sulfur to 50 parts per million (ppm). The currently applicable Euro 5 standard limits the maximum sulfur content to 10 ppm (gasoline and diesel) and was implemented in 2009. It also includes limits for the minimum octane number and cetane number for gasoline and diesel, respectively (ICCT and DieselNet, n.d.).

Similar to the EU, Canada and the United States prescribe a maximum sulfur content of 15 ppm in both gasoline and diesel. More recently, China lowered the maximum limit for fuels to 10 ppm in 2017, and India has joined the list, prescribing 10 ppm for diesel and gasoline in April 2020. Decreasing the sulfur limit from 50 ppm to 10 ppm can reduce fuel consumption by 2–3 percent in most vehicles (Marsh, Hill, and Sully 2000).

Conversely, almost all economies in the Middle East and North Africa are lagging in adopting fuel-sulfur limits comparable to the most recent (Euro 5) or to other international standards. Furthermore, even though almost all the region's economies have adopted prohibitions on the use of leaded gasoline, the only country worldwide that has *not* outlawed its use is Algeria (UNEP 2021).

Most Middle East and North Africa economies do not comply with international benchmarks for sulfur limits in diesel fuel (UNEP 2020). Only Malta (which, as an EU member, must comply with the 10 ppm limit), Morocco, and the United Arab Emirates have explicitly set sulfur limits for diesel that match international best practices (figure B3.7.1). Qatar did not set the maximum sulfur limit of diesel to 10 ppm directly, but Qatar Petroleum, the country's sole diesel supplier, announced that, as of the end of September 2020, its refinery would provide only ultra-low-sulfur diesel meeting Euro 5 specifications (*Hydrocarbon Processing* 2020). This move sets the de facto limit of diesel sulfur content to the same limit currently applicable in the EU.

Some other countries have recently lowered the sulfur limits for gasoline or are planning to do so. Oman reduced the maximum limit to 10 ppm in December 2019, becoming the first country in the Gulf Cooperation Council (GCC) to implement such strict regulations on gasoline. The United Arab Emirates reduced the sulfur limit from 100 ppm to 50 ppm in May 2017. Saudi Arabia intends to lower sulfur limits from 500 ppm to 10 ppm with the completion of its Clean Fuels Project in 2024 (Stratas Advisors 2019).

(continued)

BOX 3.7

Fuel Quality Standards in the Middle East and North Africa *(Continued)*

FIGURE B3.7.1

Diesel Sulfur Limits in the Middle East and North Africa, by Economy, 2020

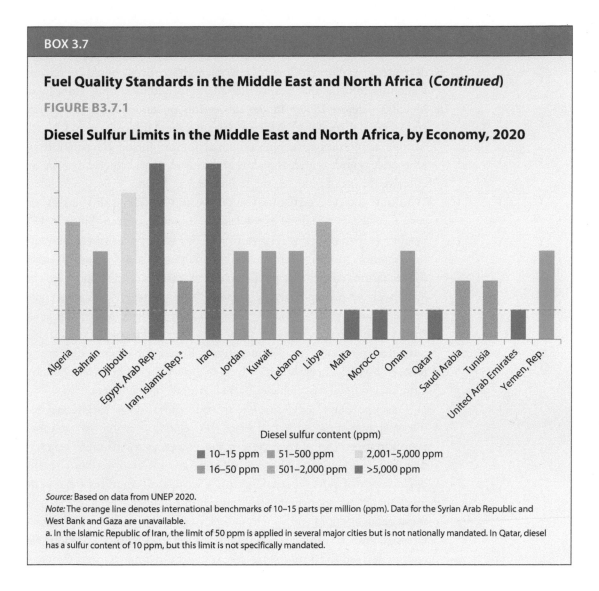

Diesel sulfur content (ppm)

- ■ 10–15 ppm ■ 51–500 ppm ▫ 2,001–5,000 ppm
- ■ 16–50 ppm ■ 501–2,000 ppm ■ >5,000 ppm

Source: Based on data from UNEP 2020.
Note: The orange line denotes international benchmarks of 10–15 parts per million (ppm). Data for the Syrian Arab Republic and West Bank and Gaza are unavailable.
a. In the Islamic Republic of Iran, the limit of 50 ppm is applied in several major cities but is not nationally mandated. In Qatar, diesel has a sulfur content of 10 ppm, but this limit is not specifically mandated.

Switching to higher-quality fuel imposes costs on fuel producers, distributors, and end consumers, all of whom may need to be supported. To ease the transition to these new standards, subsidies for technology upgrades by both producers and distributors should be considered. To nudge these actors into faster adoption of the necessary technologies, these programs could be designed such that the subsidies decline during the phase-in period for the implementation and adoption of the new regulations. This would incentivize fuel producers and distributors to avoid postponing necessary investments, hence improving air quality faster. Conversely, to incentivize end consumers to use higher-quality fuel even before standards become mandatory, a gradually increasing tax on lower-quality fuels could be a helpful tool.

Vehicle fuel switching. Increasing the share of vehicles run by alternatives to fossil fuels—for example, electric or hydrogen gas cars—can decrease air pollution. Switching to such alternatives should be accompanied by reforms that also switch energy production away from fossil fuels (as discussed later in the subsection on emissions from energy and industrial sources). However, because the Middle East and North Africa's economies have high potential for producing electricity from renewable sources, increasing the share of electric vehicles may be an effective means of reducing the region's air pollution.

Lifting import restrictions and lowering or abolishing tariffs on such vehicles, as was done in the Islamic Republic of Iran (Singh 2015), would lower their prices relative to vehicles run by traditional combustion engines. Incentives could also include purchase subsidies, subsidies for recharging vehicles at public charging stations, and support for the installation of private charging systems. Nonmonetary incentives could include allowing such vehicles to enter LEZs for free, use special lanes shared with public transportation, and use free parking services in short-term parking zones. Similar incentive schemes for electric cars are in place in several European cities such as Oslo or Stockholm, which have large shares of electric vehicles in their overall vehicle fleets.

To support an electrification of the transportation sector, the needed infrastructure must be put in place. To enable a switch to a cleaner transportation fleet, a transition of the power sector to more renewables is a prerequisite. These challenges are intertwined because the deployment of renewable energy on a large scale requires large-scale investments, as does the broader adoption of electric vehicles. Hence, these changes could and should unlock potential synergies of necessary infrastructure investments, such as installing grid storage solutions for energy derived from renewable sources and charging stations for electric vehicles. Similarly, future projects to expand public transportation systems could also be accompanied by measures to expand the use of alternative propulsion systems—for example, by introducing electric bus fleets as some of the region's cities did, including Doha, Marrakesh, and Tunis (UITP 2019).

Extending Public Transportation and Nonmotorized Options

In the Middle East and North Africa, two out of three residents live in urban areas (figure 3.20, panel a), and this figure is set to rise in the coming decades. Moreover, the region's population is highly concentrated in its largest centers, with more than one-fourth of the population living in an economy's largest city, outpaced only by Sub-Saharan Africa in this respect (figure 3.20, panel b).

FIGURE 3.20

Shares of Total Population Living in Urban Areas and Country's Largest City, by Global Region, 2018

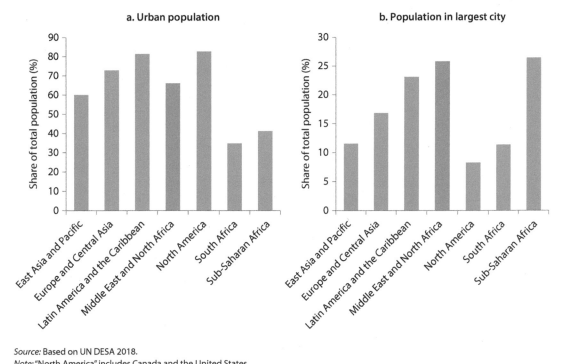

Source: Based on UN DESA 2018.
Note: "North America" includes Canada and the United States.

Extending public transportation. This high concentration in large cities makes a functioning and efficient public transportation system necessary both to reduce air pollution and to provide the local population with increased mobility and accessibility. Having convenient access to public transportation is an important part of making "cities and human settlements inclusive, safe, resilient, and sustainable," as formulated under the United Nations Sustainable Development Goal 11.[25] Expanding the extent of public transportation systems and encouraging their use over personal transportation reduces congestion, helps improve local air quality, and reduces GHG emissions (Parry et al. 2014).

However, public transportation systems are often poorly developed and poorly used in the Middle East and North Africa, even in many major cities. Extensive, efficient public transportation systems play a key role in sustainable city development. They are also of major importance to improving air quality. In most cities of the Middle East and North Africa, however, the primary mode of transportation is still the personal combustion vehicle (UITP 2019), often owing to a lack of adequate public transportation infrastructure.

In a recent global regional comparison, the Middle East and North Africa had the third highest share of people using their own cars for transportation. The study showed that the region's car share (the share of trips by car as a percentage of all motorized trips) amounted to 71.5 percent, the third highest share after North America and the Europe and Central Asia region (Fountas et al. 2020), as shown in figure 3.21. Weak transportation systems result in excessive use of often outdated personal vehicles, which in turn leads to high traffic intensities and congestion in the region's cities (Waked and Afif 2012).

Extending the public transportation system requires investments in vehicle fleets such as buses and adequate associated infrastructure like special road lanes for public transportation, tram rails, and pickup stations. These efforts should be accompanied by appropriate urban planning to ensure their integration into longer-term city expansion plans. In the procurement of public transportation fleets, it is crucial to ensure that vehicles meet emissions standards (such as those discussed earlier) or preferably rely on alternative energies such as electric or hybrid engines. Some cities in the Middle East and North Africa

FIGURE 3.21

Share of All Motorized Trips Using Personal Cars, by Global Region

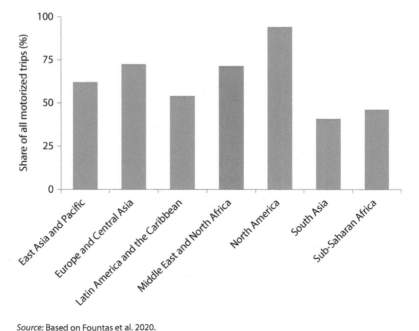

Source: Based on Fountas et al. 2020.
Note: Years of data differ between countries. For a subset of countries, the share is estimated. "North America" includes Canada and the United States.

(Alexandria, Doha, Tehran, and Tunis) have already made progress in this respect by introducing electric buses in their fleets (UITP 2019), a trend that other cities should emulate.

To induce more of the population to use public transportation, prices for these services must be affordable and provide a cost advantage over using the personal car. For people without a vehicle (that is, the poorer part of the population), prices for public transportation must not be prohibitive. Removing taxes from tickets or making them tax deductible may increase the incentive to use such services. Similarly, tax exemptions for long-term subscriptions that private companies provide to employees as part of their compensation can incentivize companies to encourage their employees to use public transit.

Evidence shows that public transportation infrastructure projects have beneficial effects on air pollution in Middle East and North Africa cities. For example, the opening of a new metro line in Cairo decreased vehicle use and at the same time reduced PM_{10} concentrations by around 3 percent (Heger, Zens, and Meisner 2019). The extension of the urban rail transit system in the city of Ahvaz in the Islamic Republic of Iran has the potential to reduce air pollutants substantially by reducing the amount of fuel used, owing to increased ridership on the new metro lines (Tabatabaiee, Abbas, and Rahman 2011). Furthermore, increasing connectivity between cities with public transportation systems (for example, through rails) can have the added benefit, if rails are used for transporting goods, of potentially reducing the number of heavy-duty vehicles used for this purpose, hence lowering their emissions.

Extending public transportation systems to be more efficient, timely, and widely accessible could also increase job prospects for the poor. Especially among people lacking access to personal vehicles, such as low-income households, the extension of the public transportation system could open up more job opportunities (Pang, Chen, and Zhang 2017). Living standards within these populations could be increased while also strengthening the local economy and potentially raising government tax revenues. Extending public transportation systems to cities' underserved areas can thus help revitalize their economies by improving connectivity.

In the Middle East and North Africa and around the world, the World Bank has supported various public transportation projects. In Morocco, for example, the Bank supported the government in strengthening the capacity of urban transportation institutions in a project started in 2015. By the end of 2020, around US$186 million in the form of loans had been disbursed, and the project had achieved most of its objectives despite the adverse effects caused by the COVID-19 crisis (World Bank 2021a). Within the course of the project, municipality-owned urban transportation centers were set up, mobility master plans were prepared, and more than 20 cities were assisted in improving their institutional capacity for urban transportation. Given these successes, additional financing was approved in December 2020, extending the project until 2024 with loans worth up to US$150 million (World Bank 2020b).

The World Bank has also recently supported the implementation of several public transportation projects in other regions, including metro lines in Colombia starting in 2018 (World Bank 2018) and bus rapid transit extensions in Peru starting in 2020 (World Bank 2019a). Advancing such projects in the Middle East and North Africa economies is important given their notoriously low use of public transportation (box 3.8).

Nonmotorized transportation options. Increasing the share of nonmotorized transportation is another important step toward reducing emissions and raising air quality. Facilitating and promoting the switch away from transportation in combustion vehicles altogether could further reduce air pollution while simultaneously inducing more physical

---

BOX 3.8

Public Transportation in Middle East and North Africa's Cities

Adoption of public transportation by large parts of the population is low in many major cities of the Middle East and North Africa. In its recent "MENA Transport Report 2019," the International Association of Public Transport compared the modal shares in some of the region's cities with those in major cities of other regions (UITP 2019). Figure B3.8.1 shows the number of trips by public transportation as a share of all motorized trips. It includes selected major cities within the Middle East and North Africa for which the modal share was reported in the past 10 years alongside the modal shares of other international cities in 2012.

As can be seen, the modal share of public transportation in the selected Middle East and North Africa cities is generally rather low relative to the international comparators. With few exceptions (Oran in Algeria as well as Tehran and Tabriz in the Islamic Republic of Iran), public transportation accounts for less than a quarter of motorized trips in the region's cities under consideration. Especially cities in the high-income countries in the Gulf Cooperation Council (GCC), where cars are a symbol of prestige, have relatively low public transportation use compared with cities of comparable income levels. Hence, inducing a switch of the modal share toward mass transit and nonmotorized forms of transportation such as cycling or walking may also need some form of cultural rethinking to make the latter socially more accepted.

(continued)

BOX 3.8

Public Transportation in Middle East and North Africa's Cities (*Continued*)

FIGURE B3.8.1

Trips by Public Transportation as a Share of Total Motorized Trips in Selected Cities Worldwide and in the Middle East and North Africa

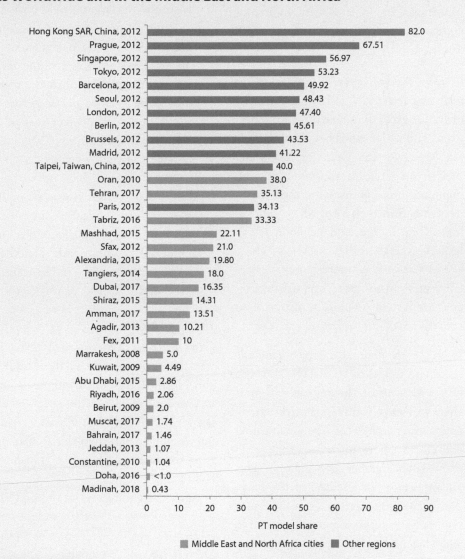

PT model share

■ Middle East and North Africa cities   ■ Other regions

Source: Adapted from UITP 2019.

Note: Data for cities outside the Middle East and North Africa region are 2012 data, taken from the 2015 International Association of Public Transport (UITP) Mobility in Cities database. Among Middle East and North Africa cities, those with modal split data older than 10 years are excluded in this benchmarking, which refers to the share of trips by public transportation (PT) in comparison to overall motorized trips. This means that trips by nonmotorized modes (for example, by walking or bike) are not considered. Owing to data limitations, the figure does not include either Cairo or Algiers, where PT was used for around two-thirds of motorized trips (according to UITP data from 2001 and 2004, respectively). Furthermore, other cities such as Marrakesh and Rabat in Morocco feature high shares of walking in their overall modal split.

(continued)

BOX 3.8

Public Transportation in Middle East and North Africa's Cities (*Continued*)

Projects in the Pipeline

Many of the region's countries have recently stepped up their game to extend their public transportation systems, and numerous projects are currently in the pipeline (UITP 2019):

- *In the Mashreq*, several countries are investing heavily. The Islamic Republic of Iran is expanding its national rail network, and major cities like Tehran, Shiraz, Mashhad, and Tabriz are extending their urban public transportation systems, including projects involving their metro networks, inner-city bus lines, and rail lines. Iraq, Jordan, and Lebanon are all looking to extend their national railway systems as well as intercity bus services connecting major cities, and urban projects for intracity public transportation systems are on the rise.

- *In the Maghreb*, notable projects include the metro extension in Algiers (already a city with one of the region's highest shares of public transportation trips). Africa's first high-speed line—connecting Tangier with Kenitra in Morocco—opened in November 2018 and represents an important cornerstone of the kingdom's 2040 rail strategy. And the Tunisian railway expansion connects Gabès and Medenine with an extension to Port Zavis near Djerba Island.

- *In the GCC*, countries are investing heavily in their public transportation systems to increase their low modal shares, with

several projects in major cities like Doha, Muscat, Kuwait, Abu Dhabi, and Dubai, among others. The Riyadh Metro system constitutes the largest single-phase metro construction project worldwide and was set to open, at least partially, in 2021. The final system will include six lines totaling about 170 kilometers of rails connecting various parts of the Saudi capital, at a total project cost of about US$24.4 billion (Smith 2020). To complement the metro system, a citywide 1,900-kilometer bus network with around 3,000 stops is also being developed. These measures constitute an important step to rebalance the current modal split, where only 2 percent of motorized trips are by public transportation (Smith 2020).

The Cairo Model

Cairo has one of the most extensive public transportation systems in the Middle East and North Africa and continues to extend it. Cairo is not part of the sample shown in figure B3.8.1 because of data constraints, but already in 2001, two-thirds of motorized trips taken were by public transportation (UITP 2019).

Since then, public transportation services have been extended further, and the ongoing metro extension is a notable example. Despite the difficult situation caused by the COVID-19 crisis, the fourth phase of line 3 was opened in August 2020, connecting the Heliopolis suburb with the Adly Mansour transit hub in the east of Cairo.

(continued)

BOX 3.8

Public Transportation in Middle East and North Africa's Cities (*Continued*)

Extensions of the line—in the east to Cairo International Airport and to the west across the Nile—are under construction. Construction of line 4 should start shortly (Burroughs 2020).

Cairo officials also seek to expand other means of public transportation to incorporate Cairo's surrounding areas. Several projects to introduce monorail or light-rail trains are under review, with a potential length of more than 150 kilometers. The introduction of special lanes to facilitate wider adoption of both buses and bicycles is in the planning stage.

---

activity by residents. As noted earlier, the comorbidities that AAP shares with low physical activity and obesity exacerbate the health effects of the individual risk factors. Therefore, making cities more walkable advances an important option for nonmotorized transportation.

Nonmotorized options are increasing in importance internationally as well as in the Middle East and North Africa. Some cities, such as Paris, have made great strides in the extension of nonmotorized alternatives by extending bicycle lanes. In the Middle East and North Africa, infrastructure that supports nonmotorized transportation is still in its infancy, but initiatives like bike-sharing services are also on the rise. There are several such services in Iranian cities (Therna, Shiraz, Mashhad, and Tabriz); in several cities in the GCC countries (Dubai, Abu Dhabi, Riyadh, Kuwait, and Kaec); and in Byblos in Lebanon, El Gouna in Egypt, and Marrakesh in Morocco (UITP 2019).

To make nonmotorized transportation options safe, it is essential to ensure that the roads are safe for users of these options and for pedestrians. A viable way to promote the wider adoption of bicycles in the region's cities would be the extension of safe cycling paths, which could come in combination with special lanes reserved for buses. However, lanes that combine buses and bicycles lead to relatively frequent close interactions between them, posing a danger to cyclists and potentially increasing the incidence of bicycle accidents. Hence, guidelines for the design of such lanes should be thoroughly explored and refined to ensure safety of all road users, especially bicyclists (De Ceunynck et al. 2017).

Furthermore, a range of potential urban geographical issues or cultural challenges facing female cyclists must be considered when designing traffic systems and policies to promote the broader adoption of cycling in the region. Increasing the walkability of cities requires not only safe pavements but also enhanced traffic management (for example, traffic lights at crosswalks). Additionally, place-based policies such as pedestrian zones (further discussed in the next subsection) are useful for ensuring pedestrians' safety and relieving congested city centers of excessive traffic.

Implementing Place-Based Policies

The air in urban areas, particularly in city centers, is often significantly more polluted than the air in surrounding suburbs or rural areas. High traffic loads due to commuters, in combination with impaired air exchange stemming from multistory buildings, are among the main factors driving the generally worse air quality in urban areas. Cities around the world have thus adopted various types of policies that restrict traffic either within an entire city or in parts of it. Such place-based policies can be effective in lowering air pollution and come in different forms (see box 3.9 for some international examples). LEZs, pollution and congestion charges, and driving restrictions are the best known of these place-based policies.

BOX 3.9

Place-Based Policies and Their Effects on Air Pollution

Since the first low emission zones (LEZs) were implemented in several cities in Sweden in the 1990s, they have become widespread around the world, especially in Europe. Several variations on these systems are described below.

London's LEZ and Ultra LEZ

Currently, the largest LEZ in operation is in the United Kingdom; it covers almost all of Greater London and applies to a wide range of commercial vehicles that do not meet certain Euro emissions standards as well as to vans, pickups, and other more-polluting vehicles. Eligibility to enter the LEZ is monitored through an automatic license plate scanning system.

The introduction of London's LEZ has substantially increased the rate of fleet turnover to vehicles meeting higher emissions standards. Furthermore, a significant, albeit not large, reduction of particulate matter (PM) concentration has been recorded within the LEZ (Ellison, Greaves, and Hensher 2013). Since April 2019, parts of central London are now declared to be an "Ultra Low Emission Zone (ULEZ)" with stricter regulations. In October 2021, the ULEZ was expanded to cover a larger area, including the North and South Circular roads (Lydall 2021).

Tehran's Inspection-Based LEZ

Tehran introduced an LEZ intended to replace the driving restrictions based on

(continued)

BOX 3.9

Place-Based Policies and Their Effects on Air Pollution (*Continued*)

license plate numbers, even though the two systems were in place simultaneously for a time. Tehran's LEZ bans the entrance of cars that have not successfully undergone the mandatory yearly inspection. However, the extension of the system to also restrict entry for cars not fulfilling certain emissions standards would be necessary to make it a "real" LEZ (Heger and Sarraf 2018). The cars that undergo inspections receive a sticker whose color indicates whether the emissions are low enough to enter the LEZ or are too high for the LEZ. The Tehran Air Quality Control Company, which assessed the implementation of the LEZ scheme, reports that the LEZ has been effective in reducing traffic emissions and consequently curbed air pollution.

Stockholm's Congestion Pricing

Stockholm introduced a congestion pricing system in August 2007 after a six-month trial period. The program was opposed by most of the municipalities in the country except those inside the affected zone. The charges vary depending on the time of the day and are collected automatically using license plate scanning technology. In 2016, the congestion tax was raised, with a focus on commuting hours, and outreach was extended. The additional revenues were earmarked for the extension of the Stockholm metro system (STA and Trafikverket 2015).

The system has had positive effects on air pollution levels and on the health of local children. The levels of NO_2 and PM_{10} fell by 15–20 percent and by 10–15

percent, respectively, and the number of acute asthma care visits of children under the age of five decreased by 50 percent (Simeonova et al. 2019). Vehicles using alternative fuels were temporarily exempted from the charges through 2008, which increased sales of such cars significantly and showcased the potential to use such systems as incentives themselves (Börjesson et al. 2012). Studies on the impact on retail revenues concluded that there were no negative effects (Daunfeldt, Rudholm, and Rämme 2009).

Beijing's Driving Restrictions

Driving restrictions for vehicles, often randomly assigned based on certain parts of their license plate number, have been implemented in numerous cities around the world. In July 2008, Beijing implemented an odd-even system that restricted cars to driving only every other day based on the last number of their license plate. This system was replaced by driving restrictions that prohibited the use of cars for one day per week, again based on license plate number. During periods of particularly severe air pollution, Beijing reverts to the odd-even system. These restrictions do not apply to electric cars.

The effects of Beijing's driving restrictions on air pollution were also positive, significantly reducing PM, with the odd-even system decreasing PM_{10} concentration levels by 18 percent and the one-day-per-week system by 21 percent (Viard and Fu 2015).

LEZs, which are widespread in Europe, restrict the most-polluting vehicles, usually allowing only vehicles with some sort of validation or certification into the zone (or charging them if they do not). For example, in Tehran, only cars that have undergone the mandatory annual inspection are allowed to enter the LEZ, with fines being imposed on noncompliers. In a similar spirit, congestion charges restrict access to certain areas of a city by charging a certain amount for vehicles to enter. Hence, vehicle owners' willingness to pay serves as a criterion for entrance into the regulated zone.

One common criticism of both types of restrictions is that they affect people differently based on their income and could affect the job prospects of those who cannot afford cleaner vehicles or the charges levied. Hence, another way to restrict driving in certain areas—practiced internationally but also in the Middle East and North Africa—is to use random assignment based on license plate numbers. One typical way of doing this is to use an odd-even scheme: every other day, only vehicles whose license plates end with an odd number may enter the restricted zones; on the alternate days, only vehicles whose plates end with even numbers may enter.

Place-based measures necessitate complementary investments, and the costs to residents must be considered in their introduction. Such investments include the development of efficient public transportation systems as alternative entrances and exits for the vehicle-restricted locations. Furthermore, an effective detection and collection system must be established to detect noncompliers. Automated license plate scanning systems are often needed to implement LEZs or congestion charges, although driving restrictions based on, for example, detecting the last number of the license plate may be simpler and can be carried out manually. Typically, penalties for noncompliance are used as economic incentives. Cities should also consider ways to ease the transition for local businesses and residents in the form of allowances for switching their cars to alternatives with lower or zero emissions or offering free public transportation if they get rid of their cars.

Reducing Vehicle Emissions: The Policy Spectrum

A broad spectrum of actions can be taken to combat air pollution from vehicles in the Middle East and North Africa. The discussion above outlined some possible routes to reducing vehicle emissions. Table 3.2 provides a comprehensive overview of such measures. For each measure, a rapid assessment of the main aspects important for policy evaluation has been carried out, which is based on international experiences and region-specific expert judgment. While not exhaustive in this regard, the selected

TABLE 3.2

Overview of Policy Options to Reduce Vehicle Emissions

Main subobjective	Measure	Timeline for implementation	Financial cost	Effectiveness
Reduce emissions from combustion vehicles	Emission control (retrofit or new fit)	medium	medium	high
	Vehicle combustion efficiency technology (retrofit or new fit)	medium	medium	medium
	Vehicle inspection and testing	short	medium	high
	Optimizing speed limits	short	low	medium
	Fuel-quality upgrade	medium	high	high
Reduce number of combustion vehicles	Vehicle scrappage	short	high	high
	Vehicle fuel switching	medium	medium	high
Reduce trips in combustion vehicles	Traffic management (for example, changing directions, diversion to less congested roads)	medium	low	medium
	Parking management	medium	low	medium
	Freight management	medium	medium	medium
Reduce demand for trips in combustion vehicles	Increasing fuel prices	short	medium	high
	Encourage modal shift (discourage driving of personal combustion vehicles)	long	high	high
	Public procurement of low-emission vehicles	long	medium	medium
	Vehicle-restricted areas	medium	medium	medium
	Low emission zones	medium	medium	medium
	Road space rationing (for example, alternate day or no-drive days)	short	low	high
	Pollution or congestion charges	medium	low	high
	Communication technology to encourage home-based work and telecommuting	short	low	low
	Information campaigns encouraging low emission travel	medium	low	low

Source: Based on World Bank data.
Note: The classifications of the qualifiers—timeline (short, medium, long); cost (low, medium, high); and effectiveness (low, medium, high)—are based on expansive literature reviews and the expert judgment of the report-writing team. (A full list of surveyed literature is available on request.)

qualifiers cover (a) the *time* that it takes to implement a given measure, (b) the financial *costs* associated with it, and (c) the expected *effectiveness*. The financial costs include both direct costs for technology upgrades and operations and maintenance costs.

Table 3.3 then provides more-detailed descriptions of the technology needed to implement the measures, market-based instruments (MBIs) that can assist in their adoption, and supporting regulations for these MBIs.

TABLE 3.3

Detailed Description of Policy Options to Reduce Vehicle Emissions

Main subobjective	Measure	Technology	Market-based instrument (MBI)	Regulations supporting MBIs
Reduce emissions from combustion vehicles	Emission control (retrofit or new fit)	End-of-pipe technology such as diesel particulate filters (DPFs), selective catalytic reduction (SCR), and exhaust gas recirculation (EGR) (for example, Euro I–VI[a] emission standards)	Subsidies for cleaner technologies; charges for emission-intensive technologies	Mandating specified emission-control standards
	Vehicle combustion efficiency technology (retrofit or new fit)	Vehicle efficiency standards (improving fuel economy)	Subsidies for cleaner cars, higher taxes on dirtier ones	Licensing of appropriate technologies
	Vehicle inspection and testing	Vehicle inspection garages (including dynamometers and so forth); training of staff	Tax-free repair services; entrance to LEZ free of charge; lower parking fees	Mandatory vehicle tests (for example, annually)
	Optimizing speed limits	Detection (gantry, cameras, and so forth) and collection system	Penalties for speeding and noncompliance	Traffic mandates
	Fuel-quality upgrade	Desulfurization; Euro 1–5[a] fuel standards, and improved fuel distribution system	Rebates for cleaner fuel producers and distributors; additional charges on dirtier ones such as diesel	Ban of low-quality fuels or mandates for higher-quality fuels (for example, sulfur below 10 ppm)
Reduce numbers of combustion vehicles	Vehicle scrappage	Scrappage technology	Car allowance rebate system (for example, "cash for clunkers")	Vehicle or technology age limits
	Vehicle fuel switching	Electric vehicles, hydro vehicles, and natural gas vehicles	Subsidies for greener vehicles; free parking services or entry into special zones; tax exemptions	Mandating cleaner vehicle fleets; entrance into LEZs or use of special lanes
	Traffic management (for example, changing directions, diversion to less-congested roads)	Street monitoring system; automated diversion technology; street planning	Penalties for noncompliance	Traffic mandates
Reduce trips in combustion vehicles	Road space rationing (for example, alternate day or no-drive days)	Detection (gantry, cameras, and so forth) and collection system	Subsidized public transportation; subsidized park-and-ride fees	Restricted admission based on license plate number
	Parking management	Parking infrastructure, collection system	Higher parking prices in areas with higher air pollution	Exempting low-emission vehicles from parking charges; offering special parking spaces only for low-emission vehicles' park-and-ride; infrastructure mandates
	Freight management	Logistical modeling and planning (for example, consolidating loading for freight activity)	Penalties for overweight vehicles	Maximum freight weight for vehicles

(continued on next page)

TABLE 3.3

Detailed Description of Policy Options to Reduce Vehicle Emissions (*continued*)

Main subobjective	Measure	Technology	Market-based instrument (MBI)	Regulations supporting MBIs
Reduce demand for trips in combustion vehicles	Increase fuel prices	Precise monitoring system at source; automatic tax collection systems; staff training for enforcement	Higher fuel prices by removing fuel subsidies, taxing fuel	Removing subsidies; introducing taxes
	Encourage modal shift (discourage driving of personal combustion vehicles)	Public transportation investment; infrastructure investments for low-emission alternatives (such as bicycles); campaigns for cycling and walking	Subsidized prices for public transportation; subsidies or tax credits for companies offering public transportation tickets to employees	Special traffic lanes for public transportation and cycling
	Public procurement of low-emission vehicles	Electric buses; gas-powered buses	Additional budgets for municipalities	Minimum standards for public transportation vehicles
	Vehicle-restricted areas	Access control of all motorized vehicles; detection (gantry, cameras, and so forth) and collection system	Penalties for noncompliance	Traffic mandates; exemptions for low-emission vehicles
	Low emission zones	Access control for certain vehicles; detection (gantry, cameras, and so forth) and collection system	Penalties for noncompliance	Traffic mandates; exemptions for low-emission vehicles
	Pollution or congestion charges	Detection (gantry, cameras, and so forth) and collection system	Usage fees (corresponding to cleanliness of vehicle technology)	Traffic mandates; exemptions for low-emission vehicles
	Communication technology to encourage home-based work and telecommuting	Information technology (Zoom, WebEx, and so forth)	Partial acquisition cost coverage for companies	Mandate for allowing home office where possible
	Information campaigns to encourage low-emission travel	Creation of campaigns; distribution of campaigns through various networks	Subsidized (government-financed) television ads, billboards, and so forth	None

Source: Based on World Bank data.

Note: LEZ = low emission zone; ppm = parts per million.

a. Euro standards for light-duty vehicles are denoted with Arabic numerals (for example, Euro 6), while standards for heavy-duty vehicles are denoted with Roman numerals (for example, Euro VI).

Although these lists provide a comprehensive overview of possible options for the reduction of vehicle emissions, the optimal policy mix is highly dependent on country and city characteristics and must be assessed in detail on a case-by-case basis.

Policies to Reduce Emissions from Energy and Industrial Sources

Industrial processes and energy generation are among the most important sources of airborne PM. These industrial processes include emissions from chemical or mechanical processes emitted not only in the manufacturing of goods (for example, in the cement industry) but also those emissions from on-site energy production through coal or oil thermal power plants on the premises of the producing firms. In many Middle East and North Africa cities, these large-scale plants are suspected to be the primary causes of air pollution and GHG emissions—accounting for more than half of CO_2 emissions in 2014 (Abbass, Kumar, and El-Gendy 2018). The region's energy mix is also heavily skewed toward fossil fuels, with more than 95 percent of its energy derived from them (Menichetti et al. 2019), contributing heavily to carbon and air pollutant emissions.

Lack of stringent regulation and monitoring systems for industrial emissions, low fossil fuel prices, and the high reliance of the region's economies on fossil fuels represent hurdles to achieving air quality improvements. Contributing to these issues, few countries have set air quality standards in the form of laws or regulations (UNEP 2017). Partly as a result (according to World Bank Enterprise Surveys), the adoption of measures to reduce the emissions of air pollutants and to raise energy efficiency in the private sector is rather low, at least in some of the region's economies.

This section reviews the main options for reducing emissions from energy and industrial sources, including

- Mandating and effectively enforcing emission controls through inspection and pollution charges;

- Charging and trading emissions;

- Raising industrial resource efficiency;

- Switching to renewables for energy production; and

- Properly pricing fossil fuels.

These measures are presented in tables 3.4 and 3.5, which contain additional measures not discussed in the text, because their aim is to present the universe of policy options available. The section also provides examples of successful adoption of selected measures by Middle East and North Africa economies as well as international best practices.

Looking at the firm perspective confirms that green management and green investment practices are limited among private firms in emerging markets, including the Middle East and North Africa. The most recent World Bank Enterprise Survey data show that economies in the southern and eastern Mediterranean (including Egypt, Jordan, Lebanon, Morocco, Tunisia, and West Bank and Gaza) have rather low adoption rates of green technology practices. Only around 1 in 10 firms mention the environment and climate change in their strategic objectives, and even fewer have a manager responsible for addressing such issues. Although around 1 in 3 firms monitored their energy consumption, less than 1 in 4 adopted some form of energy management, and less than 1 in 5 actually implemented measures to increase their energy efficiency. Similarly, adoption of waste minimization and recycling practices is not widespread among the surveyed firms (about 1 in 4 firms). Although these numbers look very low, other regions such as Europe and Central Asia show similar numbers.

Instituting Emissions Controls, Inspections, and Pollution Charges

In many economies of the Middle East and North Africa region, industrial emissions are a major polluting source but are subject to relatively lax standards and minimal oversight. For more than half of the region's economies, no regulations for industrial sources were identifiable when the last stocktaking was done (see UNEP 2015 and similar sources).

Phasing out the use of inefficient production technologies or machines that do not necessarily emit pollutants themselves but need a high energy input requires a clear legislative framework. However, all of these mandates must be accompanied by credible means of enforcement, including regular (unannounced) inspections and penalty schemes intimidating enough to ensure compliance by both public and private firms. Furthermore, this requires training of the staff performing these inspections as well as a clear codification of relevant violations and their consequences.

Internationally recognized certificates for environmental management systems of industrial sites are comparatively scarce in the Middle East and North Africa. Currently, the legal framework in many of the region's economies is opaque regarding environmental management systems. One other way to assess the state of environmental management is to study an economy's number of ISO 14001 certificates (ISO 2020), which reflect international standards for the systematic identification and management of environmental threats and are issued to firms that voluntarily adopt measures to mitigate and properly manage such threats. As of December 31, 2019, there are an average of around 300 such certificates per Middle East and North Africa economy, compared with a global average of around 1,000 certifications per country (excluding China, which alone has over 130,000 such certificates). A notable exception is the United Arab Emirates, with over 1,800 such certificates.

Using International Organization for Standardization (ISO) 14001 certificates could be a way of rewarding company efforts by introducing publicly available company ratings that can be combined with product labeling that displays the companies' performance to the end consumer. Despite some evidence that the broader adoption of such certificates may lower the emission of air pollutants (Potoski and Prakash 2005, 2013), their effects may differ across countries and are dependent on the domestic regulatory framework (Prakash and Potoski 2014; Arimura et al. 2016)—highlighting the need for strong legal requirements set by policy makers.

Supporting and incentivizing companies in their transition to newer technologies by subsidizing the retrofit or new fit of their industrial complexes would advance the blueing of the Middle East and North Africa's skies and seas. Equipping high-polluting facilities with suitable technologies, such as fume scrubbers or exhaust filters (for example, electrostatic precipitators and flue gas desulfurization), is a crucial step for reducing air pollution stemming from these sources.[26] Subsidies, credit lines, guarantees, and technical assistance can help facilitate a switch to less-polluting technologies. Positive incentives could include tax rebates or certifications of products, signaling to the consumer the environmental effort made by the respective producer (see box 3.10 for some examples).

Setting Industrial Emissions Charges and Introducing Emissions Markets

Pricing the emissions of carbon or air pollutants is a cost-effective way of reducing them. It is often the least-cost option to achieve deep cuts in emissions, especially of CO_2 (IMF 2021). Another advantage is that generated revenues can be used for increased investments in and subsidies for green infrastructure. Similarly, it is important to put a price on the energy created by burning fossil fuels. This is discussed further below in the subsection on "Properly Pricing Energy."

Taxing emissions. The basic economic theory underlying emissions taxes—that is, making something more expensive to reduce its usage—is straightforward. These taxes are often unpopular at the outset and require careful communications campaigns. Putting a price on emissions can help countries and companies to decarbonize economies including their supply chains, increase air quality by reducing air pollutants, and at the same time increase the social benefits of the economy or city.[27] It is a powerful tool to nudge companies to invest in cleaner technologies,

BOX 3.10

Successful Pollution Abatement Projects in the Middle East and North Africa

Egypt and Lebanon, with international support, have implemented pollution abatement projects that support businesses in transitioning to cleaner industrial production. The Egyptian Pollution Abatement Programme (EPAP), initiated by the Ministry of Environment with support from international organizations such as the World Bank, has been in effect since the early 1990s. It was rolled out in three phases, with the latest one launching in 2015. The project's main goal is to set up a framework that encourages cleaner industrial production by providing loans to companies for pollution-reducing investments. The third phase, EPAP3, includes a bank credit line up to €120 million to finance the pollution abatement projects of public and private enterprises; a grant facility (€20 million) to soften the terms of the granted loans; and a technical assistance program with €6 million to strengthen the capabilities of the various stakeholders (AFD, n.d.).

Similarly, the Lebanon Environmental Pollution Abatement Project (LEPAP)[a] was set up in 2014 in cooperation with international organizations such as the World Bank and the Italian Agency for Development Cooperation and provides free technical assistance to industrial enterprises in the form of national and international consultants. Furthermore, LEPAP includes a financial mechanism that provides concessional loans supported by the Banque Du Liban through commercial banks, which have interest rates close to zero. These loans are provided for a period of seven years with a two-year grace period, and LEPAP also supports companies in the preparation of technical specifications and the fulfillment of technical requirements for their projects.

Projects financed by these initiatives have been successfully expanded and scaled up. EPAP has financed over 35 subprojects to improve pollution abatement. The program has an astonishing track record for completed projects, with projects that received assistance leading to significantly reduced emissions. For example, in the course of the project's second phase, air pollutants of financed subprojects were reduced by 91 percent on average, with SO_2 emissions being reduced by 84 percent and particulate matter (PM) emission by 94 percent. Similarly, wastewater pollution of the financed projects was almost completely eliminated, with average decreases in wastewater effluents of 98 percent (World Bank 2015).

For LEPAP, as of September 2019, six industrial plants have applied for loans, and three of them have already received loans worth around US$3 million, with the other ones applying for loans amounting up to US$2 million. Additional financing is planned to increase the number of projects to 20–25 public and private enterprises and provide them with loans and technical assistance to address pollution emissions in a cost-effective manner.

a. For more information, see the LEPAP website: http://lepap.moe.gov.lb/.

adopt more efficient practices, apply end-of-pipe pollution abatement, and conserve energy inputs. The revenues can then be used to advance other development policies such as ones enhancing critical infrastructure or supporting businesses to update their production facilities.

An efficient emission taxation scheme requires an effective monitoring and administration system as well as stringent enforcement mechanisms. This effort requires technological solutions to measure emissions of GHGs and other air pollutants as well as regulatory changes that are necessary for the imposition of the tax in the existing tax framework. Taxes on carbon emissions are the most widely used form of emissions taxation, with carbon taxes present in various European countries (for example, France, Ireland, Spain, and the Nordic countries) as well as in Argentina, Japan, Mexico, South Africa, and a large number of Canadian provinces.

Trading emissions. An alternative way of limiting emissions from energy-production-related and industrial sources is the implementation of an emissions trading system (ETS). Such a trading system would allow companies causing more emissions to choose to invest in necessary technology upgrades immediately or postpone them to a more suitable time by buying carbon certificates from other, less-polluting companies. One of the first and largest carbon trading systems is the cap-and-trade program set up by the US state of California (box 3.11).

With the introduction of its national ETS, China could overtake the EU as the world's biggest carbon market; however, the initial version of China's ETS will cover only coal- and gas-fired power plants. In some countries, several carbon pricing systems coexist. For example, several European countries are part of the EU ETS and have national carbon taxes, and Germany launched an additional national ETS for heating and transportation fuels starting in 2021 leading to price hikes in the latter of around 10 percent. Middle-income countries such as Kazakhstan and Mexico have already implemented or are scheduled to implement an ETS, and Brazil, Indonesia, Pakistan, and Turkey are currently considering doing so and have initiated studies for their specific design (World Bank 2021b). This showcases the potential of such systems at a country or subregional level for the Middle East and North Africa as well. Hence, it is recommended that governments in the region consider implementing ETSs and commence assessments for their feasibility and ideal design.

Targeting PM explicitly. Recently, the Indian state of Gujarat launched a pilot program in the industrial city of Surat that, like a traditional ETS, allots a fixed number of permits to plants but specifically targets PM air pollution. It is the world's first cap-and-trade scheme that does so and, if successful, it is planned to be scaled up to cover all of Gujarat and perhaps other states in India. It also serves as an international

BOX 3.11

California's Emissions Trading System

California's cap-and-trade system places a limit on carbon pollution while granting companies the flexibility to make the lowest-cost reductions first. The ETS was launched in 2013 after California established its landmark climate law in 2006[a] and made reporting of greenhouse gas (GHG) emissions mandatory in January 2009. This reporting step was taken to determine the state's emissions in 1990, collecting data on actual emissions from historical energy consumption and production data, so as to base the cap that would be introduced on real emissions instead of projected ones. The state requires entities that emit more than 10,000 tons of GHG emissions annually to report their emissions, and those that emit more than 25,000 tons must verify their emissions with an independent third party.

Within the system, allowances to companies (permits that each allow a facility to emit 1 ton of GHG emissions) are distributed either for free, based on metrics such as output and efficiency, or sold in state-administered auctions. These allowances can then be traded on a secondary market, where companies whose GHGs are lower can sell them to ones that want to emit more GHGs. This way, a certain company can flexibly decide whether it wants to invest in emission-reducing technology or buy allowances instead. It is also possible to "bank" allowances for later use. The price floor for allowances was set at US$10 per ton in 2012 and has been increased by 5 percent plus inflation annually to keep the market stable in case of demand fluctuations.

The state of California decided to link its ETS with the one implemented in Quebec, Canada—enabling their respective ETSs to trade certificates with each other—because they share similarly stringent caps and have a closely aligned policy design. Initially covering the emissions of six GHGs by industrial and electricity sectors, the cap was expanded in 2015 to include transportation fuels and natural gas, covering around 85 percent of the state's GHG emissions since then. Furthermore, in 2016 California passed legislation to set its 2030 emission-reduction target to be 40 percent lower than its 1990 emissions. The program is continuously evaluated and improved at intervals of roughly two years. Notably, in an adaptation in 2017, California regulators explicitly included a program to further reduce local air pollution. Through this continuous updating of the regulations, legislators try to maintain regulatory certainty while providing a dynamic program that adapts to current needs.

Source: EDF 2018.
a. The Global Warming Solutions Act or Assembly Bill 32.

role model for the implementation of such schemes around the world, including the Middle East and North Africa (BBC News 2019).

International experts have endorsed the pilot program and see the potential to cut PM air pollution. Preliminary assessments attest to the program's effectiveness (Tripathi 2019) and a scientific evaluation in the form of a randomized control trial is being carried out at the time of this writing to rigorously determine its impacts (Greenstone et al. 2019).

Raising Energy Efficiency

The Middle East and North Africa region's energy use per unit of economic output is among the highest worldwide. On average, to produce output worth US$1,000 (2017 US$, purchasing power parity [PPP]), the region used energy equivalent to around 135 kilograms (kg) of oil in 2014 (figure 3.22, panel a). This is exceeded only in Sub-Saharan Africa as well as slightly so in the East Asia and Pacific region.

Even more concerning, the Middle East and North Africa was the only region worldwide whose average energy intensity increased in the

FIGURE 3.22

Energy Use Per Unit of Output and Growth Rate, by World Region

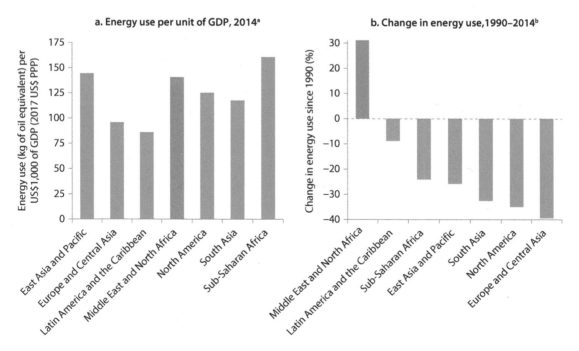

Source: Based on the World Development Indicators database.

Note: "North America" includes Canada and the United States. PPP = purchasing power parity.

a. Panel a represents a region's average energy use as the kilograms (kg) of oil needed per unit of GDP (US$1,000, 2017 US$ PPP).

b. Panel b shows the percentage change in each region's average energy use per unit of GDP in 2014 relative to its energy use per unit in 1990.

past three decades. The average energy input to produce its output has risen more than 30 percent higher than in 1990 (figure 3.22, panel b). In contrast, all other regions were able to cut their energy input to produce a given amount of output.

Electricity and other energy prices are low in many parts of the Middle East and North Africa, leading to few efforts to efficiently distribute and use energy. Reflecting this negligence, almost 15 percent of the region's produced electric power in 2014 never reached its destination (figure 3.23). Although this does not directly refer to the inefficient use of energy by industrial complexes themselves, it is a symptom of the same underlying issue: low prices that do not incentivize the economical use of energy or investments to increase energy efficiency. This makes clear that, in addition to the measures that industries should adopt (outlined below), investments in the electrical grid system and other infrastructure are needed to cut the losses of energy from source to destination.

Manufacturing firms are increasingly monitoring their energy consumption, but the adoption of energy efficiency-enhancing measures

FIGURE 3.23

Electric Power Transmission and Distribution Losses, by World Region, 2014

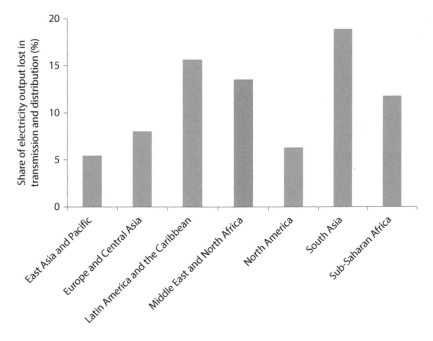

Source: Based on the World Development Indicators database.
Note: "North America" includes Canada and the United States.

is lagging. Recent data from the World Bank Enterprise Survey reveals that, on average, almost half of manufacturers in six of the region's economies—Egypt, Jordan, Lebanon, Morocco, Tunisia, and West Bank and Gaza—monitor their energy consumption, and around a third adopt some form of energy management. However, less than a quarter had explicit targets for their energy consumption, and less than 20 percent adopted concrete measures to enhance their energy efficiency.

In contrast, comparable European countries were faring better.[28] The low adoption of measures in some Middle East and North Africa economies results partly from the fact that only around 7 percent of manufacturing firms were subject to an energy performance standard, although around 22 percent were subject to an energy tax or levy, according to the Enterprise Surveys. On the upside, about 40 percent of manufacturing firms in the surveyed economies declared that they adopted improvements in the lighting systems in the past three years, and almost 30 percent had improved their heating and cooling equipment.

When firms were asked why they are not investing in proper energy management in the form of energy efficiency measures, the main reasons stated were (a) lack of prioritization relative to other forms of investment, and (b) lack of financial resources. Credit constraints have also been shown to be an impediment to the implementation of green management practices (EBRD 2019). Government regulation, as discussed in detail earlier in this chapter, is important to move the needle on green practice adoption by firms, but so is also raising awareness and customer demand. Larger rates of green practices—larger than those reported here—were found among firms that face customer pressure to act in an environmentally friendly way or that are subject to an energy tax or levy if they do not (EBRD 2019).

There is great scope for increased energy efficiency of industries. As mentioned earlier, low energy prices lead to low incentives for industry and private consumers to address inefficiencies in their energy consumption, and low prices do not encourage energy conservation (El Khoury 2012). However, increasing energy efficiency potentially represents the most cost-effective measure for reducing GHG emissions and improving air quality. The importance of raising energy efficiency to rein in excessive energy consumption has also been recognized by governments in the Middle East and North Africa. Box 3.12 provides examples from Saudi Arabia's efforts to increase energy efficiency through various programs in both the industrial and residential sectors.

Measures for raising energy efficiency include market-based instruments. For example, auction systems could incorporate a set price for each unit of energy savings, for which key market actors like utility companies have to propose their projects that generate energy savings

BOX 3.12

Saudi Arabia's Efforts to Increase Energy Efficiency

In Saudi Arabia, the National Energy Efficiency Program (NEEP) was launched in 2003 and defined eight policy objectives to increase energy efficiency (mostly electricity) by 30 percent from 2005 levels by 2030. It targeted mainly the industrial sector but included measures to increase energy efficiency in the residential sector as well. Among them were the introduction of energy audits, energy efficiency labels, and standards and labeling for appliances. Measures to reduce the demand for energy included the installation of high-efficiency air conditioners and the introduction of a construction code to improve insulations of buildings. NEEP also introduced a program to disseminate energy efficiency information and raise awareness as well as technical and managerial training through workshops and seminars.

In October 2010, NEEP was transferred to a permanent entity, the Saudi Energy Efficiency Center (SEEC), that continued the mission to "reduce energy consumption and improve energy efficiency to achieve the lowest possible energy intensity" (Fawkes 2014). It coordinates all activities related to energy consumption efficiency improvement between governmental and nongovernmental stakeholders and launched the Saudi Energy Efficiency Program (SEEP) in 2012. In addition to the foci of NEEP on the industrial and the buildings sector, SEEP also included measures targeting the transportation sector. However, unlike the previous program, SEEP's guiding principles do not include price reforms (*Arab News* 2014).

Since its inception, SEEP has partnered with foreign governments and international organizations and has introduced various regulations to raise energy efficiency in the three sectors that are its focus:

- *In the industrial sector*, SEEP established energy intensity targets based on international benchmarks for especially energy-intensive industries such as the petrochemical, cement, and steel industries. Preceding these efforts, it developed a baseline assessment of a large number of companies from various sectors and stages of production. Further actions included the issuance of standards for electric motors, with a minimum standard prescribed since 2015 and an update of the regulation in 2018.

- *In the buildings sector*, standards for thermal insulation were issued, the Saudi Building Code has been amended, and feasibility studies for moving the energy use intensity to international best practices have been constructed. Furthermore, several standards were issued and continuously enhanced for different appliances that account for a high share of energy use, such as air conditioners, water heaters, lighting products, and white goods (for example, refrigerators, freezers, and washing machines). These efforts were supplemented by clear labeling schemes.

- *In the transportation sector*, to improve energy efficiency, SEEP issued a fuel economy standard in 2014 for light-duty vehicles, implementing it in January 2016, supplemented by fuel economy labels. A second phase is being planned. It also introduced standards for tire-rolling resistance, and regulations for heavy-duty vehicles were assessed.

at a given price. Other possibilities could be to use (a) dynamic electricity prices so that retail prices follow wholesale prices that fluctuate over time and hence can incentivize energy-consumption behavior; or (b) tax-based instruments like rebates or tax credits to promote energy efficiency.

Regulatory mechanisms to push energy efficiency can come in the form of mandatory efficiency targets. These can be supplemented by measures like the assignment of a qualified energy manager to companies or groups of companies in a sector, reporting obligations for those companies or regular audits. Carrying out such inspections would require thorough training of staff and a clear communication of mandatory efficiency targets. Furthermore, the implementation of such targets may need a phasing-in period to allow companies, especially SMEs, to adjust to these requirements with investments in appropriate infrastructure.

Voluntary initiatives to increase energy efficiency should be supported by providing incentives and knowledge. A range of less-restrictive measures includes the promotion of voluntary goals for energy efficiency or the introduction of an energy savings insurance mechanism that mitigates risks for SMEs when implementing energy-efficiency measures. These measures may be confronted with less resistance and hence may be easier to implement or support. To provide support for companies that engage in such voluntary actions, governments in the Middle East and North Africa could organize conferences to induce knowledge spillovers among participants or introduce special certificates for products manufactured in participating plants.

Switching to Renewable Energies

Currently the energy mix in the Middle East and North Africa region is heavily skewed toward fossil fuel sources, primarily from oil and gas. Gas accounts for 48 percent of electricity generation, oil for 44 percent, coal for 5 percent, and renewables for only 3 percent (Menichetti et al. 2019). Even in Morocco—"a front-runner" in investing in solar energy—renewable energy accounts for only 9 percent of electric power generation. Around 12 percent of manufacturing firms surveyed in the most recent wave of the World Bank Enterprise Surveys declared that they had adopted more climate-friendly energy generation on-site, compared with about 16 percent in the Western Balkans and the Central Europe and Baltic states.

Broader adoption of renewables, both produced by dedicated power plants and produced by energy consumers themselves, requires a mix of economic, regulatory, and investment incentives. However, fossil fuel energy subsidies are a major barrier to investment in renewables.

An approach that combines removal of fossil fuel energy subsidies with investment and initial tariff subsidies could be a viable way to induce a switch in the Middle East and North Africa's energy mix. Saudi Arabia, for example, has established a regulatory and investment framework in its Vision 2030 document, which was first revealed in 2016 (Rashad 2016) and highlights the development of the Saudi solar energy sector, financed through its US$2 trillion sovereign fund (Zafar 2020b).

Support measures to increase energy from renewable sources in the Middle East and North Africa could include tax breaks for both producers and consumers of renewable energy. Or the construction of solar plants, wind parks, or hydro plants could be supported through provision of low-interest loans guaranteed by the government. Similarly, bond issuances that include tax exemptions for the interest repaid to investors can be used to raise the needed investments.

The support of supranational organizations can also be crucial—in both financial and knowledge transfer terms—for low- and middle-income countries in the region, as was the case for the Ouarzazate Solar Power Station (also called Noor Power Station) in Morocco. Substantial investments, with the support of the World Bank Group's International Finance Corporation playing a crucial role, are also under way in Egypt, the first of which is the Benban Solar Park in southern Egypt with a capacity of 1,465 megawatts (MW).

Geographically, the Middle East and North Africa is well suited to produce energy from renewable sources such as solar and wind energy. Large parts of the region are highly suitable for large-scale solar projects given their high photovoltaic power potential (map 3.2). With their vast stretches of bare land, the region's economies are also highly suitable for wind farms, and they have accelerated their efforts to increase the share of renewable sources in their energy mix. Box 3.13 provides some regional examples of economies that have initiated adoption of renewables to meet ambitious goals to increase the share of energy produced from renewable sources.

Although renewable energy sources contribute to cleaner air, air pollution can also impair the efficiency of solar energy-generating equipment such as photovoltaic panels. Similar to cloud cover, aerosol emissions affect solar radiation reaching the surface by absorbing and scattering sunlight, thereby decreasing energy yields. Moreover, the accumulation of dust (fine airborne particles) on the surface of panels physically obstructs radiation from reaching the photovoltaic cells, diminishing efficiency. These reductions can be substantial. In China, for example, the potential of photovoltaic panels between 1960 and 2015 was reduced by 11–15 percent on average because of elevated air pollution (Sweerts et al. 2019). In especially polluted eastern and northern

MAP 3.2

Photovoltaic Power Potential in the Middle East and North Africa: A Solar Resource Map

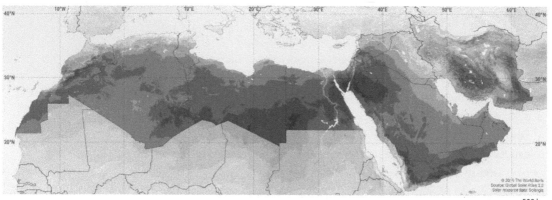

Long term average of PVOUT, period from 1994 (1999 in the East) to 2018

Daily totals:	3.6	4.0	4.4	4.8	5.2	5.6
Yearly totals:	1314	1461	1607	1753	1988	2045

KWh/kWp

⊢————⊣ 500 km

Source: Global Solar Atlas 2.0. ©World Bank.
Note: kWh = kilowatt-hour; kWp = kilowatts peak (peak power producible by a photovoltaic system or panel); PVOUT = photovoltaic power output.

BOX 3.13

Regional Examples of Investment in Renewable Energy Sources

Morocco: On the Solar Frontier

Morocco is pledging to significantly reduce greenhouse gas (GHG) emissions stemming from energy production and is investing heavily in its renewable energy sector to achieve its goal of having 52 percent of installed electricity generation capacity come from renewable sources by 2030 (Timmerberg et al. 2019). To this end, several wind farms and solar plants were constructed and are planned to be built and opened in Morocco.

The Noor-Ouarzazate complex is currently the world's largest concentrated solar power (CSP) plant, with an energy production capacity of 580 megawatts (MW), and Phase I was completed in 2016 when it was connected to the grid, providing 160 MW annually. The plant's construction was supported by a host of international organizations, including the World Bank, which provided a US$400 million loan. Phases II and III aim to increase the production capacity substantially and were commissioned in 2018 and 2019, respectively. Furthermore, under the "Integrated Wind Energy Project," five wind farms around the country, with a production capacity of 850 MW, are under construction or planned to be built in the coming years.

(continued)

BOX 3.13

Regional Examples of Investment in Renewable Energy Sources (*Continued*)

Tunisia: A Regional Pioneer in Renewable Energy Policies

Tunisia is considered a pioneer for renewable energy policies in the region. In 1995, tax exemptions for renewable energy equipment imports were introduced with Decree 95/744, and in 2005 investment subsidies for renewable energy technologies were introduced. A large share of the country's solar generation was installed under the umbrella of the Tunisian Solar Programme (PROSOL), which was originally formulated in 2005. In 2009, the government relaunched PROSOL, renamed it the Tunisia Solar Plan (TSP), and announced that the TSP would target a renewable energy penetration rate of 30 percent by 2030 with an associated law introduced in 2015 (OBG 2016). In 2016, a program was introduced to develop 1,000 MW subsequently in further projects (Enel Green Power 2017).

The GCC States: Ambitious Plans

The Gulf Cooperation Council (GCC) countries have also recognized the potential for renewable energy production because of their geographical characteristics, and most have put forward ambitious plans to increase their share of energy produced from renewable sources.

In the United Arab Emirates, plans are being finalized to build the Al Dhafra solar photovoltaic power project, which, once fully operational, will be one of the world's largest solar power plants and will replace the Noor Abu Dhabi solar power project as the largest one in the United Arab Emirates. It will offset around 2.4 million tons of carbon dioxide (CO_2) (equivalent to removing around 470,000 cars from the roads) and

provide electricity to approximately 160,000 households. It comprises 3.2 million solar panels and has a total capacity of over 2 gigawatts (GW) (Redondo 2020).

In Saudi Arabia, the Dumat Al Jandal wind farm, set to generate 400 MW of electricity, will be the biggest wind farm in the Middle East and will commence operations in 2022. It will power up to 77,000 households and displace almost 1 million tons of CO_2 (REVE 2020).

In Qatar, the government has pledged to increase its total capacity of renewables considerably in the coming years. One step to reach this goal is the construction of the Al Kharsaah Solar Power Plant, for which financial closure was reached in July 2020. Commencement is expected to start in 2022, and its capacity of 800 MW has the potential to account for around 10 percent of the country's peak electricity demand. Over its full life cycle, the project will save up to 26 million tons of CO_2 and will contribute to Qatar's commitment to host a carbon-neutral football world championship in 2022 (Verma 2020). Smaller projects in GCC countries include ones in Bahrain, Kuwait, and Oman.

All governments in the region have committed to increase the share of energy supplied from renewable sources, with Saudi Arabia being the most ambitious one by planning to meet an impressive 50 percent of its energy demand from such sources in 2030. Both Oman and the United Arab Emirates are targeting 30 percent coverage by renewables by 2030, while Qatar is committed to generate 20 percent of its energy from solar power. Kuwait wants to reach a share of 15 percent in 2030, and Bahrain aims for 15 percent in 2035.

China, the efficiency losses are even larger, with annual average reductions of 20–25 percent (Li et al. 2017). Early evidence for reduced efficiency of photovoltaic panels due to air pollution has also been found for the Islamic Republic of Iran, with yields reduced by up to 60 percent on highly polluted days (Asl-Soleimani, Farhangi, and Zabihi 2001). These findings imply that an improvement of air quality would increase the efficiency of solar electricity generation and hence its deployment, which in turn would further reduce air pollutant emissions, generating positive feedback effects (Li et al. 2017).

To increase the amount of energy derived from renewable resources and penetrate the Middle East and North Africa's energy mix, further investments in plants and supporting infrastructure are necessary. Although some of the region's economies are making progress in extracting energy from renewable sources, the penetration of the energy mix with renewables is still in the beginning stage for most of them. Increasing their shares of renewable energy will require continued investments, especially in countries that have so far rather neglected this area of development, such as the Islamic Republic of Iran, which produces less than 1 percent of its electricity from renewable sources (CMS 2017).

Furthermore, the expansion of alternative energy sources may require upgrades of the grid system to be suitable for renewable generators. Given their high potential for energy production from renewable resources, economies in the Middle East and North Africa could profit immensely from investing in this area. This is also true from the perspective of job creation because renewable energy projects were found to deliver three times more jobs per dollar than comparable ones in the fossil fuels sector (Garrett-Peltier 2018).

Increasing the share of renewables in the region's energy mix is also an important prerequisite for switching to electric vehicles. As discussed earlier, the electrification of transportation, both private and public, can be an important contributor to reducing air pollution in the region. For this to be true, though, the electricity that vehicles run on has to be derived from (relatively) clean sources. This in turn requires investments in the renewable energy system, both at the point of creation and the point of distribution, through grid storage solutions and charging stations. Hence, the potential for positive effects of a modal switch to the broader adoption of electric vehicles is heavily intertwined with the expansion of renewable energies and an orientation of the energy mix toward them.

To engage in those activities of renewable energy extraction best suited for each economy, a thorough analysis of possibilities must be conducted. As shown above, basically the entire region has positive prospects for the use of solar energy. Most of its economies are also suitable for the large-scale use of wind energy for electricity production. Hydropower, however, is a sector that has only limited potential for increasing energy production from this source. Suitable rivers (such as the Nile) are already heavily used in this respect. Furthermore, intensifying further dam building can present serious drawbacks regarding sediment blockage, leading to or exacerbating coastal erosion of coastlines downstream (as discussed in chapter 5).

Properly Pricing Energy Derived from Fossil Fuels

Electricity in the Middle East and North Africa is generated primarily from fossil fuel sources. As noted earlier, the region's energy mix is heavily skewed toward fossil fuels, with more than 95 percent of its electricity derived from these sources (Menichetti et al. 2019). The country with the largest share of its energy from renewables, Morocco, derives less than 10 percent of its energy from renewable sources.

The region's reliance on fossil fuels is primarily due to institutional factors, such as the subsidization of energy production from fossil fuels (El-Katiri 2014; Poudineh, Sen, and Fattouh 2018). To induce a switch to a more sustainable energy mix, investing in the development of renewable energies as well as adapting the institutional framework are imperative. Regulatory bodies must shift their focus away from supporting fossil fuels to address air pollution (Abbass, Kumar, and El-Gendy 2018).

Decreasing subsidies on electricity and energy more generally has been shown to improve air quality. Decreasing subsidies for electricity would increase the price of electricity, which has been shown to significantly reduce air pollution. For example, regarding pollutant emissions from industrial plants in Anhui, China, it has been found that a 1 percent increase in electricity prices leads to decreases of 1–5.8 percent in SO_2 and PM emissions concentrations for the metals and cement production sectors (Tan-Soo et al. 2019). Hence, electricity prices could be an effective policy tool for managing air pollution, which is highly relevant for the Middle East and North Africa's economies, given their unsustainable energy mix.

Reducing or removing subsidies for oil and gas for electricity production is an important step to be taken by the region's economies. Similar to the importance of fuel subsidy removal in transportation, removal of subsidies is crucial in the case of electricity production. Current regional pricing mechanisms make it hard to draw sensible cost comparisons between these energy sources, and they disadvantage the renewable energy sector. Hence, structural reforms of the regional energy market and pricing mechanisms are necessary to present renewable energy as a cost-competitive alternative to conventionally used fossil fuels (El-Katiri 2014).

Once the subsidies are removed, a next step would be to start putting a price on carbon, pollution, or both. Additional carbon pricing or trading schemes in the energy-producing sector are emerging around the world. However, the Middle East and North Africa is the only region worldwide where no country has a carbon tax, or a related emissions tax, or has implemented a trading mechanism for carbon emissions (World Bank 2021b).

Reducing Energy and Industrial Emissions: The Full Policy Spectrum

As for vehicle emissions, a broad range of measures can be taken to decrease air pollution from industrial sources and during the generation of energy. Tables 3.4 and 3.5 summarize these measures. The individual measures were assessed in terms of their expected timeline for implementation, financial costs, and expected effectiveness.

Policies to Reduce Emissions from Agricultural and Municipal Waste Burning

This section reviews the main options for reducing emissions from the burning of agricultural and municipal waste, including regulatory reform, incentive programs, awareness raising, and improving SWM.

Many Middle East and North Africa economies still practice open waste burning, in the form of both municipal waste and agricultural waste burning. The burning of municipal waste significantly increases air pollution to hazardous levels (Wiedinmyer, Yokelson, and Gullett 2014) and contributes to around 270,000 premature deaths worldwide (Kodros et al. 2016).

TABLE 3.4

Overview of Policy Options to Reduce Emissions from Industry and Energy Production

Main subobjective	Measures	Timeline for implementation	Financial cost	Effectiveness
Reduce industrial and energy emissions	Emission control (end-of-pipe technology)	medium	medium	high
	Energy efficiency	medium	high	high
	Taxes on emissions; establishment of emission markets	short	medium	medium
	Inspections and pollution charges	medium	low	high
	Remove fossil fuel subsidies	short	high	high
	Switch from coal, oil, and gas to renewables	long	medium	high
	Encourage consumers to purchase cleaner energy	medium	low	medium

Source: Based on World Bank data.
Note: The classifications of the qualifiers—timeline (short, medium, long); cost (low, medium, high); and effectiveness (low, medium, high)—are based on expansive literature reviews and the expert judgment of the report-writing team. (A full list of surveyed literature is available on request.)

TABLE 3.5

Detailed Description of Policy Options to Reduce Emissions from Industry and Energy Production

Main subobjective	Measure	Technology	Market-based instrument (MBI)	Regulations supporting MBIs
	Emission control (end-of-pipe technology)	Fume scrubbers and exhaust filters such as electrostatic precipitators and flue gas desulfurization	Subsidies for upgrading technology	Maximum pollution levels; ban on outdated technologies
	Energy efficiency	Replacing and upgrading outdated machines or parts	Subsidies for technology upgrades; lower running costs	Ban on low-efficiency technologies
	Tax emissions; establish emission markets	Precise monitoring system at source; automatic tax collection systems; emission trading system	Income for cleaner industrial companies (through trading); income for government through incurred fees	Carbon or GHG taxes
	Inspections and pollution charges	Upgrades of scrubbers, filters, waste management systems, staff training for inspections	Rewards (tax breaks and so forth) for eco-friendly behavior; penalties for overpollution	Mandated (unannounced) inspections per year; mandated maximum pollution levels
Reduce industrial and energy emissions	Remove fossil fuel subsidies	Provision of alternative energy sources; retrofitting of existing infrastructure	Savings for government; subsidies for switching	Remove subsidies
	Switch from coal, oil, and gas to renewables	Construction of solar and wind parks, hydro plants; upgrades in grid system; vocational training for laid-off employees	Government subsidies for RE technologies; feed-in bonuses; surcharges on energy from coal	Prescribing minimum share of RE for "green" energy badges
	Encourage consumers to purchase cleaner energy	Allow consumers to select their source of energy	Subsidies for using cleaner energy sources	Ban on advertisement for "dirty" energy sources
	Switch from coal, oil, and gas to renewables	Construction of solar and wind parks, hydro plants; upgrades in grid system; vocational training for laid-off employees	Government subsidies for RE technologies; feed-in bonuses; surcharges on energy from coal	Prescribing minimum share of RE energy for "green" energy badges
	Encourage consumers to purchase cleaner energy; remove fossil fuel energy subsidies	Allow consumers to select their source of energy upgrades to other energy sources	Subsidies for using cleaner energy sources; subsidies for energy switching; savings for government	Ban on advertising for "dirty" energy sources; redirect subsidies—for example, to RE, taxes on fossil fuels

Source: Based on World Bank data.
Note: GHG = greenhouse gas; RE = renewable energy.

Beirut's residential areas that experienced frequent waste burning episodes in their vicinity experienced severe increases in PM concentrations (Baalbaki et al. 2016). Translating these effects into health effects, the authors estimate that the short-term cancer risk on days when waste was burned increased twentyfold. Half the region's economies burn more than the world average on a per capita basis (figure 3.24).

Tackling the issue of municipal waste burning consists of setting up strict enforcement mechanisms and providing the means to properly dispose of waste. However, simply banning the practice combined with penalties for noncompliance may not be effective if the compliance and enforcement strategies are not prepared and effectively executed. This requires rigorous staff training and investment in infrastructure to enable authorities to effectively enforce those bans. It also requires information campaigns to inform residents about the adverse effects of their behavior.

Investing in SWM and adopting the principles of a circular economy are crucial to managing municipal waste burning. Presenting residents with a viable alternative to burning their waste crucially includes a strengthening of waste management services, which are still poorly developed in many economies of the Middle East and North Africa.[29] Hence, investment in proper waste management systems is imperative. The issue of weak SWM also relates to marine-plastic debris, where

FIGURE 3.24

Municipal Waste Burned Per Capita in the Middle East and North Africa, by Economy, 2010

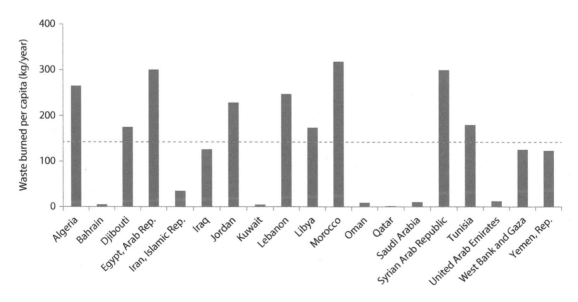

Source: Based on Wiedinmyer, Yokelson, and Gullett 2014.
Note: The orange line denotes world average. kg = kilogram.

inadequately disposed waste is one of the main culprits that lead to large flows of plastic entering the sea. Strengthening these systems and pushing for the adoption of circular economy principles therefore have the potential to address the adverse effects of poor waste management across issues. The inadequacy of the SWM system, reasons for this inadequacy, and propositions for how to resolve this issue are presented in more detail in chapter 4, where the principles of a circular economy are also illustrated. This concept is important to reduce the amount of waste that arises in the first place, which has ramifications for the amount of waste that is burned and pollutes the region's air.

The burning of agricultural waste is prevalent in some of the region's economies and contributes to poor air quality. Burning of (agricultural) waste, mostly in the form of crop residues, organic household waste, and the like, has always been a common practice in rural areas in Lebanon (Baalbaki et al. 2016). Every ton of rice straw burned produces emissions of 3 kg of PM; 60 kg of CO; 1,460 kg of CO_2; 199 kg of dust; and 2 kg of SO_2 (Rosmiza et al. 2014)—all of which can lead to various respiratory diseases and cancer. An important regional example is the "black cloud" season in autumn, when the burning of rice straws increased air pollution significantly in Cairo (Aboel Fetouh et al. 2013).

Emissions due to burning of crop residues are high in certain economies of the Middle East and North Africa, indicating the prevalence of this practice there (figure 3.25). The Islamic Republic of Iran and countries in North Africa show the highest emissions, in terms of tons of CO_2 per 100,000 people, stemming from the burning of crop residues, followed by Syria and Iraq. Countries in the GCC generally have lower emissions. Air pollution due to agriculture is a localized phenomenon; consequently, the burden associated with it is not spread out evenly across a country.

Air pollution reduces agricultural yields and in the process threatens an important source of income for low-income households. The adverse effects of rising temperatures caused by climate change on agricultural yields has been extensively investigated and validated by a host of scientific studies. However, air pollutants were also found to be detrimental to the productivity of farm plots by decreasing radiation and by changing temperature and precipitation patterns.

Emissions of black carbon (a component of PM) and ozone reduced wheat yields by around 36 percent in India from 1980 to 2010 relative to a situation absent of climate and air pollutant emission trends (Burney and Ramanathan 2014). The overwhelming fraction of losses were due to air pollutants rather than climate pollutants. Similarly, in China, elevated $PM_{2.5}$ concentrations had a significant adverse effect on average yields of wheat and corn from 2001 to 2010 (Zhou, Chen, and Tian 2018).

Addressing regional and local air pollution could have a more immediate positive effect on agricultural yields than the abatement of climate

FIGURE 3.25

Emissions (tCO₂e) from Crop Burning in the Middle East and North Africa, by Economy, 2018

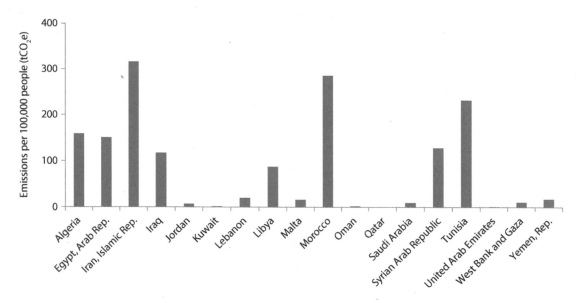

Source: Based on FAOSTAT database, Food and Agriculture Organization of the United Nations, https://www.fao.org/faostat/.
Note: tCO₂e = tons of carbon dioxide equivalent.

pollutants, where adverse effects but also gains materialize with a certain delay. Thus, such an approach could also counter some of the expected future yield losses resulting from climate change caused by the emission of climate pollutants (Burney and Ramanathan 2014).

The burning of agricultural waste in the Middle East and North Africa has to be controlled, with incentives and awareness programs for farmers being important tools. For rural areas in in the region, with their often-low population density and hence large unpopulated areas, effectively enforcing bans on agricultural waste burning can be burdensome. Bans on burning need to be complemented with incentives offering farmers alternative uses of their agricultural waste. Potential uses include animal food or fertilizer (with appropriate market structures) and biomass energy. Collection centers would facilitate disposal for farmers while also providing them with an additional source of income.

Information campaigns are also important to raise awareness among farmers about the adverse effects of crop-residue burning and the alternative options open to them. Egypt successfully implemented such measures, which helped reduce the "black cloud" phenomenon observed in the Greater Cairo region (box 3.14). Another option is "zero-tillage agriculture," which consists of plowing residues back into the soil, thereby increasing organic matter and enriching the soil (see box 3.15 for the Happy Seeder as an example).

BOX 3.14

Reducing the "Black Cloud" Phenomenon in Greater Cairo

In Egypt, rice is one of the most cultivated crops, mainly cultivated in the northeastern part of the country, primarily in the governorates of Dakahlyia, Kafr El-Sheick, and Shakyia (El-Dewany, Awad, and Zaghloul 2018). Since only around 20 percent of the crop residues resulting from these cultivation efforts have been used for further processing, the remainder was often left on the fields and later burned, contributing heavily to air pollution. This seasonal phenomenon of increased emissions after the harvesting season (typically in October and November) is called the "black cloud" and has been of great concern to policy makers in Egypt and specifically in Cairo (El-Dewany, Awad, and Zaghloul 2018; Hanafi et al. 2012). To give a sense of the magnitude of residue burning, according to observations by the National Aeronautics and Space Administration (NASA), at least 946 fires to burn leftover straw were recorded in Egypt's river Delta in 2014 (NASA 2014).

The impact of this practice on seasonal air quality in Greater Cairo was mentioned earlier and illustrated in figure 3.15, which shows the sources of $PM_{2.5}$ ($PM_{2.5}$ = particulate matter of 2.5 microns or less in diameter) in the Greater Cairo region for summer and fall 2010. The contribution of particulate matter (PM) from open burning of waste in autumn is considerably higher than in summer. This results mostly from the burning of agricultural waste, which typically takes place in October and November of each year. Hence, agricultural burning can lead to severe, seasonal burdens on air quality.

In an initial attempt to combat the uncontrolled burning of rice straw and other crop residues, the Egyptian Environmental Affairs Agency imposed a fine on waste burning ranging from LE 5,000 to LE 100,000 in 2015. Accompanying these measures, however, authorities also incentivized traders to buy the straw from farmers, paying them LE 50 (around US$3) per ton (*Egypt Today* 2019). Private companies then dry and bale the rice straw, subsequently to be used for purposes such as animal fodder, organic fertilizers, and also more unusual purposes such as in the furniture, cement, or brick industries. These companies receive a subsidy of LE 90 for each ton of rice straw that they process (El Dahan 2011). As of November 2020, the ministry had established 731 centers for collecting rice straw in six governorates, and more than 2 million tons of rice straw had been collected.

Furthermore, the ministry reported that 1,827 seminars have been held to raise farmers' awareness about the dangers of burning crop residues as well as to promote ideas about turning the residues into income-generating products (*Egypt Today* 2020). In addition, new projects have been planned to use the rice straw for the production of medium-density fiberboard. The projects are financed by Egyptian capital from the oil sector and are seen as effective solutions to support the efforts made to transform rice straw from an environmental challenge into an economic opportunity.

(continued)

BOX 3.14

Reducing the "Black Cloud" Phenomenon in Greater Cairo (*Continued*)

The potential of such efforts to reduce air pollution can be anticipated from figure B3.14.1, which shows PM_{10} (PM_{10} = particulate matter of 10 microns or less in diameter) concentrations in Greater Cairo computed as five-month moving averages from 2010 to 2016. In 2010 to 2014, there are clear seasonal patterns in air pollution. Beginning with fall each year, there are substantial increases in the concentration of PM_{10}, stemming mostly from the burning of agricultural waste in October and November, particularly rice straw. However, starting with 2015, this hump shape in autumn is much less pronounced, indicating significant decreases in the contribution of agricultural fires to PM air pollution. This exemplifies the impact that well-organized reform programs could have on improving air quality by positively incentivizing the alternative use of crop residues.

FIGURE B3.14.1

PM_{10} Concentrations (Five-Month Moving Average) in Greater Cairo, the Arab Republic of Egypt, 2010–16

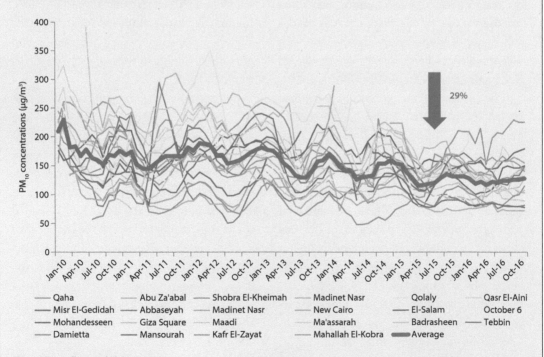

Source: Heger, Zens, and Meisner 2019.
Note: PM_{10} = particulate matter measuring 10 microns or less in diameter; μg/m³ = micrograms per cubic meter of air.

Especially for smaller farms, the investment costs of these technologies may be prohibitive. Municipalities could therefore procure them and rent them to farmers at a subsidized price or provide them without charge, as was done in India (DA&FW 2018). In Morocco, the World Bank supported a project, the "Plan Maroc Vert," which introduced such technologies in a bid to increase efficiency and reduce burning in the agriculture sector (World Bank 2014).

Raising Awareness

Raising awareness about air pollution and the damage it does is important and increases the demand for cleaner air. To effectively tackle the

---

BOX 3.15

Reusing Crop Residues as Fertilizer with the Happy Seeder

The Happy Seeder is an agricultural device that cuts crop residues, sows seeds into the soil, and deposits the sown crop residues over the area with the sown seeds as a natural fertilizer. By recycling crop residues in this way, farmers have less incentive to burn them, and stubble burning is reduced.

The environmental and economic effects of a range of in situ management practices were assessed in India, and the Happy Seeder was associated with the largest potential of reducing air pollution caused by burning stubbles. Use of the Happy Seeder instead of burning would reduce greenhouse gas (GHG) emissions by more than 78 percent. This would significantly reduce agriculture's contribution to overall GHG emissions in India and lower social costs in terms of particulate air pollution (Shyamsundar et al. 2019).

Systems using technology like the Happy Seeder can improve profits by 10–20 percent compared with farming practices involving burning. The higher profits stem from slightly higher yields and lower input costs for land preparation (Shyamsundar et al. 2019). This investigation shows the potential to gain air quality while simultaneously enhancing economic profits for farmers using innovative technologies like the Happy Seeder, and one of the authors stresses the potential for scaling up the adoption of such programs (CIMMYT 2019).

The Indian government has substantially increased subsidies for the in situ residue management (DA&FW 2018) and provides needed tools such as the Happy Seeder at low costs to farmers in an attempt to combat agricultural fires at the end of the harvesting season. With the scheme, which involves subsidies amounting to US$75 million, farms making up a total area of around 0.8 million hectares have been able to use the Happy Seeder technology in northwestern states of India. Scaling up the figures, it is estimated that direct farmer benefits amounted to US$131 million within one year.

problem of air pollution regarding its adverse effects on human life, it is imperative to raise residents' knowledge and awareness. Although most of the measures are based on the premise that governments want to tackle the issue of air pollution, it is also crucial to nurture the general population's demand for such interventions. Furthermore, the dissemination of such information gives residents a better understanding of why some measures to curb air pollution are taken. This lends more legitimacy and credibility to the governments' actions while simultaneously giving residents a way of monitoring the progress and the means to have some kind of checks and balances over government actions in this respect.

Properly informing residents about the current state of the air is an important step to take, but it necessitates the expansion of existing ground monitoring systems. To raise awareness about air pollution and its adverse effects, regular (day-to-day or real-time) monitoring of air pollutants and dissemination of this information to the public is an important precondition. As shown in the previous section, Middle East and North Africa economies often lack the appropriate infrastructure, in the form of ground monitoring stations. Monitoring stations to measure $PM_{2.5}$ concentrations are especially scarce in the region, hampering the dissemination of information about the most harmful pollutant not only to the public but also to researchers who conduct scientific studies based upon this information.

The region's economies have been starting initiatives to raise awareness about environmental issues in general as well as about air pollution in particular. Awareness on air pollution, its sources, and impacts has been considered low in the region. For example, through the subsidization of energy, raising energy efficiency has been perceived as rather unimportant by the general population given the low prices (El Khoury 2012). However, some countries in the Middle East and North Africa have initiated several campaigns to raise awareness about the various sources of air pollution (box 3.16).

Raising awareness about polluted air necessitates well-planned communication strategies, which can help build a broad base of support for abatement policies. To reach the general population with information about air pollution, it is crucial to broadly disseminate information about current trends in air quality as well as the potential effects of low quality; this information should be clear and easily understood. This can come in various forms such as introducing a traffic light system that indicates varying degrees of air pollution, as is done in Abu Dhabi (discussed earlier in box 3.2). To reach a broad base of residents effectively, information should also be spread via different channels (for example, newspapers, billboards, TV, radio, and social media) together with material

BOX 3.16

Public Awareness Programs on Air Pollution in the Middle East and North Africa

Egypt

During its project to tackle the black-cloud phenomenon caused by burning of crop residues in the Greater Cairo region (discussed in box 3.14), Egypt has recognized that raising awareness about the problem is crucial. The responsible authorities have held more than 1,800 seminars to inform farmers about the adverse effects associated with the practice of agricultural waste burning and how crop residues can be turned into products generating additional income for them (*Egypt Today* 2020). A recently approved US$200 million World Bank project includes awareness campaigns to better inform the public about the links between air pollution and related diseases. This is seen as crucial to reach a broader audience and maximize the impact of related measures.

Islamic Republic of Iran

In Tehran, the Islamic Republic of Iran, authorities report daily measures of air quality for various locations in the city and disseminate them via different channels, including billboards, mobile phone apps, and websites[a] as well as through social media platforms (Heger and Sarraf 2018). The reports provide information on concentrations of $PM_{2.5}$, NO_2, and SO_2 on an hourly basis, along with a forecast for the next three days and educational information to increase awareness of the definitions of air quality.

Morocco

In Morocco, the Qualit'Air program of the Fondation Mohammed VI pour la Protection de L'Environment launched its first driver-awareness campaigns in 2005 to highlight the air pollution caused by vehicles. It has since launched numerous initiatives in cooperation with private companies to educate drivers about global warming and air pollution in a bid to reduce their greenhouse gas (GHG) emissions through the development of voluntary carbon-offsetting programs. In 2015, the foundation designed a platform to raise awareness among children about air pollution and global warming and also electrified rural schools with clean energy through the installation of solar panels (FM6E 2016).

Bahrain

In 2019, "Beating Air Pollution" was the theme of the United Nations World Environment Day. The "Mask Challenge" that aimed to promote awareness about air pollution has been endorsed by large parts of the population in Bahrain. Companies such as Aluminum Bahrain have publicly announced their support for efforts to tackle air pollution and protect the environment. The Indian embassy in Bahrain also organized a cycling event with the theme "Fight Air Pollution" in cooperation with the United Nations Environment Programme (UNEP 2019).

(continued)

BOX 3.16

Public Awareness Programs on Air Pollution in the Middle East and North Africa (*Continued*)

United Arab Emirates

Abu Dhabi in the United Arab Emirates has developed a color-coded air quality index for public communications, together with guidelines targeted to particular segments of the community that might be vulnerable when pollution rises above certain levels.[b] It also provides advice on how to contribute to improving air quality, mostly through careful use of energy at home (including use of air conditioning) and with transportation (including vehicle maintenance, use

of cleaner fuels, and greater use of public transportation or carpooling).

Policies to reduce energy use include the reduction of fuel subsidies and encouragement of switching to cleaner vehicles, new transportation and industry technologies, public awareness campaigns, and research that includes collaboration between the Environmental Agency Abu Dhabi (EAD), Health Authority – Abu Dhabi (HAAD), academia, and the private sector (Mohamed 2017).

a. For example, see the "Tehran Air Pollution: Real-Time Air Quality Index (AQI)," World Air Quality Index Project website: https://aqicn.org/city/tehran/.
b. For more information, see the Environment Agency – Abu Dhabi (EAD) Air Quality Monitoring System platform: https://www.adairquality.ae/#LiveData.

that provides guidance for how individual actions can help reduce air pollution. Furthermore, clear communication campaigns can help foster a broad base of public support for some of the measures undertaken by the government to curb air pollution, reducing backlash against policies that would otherwise be met with discontent.

Awareness programs should identify and explain the various sources of air pollution and the actions an individual can take to combat it. As elaborated in the previous section, air pollutants can stem from different sources, and the general population often lacks knowledge about them. Moreover, even when the public has a sense of where the pollution is coming from, there may be little awareness about what actions an individual can take. Disseminating information about how such individual actions can contribute to increasing air quality is important to give people the sense that each one of them can be part of the solution. Such individual actions include the switch from a personal vehicle to nonmotorized option or public transportation, lowering energy consumption (which in the Middle East and North Africa is mainly derived from fossil fuels) or consumption of goods that are less energy-intensive.

Furthermore, the proper insulation of houses and installation of appliances such as air conditioners with higher efficiency levels would lead to lower energy consumption and leakage in the residential sector, an important point because air conditioning accounts for a large share of consumed energy.[30]

Proper environmental education early on is an important factor in increasing residents' engagement with the issues of environmental degradation and air pollution. The integration of environmental issues in the agenda of schools, universities, and other community initiatives in the education sector is important to increase knowledge about environmental issues. This raises awareness for the adverse effects that individual actions also can have on the environment in general and air pollution in particular. There has been progress in integrating the environmental agenda in university curricula in the Arab world (Saab, Badran, and Sadik 2019). However, school curricula, which have the potential to reach a much broader audience, are often outdated, neglecting the effect of climate change. Educating residents about the dangers posed by air pollution and the possible ways to avoid it, can be crucial to effectively address this problem in an inclusive manner.

To raise public awareness about air pollution and its impacts, more initiatives are necessary. Recent efforts by several cities and economies in the Middle East and North Africa are a step in the right direction (box 3.16); however, they should be reinforced to expose air pollution as the threat that it is to the health of the region's residents. Making people aware of the negative effects of air pollution and how they themselves can contribute to the mitigation of it is important to achieve a broad mobilization and nudge people to change their behavior. As already mentioned, the possibilities to change behavioral patterns is an option to reduce air pollution, cutting across the different sectors and activities that are sources for the degradation of air quality. Governments in the Middle East and North Africa could support initiatives in this area by providing financial resources for them to garner supporters and attention.

Increasing Energy Efficiency in the Residential Sector

The built environment accounts for a substantial share of energy use in the region's economies. In Egypt, the residential sector accounts for 18 percent of energy consumption and 51 percent of electricity consumption, primarily in lighting and cooling. In Saudi Arabia, cooling accounts for 70 percent of residential electricity consumption, representing around 40 percent of total annual electricity consumption (El Khoury 2012). The greater use of more efficient cooling and mechanical systems, better insulation of buildings, and installation of light-emitting diode (LED) lightbulbs instead of traditional ones can have a large cumulative effect on energy consumption. Hence, improving energy efficiency in buildings can play an important role in reducing urban air pollution as well as in reducing energy costs for households.

There is scope for increased energy efficiency, but low energy prices do not provide incentives for energy savings. Traditionally, energy subsidies have formed part of the "social contract" in the GCC countries as well as in energy-importing Middle East and North Africa economies, even though better-off residents generally benefited the most. Successful change has been politically challenging and requires careful communications campaigns and targeted social protection measures (Coady, Flamini, and Sears 2015). The reduction in global oil prices in 2014 provided an opportunity for subsidy reduction in many of the region's economies (Krane and Monaldi 2017).

A recent study estimated the potential for increasing energy efficiency to be an average of 20 percent of primary energy supply in the region's economies by 2025 (World Bank 2016), with Saudi Arabia alone accounting for over one-third of the potential. There is scope for energy savings in the transportation sector through improved, lower-carbon-emitting public transportation systems and traffic management. All of these improvements will help to improve urban air quality.

Middle East and North Africa economies have recognized this potential and have taken measures to increase energy efficiency in the residential sector. The importance of the residential sector in energy consumption and its inefficiency is being recognized by an increasing number of the region's governments, and some economies are now making progress toward improving building codes and emissions standards for home appliances to increase energy efficiency in the residential sector (World Bank 2016). Some countries have combined reforms with programs to improve services, as in the case of Egypt, which has supported a large-scale program to help households switch from using liquefied petroleum gas (LPG) to grid-connected natural gas.[31]

Morocco pushed for higher efficiency standards as well as thermal efficiency regulations for buildings through its 2014 adopted Réglementation Thermique de Construction (RTCM). Although the regulation is only partially applied, extending its scope and application is the subject of major efforts. Because of its commitment to raising energy efficiency, Morocco received funding totaling €20 million from a program financed by European governments to support the integration of energy-efficiency measures into 12,000 homes during 2021 (*Econostrum* 2020). Support programs to raise efficiency both at the building stage of new dwellings and through retrofitting of existing ones is an important precursor that could also form part of COVID-19 recovery strategies. Compared with Morocco's efforts, the reforms to raise residential energy efficiency were less far-reaching in the GCC countries and the Islamic Republic of Iran.

Labeling energy-efficient appliances can help consumers make more informed choices and nudge producers to build more energy-efficient ones. Raising awareness of the importance of energy efficiency to tackle air pollution is essential. One approach to promoting energy efficiency is to introduce obligatory tests and labeling schemes, such as the ones that were implemented or are under way in the GCC countries regarding air conditioners (Andreula 2019). These measures have the dual advantage of raising awareness among consumers about energy inefficiency (via the labeling schemes) and incentivizing producers to invest in the production and distribution of more-efficient appliances (because of demand shifts).

Greening Cities

Greening cities—for example, by increasing the number of public parks and the use of vegetated building roofs and walls—is good for air quality. Increasing the share of vegetation in cities has been found to significantly enhance air quality; its effects have varied with the types of vegetation used. Because some forms of vegetation (for example broad-leaved trees) can trap air pollutants stemming from vehicles, careful impact evaluation of green infrastructure should be performed before its installation to improve the impacts on air quality.

The link between air quality and urban vegetation has been investigated heavily. On open roads, the use of wide and tall vegetation with low porosity leads to downwind pollutant reductions, and low-level green infrastructure such as hedges can improve air quality conditions. Green walls and roofs on buildings (such as cultivation of greenery on them) are also an effective means to reduce air pollution (Abhijith et al. 2017). The potential of greener cities for air pollution abatement has also received some attention in the Middle East and North Africa, with cities increasing their share of green areas.

In addition to decreasing air pollution, green spaces positively affect the general livability of cities and can contribute to mitigating the effects of excessive temperatures. City governments could support the spread of such measures by introducing regulations mandating a minimum share of green area per district or by granting subsidies for newly built buildings that include green infrastructure. Box 3.17 provides an example of the potential for greener spaces in Cairo.

BOX 3.17

Green Space in Cairo, the Arab Republic of Egypt

There is increasing understanding of the role of green space and "green infrastructure" in mitigating urban heat island effects and in improving broader social well-being, health, and quality of life for residents as well as improving air quality. Cairo suffers from a shortage of green space, which has been exacerbated in recent decades by luxury developments and the enclosure of public spaces along the Nile. Three policy bodies and one executive-level government body guide green space provision: the Ministry for Planning and Administrative Reform, the Environmental Affairs Agency, the Organization for Urban Harmony, and the Cairo Cleaning and Beautification Agency.

Private sector organizations provide 67 percent of Cairo's green space and charge entrance fees, membership fees, or both. These spaces are used by only 30 percent of the urban population and include gated compounds and sporting clubs, smaller green areas, and play spaces. The remaining 70 percent use public spaces that are, however, unevenly distributed. The distance to the nearest green space is on average 700 meters. The population groups with the worst access to green space (more than 1,350 meters away, on average, from even the smaller and "pocket" green spaces) are those in informal settlements. Surveys have shown that women feel safer in green spaces where they can be with other women and family members, even where these spaces are small. Surveys have also indicated the

potential of religion in supporting broader green and sustainability awareness.

There is an argument for an integrated approach to green-space provision, adapted to local neighborhoods. This approach would involve public-private partnerships and effective community involvement and encompass smaller "pocket" areas and street greening, playgrounds, local sports facilities, and larger green spaces. There is scope also to convert wasteland, as illustrated by the Al-Azhar park in Eastern Cairo, a 30-hectare park created in 2005 with the assistance of the Aga-Khan Foundation and other donors. The foundation also helped restore historic monuments and supported development activities for neighboring communities.

There is scope for expanding initiatives that have been piloted, such as green roofs and vertical gardens, hillside greening on areas not suitable for construction, converting unused military and government land (nearly 840 hectares), "greening" the riverbanks and cemeteries (500 hectares), and greening unused industrial land and airport and railway premises. (The former Imbaba airport premises, for example, have been converted into a local park, overcoming arguments for developing the area for high-density housing.) Although both private and public green areas have a role in green infrastructure provision, the biggest challenge will remain the provision of accessible green space for Cairo's poorest residents.

Source: Kafafy 2010.

The Al Shaheed Park in Kuwait is another prominent regional example where the main motivation was to protect the city from sandstorms and reduce air pollution. It is Kuwait's largest park, with a green roof area of almost 20,000 square meters, and an artificial lake serves as a water reservoir during the hot season. It accommodates several amenities, and its continuous park character, preserved by equipping most buildings with an accessible green roof, give inhabitants a safe retreat from the hot climate (ZinCo GmbH 2020).

Roofs of buildings can also be used for energy production by installing solar panels, which support the overall change of the energy mix in the region's cities as well as raising awareness for the possibilities of individual energy production in the population. Both local and federal governments could support such measures by granting subsidies for new buildings that include the integration of photovoltaic solutions in their design and construction. Furthermore, to incentivize the adoption of such solutions by companies but also private households, implementing efficient solutions for the feed-in of energy produced in this way in the general electricity system as well as establishing rates for feeding in energy are a necessary prerequisite to make such investments more attractive.

Another possibility for public entities in this regard could be to act as a role model by mandating a minimum number of solar panels or similar technologies on government buildings. In the United Arab Emirates, recent projects like ones by SirajPower[32]—consisting of installing solar panels covering a roof area of around 70,000 square meters on more than 100 buildings in the Jebel Ali Free Zone East and West—highlight the potential for such efforts. Displacing about 7,500 metric tons of CO_2 per year, which corresponds to more than 125,000 trees being planted, this project has a capacity of 6.75 MW (*Construction Week* 2020). Expanding such efforts on a larger scale can effectively help spread the use of renewable energy sources apart from installing large solar or wind parks and can raise awareness of the general population by making the use of renewable sources more tangible.

Tables 3.6 and 3.7 summarize the plethora of options to tackle air pollution besides the emissions stemming from vehicles and industrial sources. Table 3.6 assesses some main aspects of the measures regarding their implementation, and table 3.7 provides closer descriptions of them.

TABLE 3.6

Overview of Policy Options to Reduce Emissions from Waste Burning and for Other Objectives

Main subobjective	Measures	Timeline for implementation	Financial cost	Effectiveness
Reduce waste burning	Stop agricultural waste burning	short	low	high
	Improve SWM (including improved collection and recycling)	medium	high	high
	Stop household waste burning	short	low	high
Other policies	Promote individual green energy production	medium	medium	high
	Promote greening of urban areas	medium	medium	medium
	Raise energy efficiency in the residential sector	medium	medium	high
	Provide information and awareness-raising campaigns (for example, for alternative energy sources)	short	low	medium

Source: Based on World Bank data.
Note: The classifications of the qualifiers—timeline (short, medium, long); cost (low, medium, high); and effectiveness (low, medium, high)—are based on expansive literature reviews and the expert judgment of the report-writing team. (A full list of surveyed literature is available on request.) SWM = solid waste management.

TABLE 3.7

Detailed Description of Policy Options to Reduce Emissions from Waste Burning and for Other Objectives

Main subobjective	Measures	Technology	Market-based instrument (MBI)	Regulations supporting MBIs
Reduce waste burning	Stop agricultural waste burning	"Happy Seeder"; biomass energy plants	Creation of a market for crop residues (for example, for waste-to-energy), penalties for noncompliance	Ban on burning agricultural waste
	Improve SWM (including improved collection and recycling)	Extended recycling capabilities, enhanced pickup services	Subsidies recycling; charges on landfills; open dumps	Ban on open dumps and landfills
	Stop household waste burning	Enhanced waste pick-up system; awareness programs	Subsidies for waste pickup services; penalties for noncompliance	Ban on open waste burning by private individuals
Other policies	Promote individual green energy production	Solar panels on buildings, windmills, feed-in system	Subsidies for solar panels, windmills; feed-in system for unused energy	Minimum number of solar panels for new (government) buildings
	Promote greening of urban areas	Public space investments, for example, in parks; greening of rooftops and buildings' walls	Subsidies for green roofs and walls	Minimum amount of green space per urban district
	Raise energy efficiency in the residential sector	More efficient cooling systems; installation of LED lightbulbs; better insulation for new dwellings	Charges on inefficient appliances; raise prices through removal of electricity subsidies	Ban on most inefficient appliances; introduction of residential building codes
	Provide information and awareness-raising campaigns (for example, for alternative energy sources)	Creation and administration of awareness campaigns	Subsidies for creation and promotion of campaigns	None

Source: Based on World Bank data.
Note: LED = light-emitting diode; SWM = solid waste management.

NOTES

1. The GRID framework, as laid out in World Bank and IMF (2021), refers to a set of integrated, longer-horizon strategies to repair the structural damage caused by the COVID-19 pandemic and accelerate climate change mitigation and adaptation efforts while also restoring momentum on the twin goals of poverty reduction and shared prosperity. A GRID growth path would have fewer emissions, less environmental degradation, and stronger ecosystems while at the same time boosting resilience and inclusion, if managed properly.
2. For more information about the Gallup World Poll and access to its data, see http://worldview.gallup.com.
3. "Ambient (Outdoor) Air Pollution." Online fact sheet, World Health Organization (updated September 22, 2021): https://www.who.int/news-room/fact-sheets/detail/ambient-(outdoor)-air-quality-and-health.
4. The data and research results presented in this section were obtained using the Global Burden of Disease (GBD) tool of the IHME's Global Health Data Exchange (http://ghdx.healthdata.org/gbd-results-tool). It should be noted that the IHME data represent estimates for the various health effects of AAP based on global models. The figures and numbers presented in this section are derived from these global models and hence are not based on direct observation. Although it would be desirable to base the analysis on directly observed figures, data limitations and lack of comprehensive studies (discussed later in this chapter, within the policy review section) necessitated the use of estimated data as a second-best option.
5. As noted earlier, this report focuses on ambient (outdoor) air pollution rather than household (indoor) air pollution. In most of the Middle East and North Africa economies—except for some of the poorest (namely, Djibouti and the Republic of Yemen) and in certain rural areas of other economies—indoor air pollution is not a big concern.
6. Air pollution has transboundary health impacts, because it is transported across national borders (see, for example, Anenberg et al. 2014; Zhang et al. 2017). This makes coordination at a supranational level crucial to stem the challenge presented by air pollution.
7. The link between air pollution and diabetes prevalence is discussed at length in Rajagopalan and Brook (2012).
8. As figure 3.8 shows, Malta has far lower death rates due to AAP than all other countries in the region. This is mainly due to its unique geographical location, being the only island. Furthermore, its membership in the European Union (EU) and hence its obligation to adhere to the EU's stricter regulations sets it apart from the other economies in this respect.
9. The higher relative effect for wasting stems from the lower prevalence of children suffering from it relative to stunting. In the sample under investigation, almost 20 percent of the children were stunted (having a low height for a certain age), whereas 8 percent of the children fit the definition of wasted (having low weight for a certain height).
10. See Barnett et al. (2006) for Australia and New Zealand, and Samet et al. (2000) for the United States.
11. Conversely, air pollution increases obesity levels by reducing physical exercise (An et al. 2018; Deschenes et al. 2020). Of course, other factors driving the comorbidity risk include smoking and unhealthy diets.

12. For China, see Chen et al. (2020); for the United States, see Son et al. (2020). In addition, Venter et al. (2020) investigated levels of different pollutants in 27 countries worldwide using remote-sensing techniques validated with air quality monitoring stations during the spring 2020 lockdown. They found that air pollution was reduced significantly, with NO_2 levels declining by 29 percent on average, ozone (O_3) by 11 percent, and $PM_{2.5}$ by 9 percent. Using exposure-response functions, they estimated that 7,400 deaths and 6,600 pediatric asthma cases were avoided in the two weeks following the lockdown.

13. Kodjak (2015) provides benefit-cost analyses on several emission standards that are or will be implemented and shows that benefits outweigh costs by a factor of 1.4 to 16, depending on the emission standard considered. Blumberg (2004) conducted a cost-benefit analysis on the introduction of an ultra-low-sulfur fuels policy in Mexico City and projected that annual net benefits exceed US\$9 billion. Li, Lu, and Wang (2020) evaluated the recent enforcement of high-quality gasoline standards in China and determined that they reduced air pollutants by 12.9 percent on average; the net benefit of the measures is about US\$26 billion.

14. Wang et al. (2014) provide cost-benefit calculations for several such policies for which benefits significantly outweigh costs.

15. Coady et al. (2017) estimate that phasing out fossil fuel subsidies would have reduced global carbon emissions by 21 percent and deaths related to fossil-fuel-induced air pollution by 55 percent in 2015 while also raising tax revenues by 4 percent and social welfare by 2.2 percent of global GDP. In 2015, the removal of such subsidies would have raised tax revenues by an estimated 3.8 percent of global GDP, and net economic benefits (that is, environmental benefits less economic costs) would have amounted to 1.7 percent of global GDP (Coady et al. 2019).

16. To set National Ambient Air Quality Standards (NAAQS), the US Environmental Protection Agency (EPA) designated six "criteria air pollutants," which are common air pollutants that can harm human health and the environment and cause property damage. These pollutants include PM, photochemical oxidants (including ozone), CO, sulfur oxides, NO_X, and lead. The EPA calls them "criteria" air pollutants "because it sets NAAQS for them based on the criteria, which are characterizations of the latest scientific information regarding their effects on health or welfare" ("Criteria Air Pollutants," US EPA website: https://www.epa.gov/criteria-air-pollutants).

17. For example, about 40 percent of subsidized fuel is smuggled out of Libya.

18. These countries include the GCC countries, Algeria, the Islamic Republic of Iran, Iraq, and Libya. Syria and the Republic of Yemen are also exporting countries but to a lower degree, and trade relations have been rocked by recent unrest.

19. Source apportionment studies are bottom-up approaches to understanding air pollution concentrations, whereas emissions inventory and dispersion modeling are top-down approaches.

20. In Ahvaz, Islamic Republic of Iran, industrial processes accounted for around 38 percent of PM air pollution from anthropogenic sources, followed closely by traffic emissions (around one-third) and waste burning (10 percent), with the remainder from unspecified anthropogenic sources in 2010–11

(Sowlat et al. 2013). In Kuwait City, traffic accounted for about two-thirds of anthropogenic $PM_{2.5}$ emissions in 2004–05 and industry for about one-third (Alolayan et al. 2013). In Jeddah, Saudi Arabia, industry was the dominant source of $PM_{2.5}$ and PM_{10} in 2011, accounting for almost 60 percent of PM_{10} emissions and close to 50 percent of $PM_{2.5}$ emissions from human sources (Khodeir et al. 2012).

21. Information on Egypt's emissions inventories comes from the World Bank's work with the Ministry of Environment, Arab Republic of Egypt.

22. Some reports are already available on how to best reduce emissions in Greater Cairo (Larsen 2019), Tehran (Heger and Sarraf 2018), and Riyadh (Heger et al. 2019). More and better studies on air pollution sources are a precursor to specific analyses that allow for detailed policy recommendations.

23. The IEA fuel subsidy estimates for the countries available (specified in the paragraph) are derived using the price-gap methodology. The estimates included in the paragraph capture subsidies only on transport oil. For the IEA energy subsidies database, see https://www.iea.org/topics/energy-subsidies.

24. The economic reforms in Egypt received support in the form of a US$1.15 billion development policy financing loan from the World Bank and US$12 billion in the form of an Extended Agreement from the International Monetary Fund.

25. For more information about Sustainable Development Goal (SDG) 11 on Sustainable Cities and Communities, including its specific targets and indicators, see the United Nations' SDG 11 Knowledge Base page: https://sdgs.un.org/goals/goal11.

26. For example, evaluating the effectiveness of regulations to decrease industrial emissions set forth in China's 11th Five-Year Plan (2006–2010), Zhang et al. (2015) show that end-of-pipe facilities (like flue-gas desulfurization) and increased technology efficiency significantly reduced major industrial pollutant emissions.

27. The effectiveness of emissions taxes in increasing air quality has been shown in the case of the Chinese "pollution tax" imposed in 2018 (Hu et al. 2018). The tax policy was generally successful in reducing levels of air pollutants like SO_2, NO_X, PM_{10}, and $PM_{2.5}$. However, Hu et al. (2018) stress that the significant effects were only in regions with large economic scale and in sectors with high emission intensity. Hence, it could prove useful to have different tax rates for different industries and to give administrative units some discretion in their implementation of the taxes.

28. In Western Balkan, Central European, and Baltic countries—comparable in income and development to middle-income Middle East and North Africa countries—more than 60 percent of manufacturing firms were monitoring energy consumption, and a third were forming explicit targets for it, according to the World Bank Enterprise Surveys. In addition, almost 40 percent of manufacturers in Central Europe and the Baltic states have adopted specific measures to rein in excessive energy consumption. In the Western Balkans, this figure was lower (30 percent), but that share still exceeds that of the Middle East and North Africa surveyed group by around 10 percentage points.

29. See Zafar (2020a). This will also be discussed in more detail in chapter 4 of this report.

30. For example, the cooling of buildings accounts for around 70 percent of electricity consumption in Saudi Arabia (El Khoury 2012).
31. Egypt Household Natural Gas Connection Project, appraised in 2014. Egypt is also working to reform broader energy markets.
32. For more information, see the SirajPower website: http://www.sirajpower .com/.

REFERENCES

Abbass, R. A., P. Kumar, and A. El-Gendy. 2018. "An Overview of Monitoring and Reduction Strategies for Health and Climate Change Related Emissions in the Middle East and North Africa Region." *Atmospheric Environment* 175: 33–43.

Abdallah, C., C. Afif, S. Sauvage, A. Borbon, T. Salameh, A. Kfoury, T. Leonardis, et al. 2020. "Determination of Gaseous and Particulate Emission Factors from Road Transport in a Middle Eastern Capital." *Transportation Research Part D: Transport and Environment* 83: 102361.

Abhijith, K. V., P. Kumar, J. Gallagher, A. McNabola, R. Baldauf, F. Pilla, B. Broderick, S. Di Sabatino, and B. Pulvirenti. 2017. "Air Pollution Abatement Performances of Green Infrastructure in Open Road and Built-Up Street Canyon Environments–A Review." *Atmospheric Environment* 162: 71–86.

Aboel Fetouh, Y., H. El Askary, M. El Raey, M. Allali, W. A. Sprigg, and M. Kafatos. 2013. "Annual Patterns of Atmospheric Pollutions and Episodes over Cairo, Egypt." *Advances in Meteorology* 2013 (Special Issue): 984853.

AFD (Agence Française de Développement). n.d. "Egyptian Pollution Abatement Project (EPAP)." Project sheet, AFD, Paris. https://www.afd.fr/en/carte-des -projets/industrial-pollution-abatement.

Ahmed, H. O. 2020. "Algeria Plans Fuel Price Rise, Sees Economy Contracting 2.6% This Year." Reuters, May 15. https://www.reuters.com/article /algeria-economy-idUSL8N2CX5LJ.

Al-Hemoud, A., L. Al-Awadi, M. Al-Rashidi, K. A. Rahman, A. Al-Khayat, and W. Behbehani. 2017. "Comparison of Indoor Air Quality in Schools: Urban vs. Industrial 'Oil & Gas' Zones in Kuwait." *Building and Environment* 122: 50–60.

Alimohammadi, H., S. Fakhri, H. Derakhshanfar, S.-M. Hosseini-Zijoud, S. Safari, and H. R. Hatamabadi. 2016. "The Effects of Air Pollution on Ischemic Stroke Admission Rate." *Chonnam Medical Journal* 52 (1): 53–58.

Alolayan, M. A., K. W. Brown, J. S. Evans, W. S. Bouhamra, and P. Koutrakis. 2013. "Source Apportionment of Fine Particles in Kuwait City." *Science of the Total Environment* 448: 14–25.

Amini, H., N. T. T. Nhung, C. Schindler, M. Yunesian, V. Hosseini, M. Shamsipour, M. S. Hassanvand, et al. 2019. "Short-Term Associations Between Daily Mortality and Ambient Particulate Matter, Nitrogen Dioxide, and the Air Quality Index in a Middle Eastern Megacity." *Environmental Pollution* 254 (Part B): 113–21.

An, R., M. Ji, H. Yan, and C. Guan. 2018. "Impact of Ambient Air Pollution on Obesity: A Systematic Review." *International Journal of Obesity* 42 (6): 1112–26.

Andersson, J. J. 2019. "Carbon Taxes and CO2 Emissions: Sweden as a Case Study." *American Economic Journal: Economic Policy* 11 (4): 1–30.

Andreula, E. 2019. "Sultanate of Oman—Energy Efficiency and Labeling Requirements for Air Conditioners." News item, UL LLC, Northbrook, IL. https://www.ul.com/news/sultanate-oman-energy-efficiency-and-labeling-requirements-air-conditioners-0.

Anenberg, S. C., J. J. West, H. Yu, M. Chin, M. Schulz, D. Bergmann, I. Bey, et al. 2014. "Impacts of Intercontinental Transport of Anthropogenic Fine Particulate Matter on Human Mortality." *Air Quality, Atmosphere & Health* 7 (3): 369–79.

AP (Associated Press). 2020. "Syria Reduces Fuel Subsidies as Economic Crisis Deepens." Al Arabiya News, May 9. https://english.alarabiya.net/business/energy/2020/05/09/Syria-reduces-fuel-subsidies-as-economic-crisis-deepens.

Arab News. 2014. "Energy Efficiency Vital for KSA's Economic and Social Development." Statement by Prince Abdulaziz bin Salman, *Arab News,* July 20.

Arimura, T. H., N. Darnall, R. Ganguli, and H. Katayama. 2016. "The Effect of ISO 14001 on Environmental Performance: Resolving Equivocal Findings." *Journal of Environmental Management* 166 : 556–66.

Arlinghaus, J., and K. van Dender. 2017. "The Environmental Tax and Subsidy Reform in Mexico." Taxation Working Papers No. 31, Organisation for Economic Co-operation and Development, Paris.

Asare, J., and M. Reguant. 2020. "Low Oil Prices during COVID-19 and the Case for Removing Fuel Subsidies." Policy brief, International Growth Centre, London.

Asl-Soleimani, E., S. Farhangi, and M. S. Zabihi. 2001. "The Effect of Tilt Angle, Air Pollution on Performance of Photovoltaic Systems in Tehran." *Renewable Energy* 24 (3–4): 459–68.

Atalayar. 2020. "Syrian Government Implements a More than 100% Increase in Diesel Fuel." *Atalayar,* October 22. https://atalayar.com/en/content/syrian-government-implements-more-100-increase-diesel-fuel.

Atlas Magazine. 2020a. "Algeria: Introduction of a Pollution Tax." *Atlas Magazine,* July 1. https://www.atlas-mag.net/en/article/introducing-a-pollution-tax.

Atlas Magazine. 2020b. "Towards the Abolishment of the Pollution Tax in Algeria." *Atlas Magazine,* October 14. https://www.atlas-mag.net/en/article/towards-the-abolishment-of-the-pollution-tax-in-algeria.

Baalbaki, R., R. El Hage, J. Nassar, J. Gerard, N. B. Saliba, R. Zaarour, M. Abboud, et al. 2016. "Exposure to Atmospheric PMs, PAHs, PCDD/Fs and Metals Near an Open Air Waste Burning Site in Beirut." *Lebanese Science Journal* 17 (2): 91–103.

Barnett, A. G., G. M. Williams, J. Schwartz, T. L. Best, A. H. Neller, A. L. Petroeschevsky, and R. W. Simpson. 2006. "The Effects of Air Pollution on Hospitalizations for Cardiovascular Disease in Elderly People in Australian and New Zealand Cities." *Environmental Health Perspectives* 114 (7): 1018–23.

BBC News. 2019. "India Air Pollution: Will Gujarat's 'Cap and Trade' Programme Work?" *BBC News*, July 19. https://www.bbc.com/news/world-asia-india-48744163.

Benes, K., A. Cheon, J. Urpelainen, and J. Yang. 2015. "Low Oil Prices: An Opportunity for Fuel Subsidy Reform." Working Paper 364, Columbia University, New York.

Blumberg, K. 2004. "Benefit-Cost Analysis of Ultralow Sulphur Fuels for Mexico." Unpublished manuscript, International Council on Clean Transportation.

Boogaard, H., K. Walker, and A. J. Cohen. 2019. "Air Pollution: The Emergence of a Major Global Health Risk Factor." *International Health* 11 (6): 417–21.

Börjesson, M., J. Eliasson, M. B. Hugosson, and K. Brundell-Freij. 2012. "The Stockholm Congestion Charges—5 Years On. Effects, Acceptability and Lessons Learnt." *Transport Policy* 20: 1–12.

Breisinger, C., A. Mukashov, M. Raouf, and M. Wiebelt. 2019. "Energy Subsidy Reform for Growth and Equity in Egypt: The Approach Matters." *Energy Policy* 129: 661–71.

Broomandi, P., F. Karaca, A. Nikfal, A. Jahanbakhshi, M. Tamjidi, and J. R. Kim. 2020. "Impact of COVID-19 Event on the Air Quality in Iran." *Aerosol and Air Quality Research* 20 (8): 1793–804.

Burney, J., and V. Ramanathan. 2014. "Recent Climate and Air Pollution Impacts on Indian Agriculture." *Proceedings of the National Academy of Sciences* 111 (46): 16319–24.

Burroughs, D. 2020. "Cairo Metro Line 4 Phase 1 Systems Contract Awarded." *IRJ International Railway Journal*, November 29.

Chekir, N., and Y. Ben Salem. 2020. "What Is the Relationship between the Coronavirus Crisis and Air Pollution in Tunisia?" *Euro-Mediterranean Journal for Environmental Integration* 6 (1): 1–9.

Chen, K., M. Wang, C. Huang, P. L. Kinney, and P. T. Anastas. 2020. "Air Pollution Reduction and Mortality Benefit during the COVID-19 Outbreak in China." *The Lancet Planetary Health* 4 (6): e210–e212.

CIMMYT (International Maize and Wheat Improvement Center). 2019. "Happy Seeder Can Reduce Air Pollution and Greenhouse Gas Emissions While Making Profits for Farmers." Press release, August 8.

CMS (CMS Cameron McKenna LLP). 2017. "Renewable Energy in Iran." Paper, CMS Cameron McKenna LLP, London.

Coady, D. P., V. Flamini, and L. Sears. 2015. "The Unequal Benefits of Fuel Subsidies Revisited: Evidence for Developing Countries." Working Paper No. 15/250, International Monetary Fund, Washington, DC.

Coady, D., I. Parry, N.-P. Le, and B. Shang. 2019. "Global Fossil Fuel Subsidies Remain Large: An Update Based on Country-Level Estimates." Working Paper No. 19/89, International Monetary Fund, Washington, DC.

Coady, D., I. Parry, L. Sears, and B. Shang. 2017. "How Large are Global Fossil Fuel Subsidies?" *World Development* 91: 11–27.

Cockayne, J., and A. Calik. 2020. "Tunisia Moves to Eliminate Fuel Subsidies." MEES (Middle East Economic Survey) newsletter, Issue 63/16, April 17.

Cohen, A. J., M. Brauer, R. Burnett, H. R. Anderson, J. Frostad, K. Estep, K. Balakrishnan, B. Brunekeef, et al. 2017. "Estimates and 25-Year Trends of the Global Burden of Disease Attributable to Ambient Air Pollution: An Analysis of Data from the Global Burden of Diseases Study 2015." *The Lancet* 389 (10082): 1907–18.

Coker, E. S., L. Cavalli, E. Fabrizi, G. Guastella, E. Lippo, M. L. Parisi, N. Pontarollo, M. Rizzati, A. Varacca, and S. Vergalli. 2020. "The Effects of Air Pollution on COVID-19 Related Mortality in Northern Italy." *Environmental and Resource Economics* 76 (4): 611–34.

Cole, M. A., C. Ozgen, and E. Strobl. 2020. "Air Pollution Exposure and COVID-19 in Dutch Municipalities." *Environmental and Resource Economics* 76 (4): 581–610.

Construction Week. 2020. "Solar Panels on Residential Buildings Are Financially Viable." *Construction Week Online*, February 13.

Crippa, M., G. Janssens-Maenhout, D. Guizzardi, and S. Galmarini. 2016. "EU Effect: Exporting Emission Standards for Vehicles through the Global Market Economy." *Journal of Environmental Management* 183: 959–71.

DA&FW (Department of Agriculture and Farmers' Welfare, Government of India). 2018. "Cabinet Approves Promotion of Agricultural Mechanisation for In-Situ Management of Crop Residue in the States of Punjab, Haryana, Uttar Pradesh and NCT of Delhi." Press release, March 7.

Daunfeldt, S.-O., N. Rudholm, and U. Rämme. 2009. "Congestion Charges and Retail Revenues: "Results from the Stockholm Road Pricing Trial." *Transportation Research Part A: Policy and Practice* 43 (3): 306–09.

De Ceunynck, T., B. Dorleman, S. Daniels, A. Laureshyn, T. Brijs, E. Hermans, and G. Wets. 2017. "Sharing Is (S)caring? Interactions between Buses and Bicyclists on Bus Lanes Shared with Bicyclists." *Transportation Research Part F: Traffic Psychology and Behaviour* 46 (Part B): 301–15.

Deloitte. 2016. "Taxation and Investment Mexico 2016: Reach, Relevance and Reliability." Report, Deloitte Touche Tohmatsu Limited, London. https://www2.deloitte.com/content/dam/Deloitte/cn/Documents/international-business-support/deloitte-cn-ibs-mexico-tax-invest-en-2016.pdf.

Deloitte. 2019. "2020 Budget Bill Proposal Presented by the Government, Centre Party and Liberal Party." tax@hand article, Deloitte Sweden, Stockholm.

Deschenes, O., H. Wang, S. Wang, and P. Zhang. 2020. "The Effect of Air Pollution on Body Weight and Obesity: Evidence from China." *Journal of Development Economics* 145 (2): 102461.

Dominici, F., R. D. Peng, M. L. Bell, L. Pham, A. McDermott, S. L. Zeger, and J. M. Samet. 2006. "Fine Particulate Air Pollution and Hospital Admission for Cardiovascular and Respiratory Diseases." *JAMA* 295 (10): 1127–34.

Dzair Daily. 2020. "Algeria: Gasoline, Diesel .. The New Fuel Prices in Detail." *Dzair Daily*, June 6.

EBRD (European Bank for Reconstruction and Development). 2019. *Transition Report 2019–20: Better Governance, Better Economies*. London: EBRD.

EC (European Commission). 2017. "Financing Labour Tax Wedge Cuts." Technical Services Note to the Eurogroup, Directorate General Economic and Financial Affairs, EC, Brussels.

Econostrum. 2020. "Morocco Boosts Home Energy Efficiency Measures." *Econostrum*, January 24.

Eghtesad Online. 2019. "Iran to Enforce Green Tax." Eghtesad Online, August 14. https://www.en.eghtesadonline.com/Section-economy-4/29862 -iran-to-enforce-green-tax.

Egypt Today. 2019. "No Burning Anymore: Rice Straw Collected for Recycling." *Egypt Today*, September 8.

Egypt Today. 2020. "Egypt Recycles 2M Tons of Rice Straw, Produces Fertilizers." *Egypt Today*, November 2.

El Dahan, M. 2011. "Cashing In on Egypt's Black Cloud." Reuters, November 9. https://www.reuters.com/article/egypt-rice/feature-cashing-in-on-egypts -black-cloud-idUKL5E7LU0B320111109.

El-Dewany, C., F. Awad, and A. Zaghloul. 2018. "Utilization of Rice Straw as a Low-Cost Natural By-Product in Agriculture." *International Journal of Environmental Pollution and Environmental Modelling* 1 (4): 91–102.

El-Katiri, L. 2014. "A Roadmap for Renewable Energy in the Middle East and North Africa." Working Paper, Oxford Institute for Energy Studies.

El Khoury, G. 2012. "Carbon Footprint of Electricity in the Middle East." *Carboun Journal*, Article no. 29.

Ellison, R. B., S. P. Greaves, and D. A. Hensher. 2013. "Five Years of London's Low Emission Zone: Effects on Vehicle Fleet Composition and Air Quality." *Transportation Research Part D: Transport and Environment* 23: 25–33.

Enache, C. 2020. "Countries Eye Environmental Taxation." Tax Foundation (blog), September 29. https://taxfoundation.org/countries-eye-environmental -taxation/.

Enel Green Power. 2017. "Tunisia Renewable Energy Framework: An IPP Standpoint." PowerPoint presentation, Enel Green Power Corp., Rome. https://atainsights.com/wp-content/uploads/2017/07/Federicca.pdf.

EPA (US Environmental Protection Agency). 2001. "National Air Quality and Emissions Trends Report, 1999." Annual report on air pollution trends, Report No. EPA 454/R-01-004, Office of Air Quality Planning and Standards, EPA, Research Triangle Park, NC.

EY (Ernst & Young Global Ltd.). 2020. "Colombia Updates Regulations on Tax Incentives for Investments in Renewable Energy Sources." Global Tax Alert, EY, London.

Farrow, A., K. A. Miller, and L. Myllyvirta. 2020. "Toxic Air: The Price of Fossil Fuels." Report, Greenpeace Middle East and North Africa, Beirut, Lebanon.

Fassihi, F. 2010. "Iran Tightens Security as Subsidy Cuts Loom." *Wall Street Journal*, November 4.

Fattouh, B., and L. El-Katiri. 2013. "Energy Subsidies in the Middle East and North Africa." *Energy Strategy Reviews* 2 (1): 108–15.

Fattouh, B., and L. El-Katiri. 2015. "A Brief Political Economy of Energy Subsidies in the Middle East and North Africa." Working paper, Oxford Institute for Energy Studies.

Fawkes, S. 2014. "Energy Efficiency in Saudi Arabia." *Only Eleven Percent* (blog), October 22. https://www.onlyelevenpercent.com/energy-efficiency-saudi -arabia/.

Financial Tribune. 2018. "Mobile Units Will Ease Technical Inspection of Vehicles in Tehran." *Financial Tribune*, January 29.

FM6E (Mohammed VI Foundation for the Protection of the Environment). 2016. "Qualit'Air 2016." Brochure, FM6E, Rabat, Morocco.

Fountas, G., Y.-Y. Sun, O. Akizu-Gardoki, and F. Pomponi. 2020. "How Do People Move Around? National Data on Transport Modal Shares for 131 Countries." *World* 1 (1): 34–43.

Frondel, M., and C. Vance. 2018. "Drivers' Response to Fuel Taxes and Efficiency Standards: Evidence from Germany." *Transportation* 45 (3): 989–1001.

Garrett-Peltier, H. 2017. "Green versus Brown: Comparing the Employment Impacts of Energy Efficiency, Renewable Energy, and Fossil Fuels Using an Input-Output Model." *Economic Modelling* 61: 439–47.

Gasana, J., D. Dillikar, A. Mendy, E. Forno, and E. Ramos Vieira. 2012. "Motor Vehicle Air Pollution and Asthma in Children: A Meta-Analysis." *Environmental Research* 117: 36–45.

Gatti, R., D. F. Angel-Urdinola, J. Silva, and A. Bodor. 2014. "Striving for Better Jobs: The Challenge of Informality in the Middle East and North Africa." MENA Knowledge and Learning Quick Notes Series, No. 49, World Bank, Washington, DC.

GBD (Global Burden of Disease 2017 Risk Factor Collaborators). 2018. "Global, Regional, and National Comparative Risk Assessment of 84 Behavioural, Environmental and Occupational, and Metabolic Risks or Clusters of Risks for 195 Countries and Territories, 1990–2017: A Systematic Analysis for the Global Burden of Disease Study 2017." *The Lancet* 392 (10159): 1923–94.

GBD (Global Burden of Disease 2019 Risk Factor Collaborators). 2020. "Global Burden of 87 Risk Factors in 204 Countries and Territories, 1990–2019: A Systematic Analysis for the Global Burden of Disease Study 2019." *The Lancet 396* (10258): 1223–49.

Ghadiri, Z., Y. Rashidi, and P. Broomandi. 2017. "Evaluation Euro IV of Effectiveness in Transportation Systems of Tehran on Air Quality: Application of IVE Model." *Pollution* 3 (4): 639–53.

Ghorani-Azam, A., B. Riahi-Zanjani, and M. Balali-Mood. 2016. "Effects of Air Pollution on Human Health and Practical Measures for Prevention in Iran." *Journal of Research in Medical Sciences* 21: 65.

GIZ (German Agency for International Cooperation). 2019. "International Fuel Prices 2018/19." Biennial study, GIZ, Bonn.

Graff Zivin, J., and M. Neidell. 2012. "The Impact of Pollution on Worker Productivity." *American Economic Review* 102 (7): 3652–73.

Greenstone, M., and R. Hanna. 2014. "Environmental Regulations, Air and Water Pollution, and Infant Mortality in India." *American Economic Review* 104 (10): 3038–72.

Greenstone, M., R. Pande, N. Ryan, and A. Sudarshan. 2019. "The Impact of an Emissions Trading Scheme on Economic Growth and Air Quality in India." Impact evaluation report, Abdul Latif Jameel Poverty Action Lab (J-PAL), Cambridge, MA.

Gupta, A., H. Bherwani, S. Gautam, S. Anjum, K. Musugu, N. Kumar, A. Anshul, and R. Kumar. 2020. "Air Pollution Aggravating COVID-19 Lethality? Exploration in Asian Cities Using Statistical Models." *Environment, Development and Sustainability* 23: 6408–17.

Hajat, A., C. Hsia, and M. S. O'Neill. 2015. "Socioeconomic Disparities and Air Pollution Exposure: A Global Review." *Current Environmental Health Reports* 2 (4): 440–50.

Hanafi, E. M., H. H. El Khadrawy, W. M. Ahmed, and M. M. Zaabal. 2012. "Some Observations on Rice Straw with Emphasis on Updates of its Management." *World Applied Sciences Journal* 16 (3): 354–61.

Hanna, R., and P. Oliva. 2015. "The Effect of Pollution on Labor Supply: Evidence from a Natural Experiment in Mexico City." *Journal of Public Economics* 122: 68–79.

He, J., H. Liu, and A. Salvo. 2019. "Severe Air Pollution and Labor Productivity: Evidence from Industrial Towns in China." *American Economic Journal: Applied Economics* 11 (1): 173–201.

Heft-Neal, S., M. Heger, M. Burke, and V. Rathi. Forthcoming. "Global Stunting Impacts from Prenatal Pollution Exposure: Evidence from a Million Children." Working paper, World Bank, Washington, DC.

Heger, M., M. Amann, J. J. Lee, and L. Croitoru. 2019. "Supporting Air Quality Management in Saudi Arabia: Health and Economic Benefits from Improving Air Quality & A Roadmap of How to Get Better Air Quality Management." Unpublished manuscript, World Bank, Washington, DC.

Heger, M., and M. Sarraf. 2018. "Air Pollution in Tehran: Health Costs, Sources, and Policies." Discussion Paper, Environment and Natural Resources Global Practice, World Bank, Washington, DC.

Heger, M., D. Wheeler, G. Zens, and C. Meisner. 2019. "Motor Vehicle Density and Air Pollution in Greater Cairo: Fuel Subsidy Removal & Metro Line Extension and Their Effect on Congestion and Pollution." Environmental study, World Bank, Washington, DC.

Heger, M., G. Zens, and C. Meisner. 2019. "Particulate Matter, Ambient Air Pollution, and Respiratory Disease in Egypt." Environmental study, World Bank, Washington, DC.

Hu, X., Y. Liu, L. Yang, Q. Shi, W. Zhang, and C. Zhong. 2018. "SO_2 Emission Reduction Decomposition of Environmental Tax Based on Different Consumption Tax Refunds." *Journal of Cleaner Production* 186: 997–1010.

Hydrocarbon Processing. 2020. "Qatar Petroleum to Supply Ultra Low Sulfur Diesel to the Local Market from Its Refinery in Mesaieed." *Hydrocarbon Processing*, September 30.

ICCT and DieselNet (International Council on Clean Transportation and DieselNet). n.d. "EU: Fuels: Diesel and Gasoline." Article, TransportPolicy.net.

IMF (International Monetary Fund). 2017. "If Not Now, When? Energy Price Reform in Arab Countries." Proceedings from the Annual Meeting of Arab Ministers of Finance, Rabat, Morocco, April 2017.

IMF (International Monetary Fund). 2021. "Reaching Net Zero Emissions." G-20 Background Note, IMF, Washington, DC.

ISO (International Organization for Standardization). 2020. "ISO Survey 2020." Annual survey of valid certifications to ISO management standards, by country. [Available as Excel datasets.] ISO, Geneva.

Jonsson, S., A. Ydstedt, and E. Asen. 2020. "Looking Back on 30 Years of Carbon Taxes in Sweden." Fiscal Fact No. 727, Tax Foundation, Washington, DC.

Kafafy, N. A.-A. 2010. "Dynamics of Urban Green Space in an Arid City: The Case of Cairo-Egypt." Doctoral dissertation, Cardiff University, United Kingdom.

Karagulian, F., C. A. Belis, C. Francisco, C. Dora, A. M. Prüss-Ustün, S. Bonjour, H. Adair-Rohani, and M. Amann. 2015. "Contributions to Cities' Ambient Particulate Matter (PM): A Systematic Review of Local Source Contributions at Global Level." *Atmospheric Environment* 120: 475–83.

Khalid, T. 2020. "Saudi Aramco Reduces Fuel Prices in May 2020." Al Arabiya, May 11.

Khalilzadeh, S., Z. Khalilzadeh, H. Emami, and M. R. Masjedi. 2009. "The Relation between Air Pollution and Cardiorespiratory Admissions in Tehran." *Tanaffos* 8 (1): 35–40.

Khatibi, S. R., S. M. Karimi, M. Moradi-Lakeh, M. Kermani, and S. A. Motevalian. 2020. "Fossil Energy Price and Outdoor Air Pollution: Predictions from a QUAIDS Model." *Biofuel Research Journal* 7 (3): 1205–16.

Kheiravar, K. H. 2019. "Economic and Econometric Analyses of the World Petroleum Industry, Energy Subsidies, and Air Pollution." Doctoral thesis, University of California, Davis.

Khodeir, M., M. Shamy, M. Alghamdi, M. Zhong, H. Sun, M. Costa, L.-C. Chen, and P. Maciejczyk. 2012. "Source Apportionment and Elemental Composition of PM2.5 and PM10 in Jeddah City, Saudi Arabia." *Atmospheric Pollution Research* 3 (3): 331–40.

Khomsi, K., H. Najmi, H. Amghar, Y. Chelhaoui, and Z. Souhaili. 2020. "COVID-19 National Lockdown in Morocco: Impacts on Air Quality and Public Health." *One Health* 11: 100200.

Kodjak, D. 2015. "Policies to Reduce Fuel Consumption, Air Pollution, and Carbon Emissions from Vehicles in G20 Nations." International Council on Clean Transportation.

Kodros, J. K., C. Wiedinmyer, B. Ford, R. Cucinotta, R. Gan, S. Magzamen, and J. R. Pierce. 2016. "Global Burden of Mortalities Due to Chronic Exposure to Ambient PM2.5 From Open Combustion of Domestic Waste." *Environmental Research Letters* 11 (12): 124022.

Krane, J., and F. Monaldi. 2017. "Oil Prices, Political Instability, and Energy Subsidy Reform in MENA Oil Exporters." Center for Energy Studies, Baker III Institute for Public Policy, Rice University.

Łapko, A., A. Panasiuk, R. Strulak-Wójcikiewicz, and M. Landowski. 2020. "The State of Air Pollution as a Factor Determining the Assessment of a City's Tourist Attractiveness—Based on the Opinions of Polish Respondents." *Sustainability* 12 (4): 1466.

Larsen, B. 2019. "Egypt: Cost of Environmental Degradation: Air and Water Pollution." Environmental study, World Bank, Washington, DC.

Li, P., Y. Lu, and J. Wang. 2020. "The Effects of Fuel Standards on Air Pollution: Evidence from China." *Journal of Development Economics* 146: 102488.

Li, X., F. Wagner, W. Peng, J. Yang, and D. L. Mauzerall. 2017. "Reduction of Solar Photovoltaic Resources due to Air Pollution in China." *Proceedings of the National Academy of Sciences* 114 (45): 11867–72.

Lozano-Gracia, N., and M. E. Soppelsa. 2019. "Pollution and City Competitiveness: A Descriptive Analysis." Policy Research Working Paper 8740, World Bank, Washington, DC.

Lydall, Ross. 2021. "When Is the London Ulez Expanding, Where Does It Expand to and Who Will Have to Pay?" *Evening Standard*, October 25.

Mansourian, M., S. H. Javanmard, P. Poursafa, and R. Kelishadi. 2010. "Air Pollution and Hospitalization for Respiratory Diseases among Children in Isfahan, Iran." *Ghana Medical Journal* 44 (4): 138–43.

Markandya, A., J. Sampedro, S. J. Smith, R. Van Dingenen, C. Pizarro-Irizar, I. Arto, and M. González-Eguino. 2018. "Health Co-Benefits from Air Pollution and Mitigation Costs of the Paris Agreement: A Modelling Study." *Lancet Planetary Health* 2 (3): e126–e133.

Marsh, G., N. Hill, and J. Sully. 2000. "Consultation on the Need to Reduce the Sulphur Content of Petrol and Diesel Fuels Below 50 PPM: A Policy Makers Summary." Report for the Directorate General for Environment, European Commission, prepared by AEA Technology, Abingdon, UK.

Menichetti, E., A. El Gharras, B. Duhamel, and S. Karbuz. 2019. "The MENA Region in the Global Energy Markets." In *Foreign Policy Review* Special Issue, "MENARA: Middle East and North Africa Regional Architecture": 75–119. Institute for Foreign Affairs and Trade, Budapest.

Miranda, M. L., S. E. Edwards, M. H. Keating, and C. J. Paul. 2011. "Making the Environmental Justice Grade: The Relative Burden of Air Pollution Exposure in the United States." *International Journal of Environmental Research and Public Health* 8 (6): 1755–71.

MOCCAE (Ministry of Climate Change and Environment, United Arab Emirates). 2019. "UAE National Emissions Inventory Project: Final Results, 2019." MOCCAE, Dubai, UAE.

Mohamed, R. 2017. "Abu Dhabi State of the Environment Report 2017: Air Quality." Report, Environment Agency – Abu Dhabi (EAD). https://www.soe.ae/wp-content/uploads/2017/10/environmental-report-air-quality.pdf.

Moisio, M., H. van Soest, N. Forsell, L. Nascimento, G. de Vivero, S. Gonzales, F. Hans, et al. 2020. "Overview of Recently Adopted Mitigation Policies and Climate-Relevant Policy Responses to COVID-19: 2020 Update." New Climate Institute, PBL Netherlands Environmental Assessment Agency, International Institute for Applied Systems Analysis.

Mostafa, M. K., G. Gamal, and A. Wafiq. 2020. "The Impact of COVID 19 on Air Pollution Levels and Other Environmental Indicators – A Case Study of Egypt." *Journal of Environmental Management* 277: 111496.

Myllyvirta, L. 2020. "Quantifying the Economic Costs of Air Pollution from Fossil Fuels." Briefing paper, CREA (Centre for Research on Energy and Clean Air), Helsinki.

NASA (National Aeronautics and Space Administration). 2014. "Fires in the Egypt River Delta." Satellite image, NASA, Washington, DC. https://www.nasa.gov/content/goddard/fires-in-the-egypt-river-delta/#.V07aa8!BvqD.

OBG (Oxford Business Group). 2016. "Government Promoting Renewables in Tunisian Energy Sector." In *The Report: Tunisia 2016*, edited by Robert Tashima. London: OBG.

OECD (Organisation for Economic Co-operation and Development). 2016. *The Economic Consequences of Outdoor Air Pollution*. Paris: OECD Publishing.

OECD (Organisation for Economic Co-operation and Development). 2020. "Colombia." In *Taxation in Agriculture*, 122–26. Paris: OECD Publishing.

Otmani, A., A. Benchrif, M. Tahri, M. Bounakhla, M. El Bouch, and M. Krombi. 2020. "Impact of Covid-19 Lockdown on PM_{10}, SO_2 and NO_2 Concentrations in Salé City (Morocco)." *Science of The Total Environment* 735: 139541.

Pandey, M. P., and J. S. Nathwani. 2003. "Canada Wide Standard for Particulate Matter and Ozone: Cost-Benefit Analysis Using a Life Quality Index." *Risk Analysis* 23 (1): 55–67.

Pang, M.-B., M.-L. Chen, and N. Zhang. 2017. "Scheduling Optimization of Intelligent Public Transport System Based on MAST." *Journal of Transportation Systems Engineering and Information Technology* 17 (1): 1009–6744.

Parry, I. W. H., D. Heine, E. Lis, and S. Li. 2014. *Getting Energy Prices Right: From Principle to Practice*. Washington, DC: International Monetary Fund.

Pope, C. A. III, and D. W. Dockery. 2006. "Health Effects of Fine Particulate Air Pollution: Lines that Connect." *Journal of the Air & Waste Management Association* 56 (6): 709–42.

Potoski, M., and A. Prakash. 2005. "Green Clubs and Voluntary Governance: ISO 14001 and Firms' Regulatory Compliance." *American Journal of Political Science* 49 (2): 235–48.

Potoski, M., and A. Prakash. 2013. "Do Voluntary Programs Reduce Pollution? Examining ISO 14001's Effectiveness Across Countries." *Policy Studies Journal* 41 (2): 273–94.

Poudineh, R., A. Sen, and B. Fattouh. 2018. "Advancing Renewable Energy in Resource-Rich Economies of the MENA." *Renewable Energy* 123: 135–49.

Prakash, A., and M. Potoski. 2014. "Global Private Regimes, Domestic Public Law: ISO 14001 and Pollution Reduction." *Comparative Political Studies* 47 (3): 369–94.

Psaledakis, D. 2020. "U.S. Imposes Libya-Related Sanctions on Individuals, Company." Reuters, August 6.

Rajagopalan, S., and R. D. Brook. 2012. "Air Pollution and Type 2 Diabetes: Mechanistic Insights." *Diabetes* 61 (12): 3037–45.

Rashad, M. 2016. "Saudis Await Prince's Vision of Future with Hope and Concern." Reuters, April 24.

Rayess, R. 2020. "More Difficult Times to Come for Lebanon When Food, Fuel Subsidies Lifted." Al Arabiya News, November 24.

Redondo, R. 2020. "Al-Dhafra Solar Plant to Become World's Largest." Atalayar, April 29.

Reuters. 2020. "Fuel or Flour? Lebanon to Ration $2bn in Subsidies, PM Says." Al Jazeera, December 29.

REVE (Wind Energy and Electric Vehicle Magazine). 2020. "Wind Energy in Saudi Arabia, Wind Turbines Arrive for Wind Farm." REVE, July 29.

Rhys-Tyler, G. A., W. Legassick, and M. C. Bell. 2011. "The Significance of Vehicle Emissions Standards for Levels of Exhaust Pollution from Light Vehicles in an Urban Area." *Atmospheric Environment* 45 (19): 3286–93.

Rosmiza, M. Z., W. P. Davies, Rosniza C. R. Aznie, M. Mazdi, M. J. Jabil, Wan W. Y. Toren, and Che C. M. Rosmawati. 2014. "Farmers' Participation in Rice Straw-Utilisation in the MADA Region of Kedah, Malaysia." *Mediterranean Journal of Social Sciences* 5 (23): 229–37.

Ross, M. L., C. Hazlett, and P. Mahdavi. 2017. "Global Progress and Backsliding on Gasoline Taxes and Subsidies." *Nature Energy* 2 (1): 1–6.

Saab, N., A. Badran, and A.-K. Sadik, eds. 2019. "Environmental Education for Sustainable Development in Arab Countries." Annual report of the Arab Forum for Environment and Development (AFED), Beirut, Lebanon.

Saab, N., and R. R. Habib. 2020. "Health and the Environment in Arab Countries." AFED report series. Beirut: Arab Forum for Environment and Development.

Saidi, N. 2019. "How Carbon Taxes Can Boost State Coffers and Clean the Environment." *The National*, October 7.

Sajjad, F., U. Noreen, and K. Zaman. 2014. "Climate Change and Air Pollution Jointly Creating Nightmare for Tourism Industry." *Environmental Science and Pollution Research* 21: 12403–18.

Samet, J. M., F. Dominici, F. C. Curriero, I. Coursac, and S. L. Zeger. 2000. "Fine Particulate Air Pollution and Mortality in 20 US Cities, 1987–1994." *New England Journal of Medicine* 343 (24): 1742–49.

Sanderson, Daniel. 2018. "Abu Dhabi Pollution Map Paves Way for Emissions Crackdown." *The National*, December 12.

Scovronick, N., M. Budolfson, F. Dennig, F. Errickson, M. Fleurbaey, W. Peng, R. H. Socolow, D. Spears, and F. Wagner. 2019. "The Impact of Human Health Co-Benefits on Evaluations of Global Climate Policy." *Nature Communications* 10 (1): 1–12.

Sekmoudi, I., K. Khomsi, S. Faieq, and L. Idrissi. 2020. "Covid-19 Lockdown Improves Air Quality in Morocco." arXiv preprint arXiv:2007.05417.

Shaddick, G., M. L. Thomas, A. Green, M. Brauer, A. van Donkelaar, R. Burnett, H. H. Chang, et al. 2018. "Data Integration Model for Air Quality: A Hierarchical Approach to the Global Estimation of Exposures to Ambient Air Pollution." *Journal of the Royal Statistical Society: Series C (Applied Statistics)* 67 (1): 231–53.

Shaheen, A., R. Wu, and M. Aldabash. 2020. "Long-Term AOD Trend Assessment over the Eastern Mediterranean Region: A Comparative Study Including a New Merged Aerosol Product." *Atmospheric Environment* 238: 117736.

Shyamsundar, P., N. P. Springer, H. Tallis, S. Polasky, M. L. Jat, H. S. Sidhu, P. P. Krishnapriya, et al. 2019. "Fields on Fire: Alternatives to Crop Residue Burning in India." *Science* 365 (6453): 536–38.

Simeonova, E., J. Currie, P. Nilsson, and R. Walker. 2019. "Congestion Pricing, Air Pollution, and Children's Health." *Journal of Human Resources*: 0218–9363R2.

Singh, S. 2015. "Iran Automotive Industry: Can American Car Manufacturers Overcome Chinese Resistance?" *Forbes*, August 11.

Smith, Kevin. 2020. "Riyadh Metro Testing Progresses with First Lines on Course to Open in 2021." *IRJ International Railway Journal*, December 10.

Son, J.-Y., K. C. Fong, S. Heo, H. Kim, C. C. Lim, and M. L. Bell. 2020. "Reductions in Mortality Resulting from Reduced Air Pollution Levels Due to COVID-19 Mitigation Measures." *Science of the Total Environment* 744: 141012.

Sowlat, M. H., K. Naddafi, M. Yunesian, P. L. Jackson, S. Lotfi, and A. Shahsavani. 2013. "PM_{10} Source Apportionment in Ahvaz, Iran, Using Positive Matrix Factorization." *Clean Soil, Air, Water* 41 (12): 1143–51.

STA and Trafikverket (Swedish Transport Agency and Swedish Transport Administration). 2015. "Changes in Stockholm's Congestion Tax." Brochure, STA, Norrköping; and Trafikverket, Borlänge, Sweden.

Stratas Advisors. 2019. "The Middle East Stays on Course in Sulfur Reduction Plans." Excerpt from a report by Stratas Advisors, Houston.

Sun, Q., P. Yue, J. A. Deiuliis, C. N. Lumeng, T. Kampfrath, M. B. Mikolaj, Y. Cai, et al. 2009. "Ambient Air Pollution Exaggerates Adipose Inflammation and Insulin Resistance in a Mouse Model of Diet-Induced Obesity." *Circulation* 119 (4): 538–46.

Sweerts, B., S. Pfenninger, S. Yang, D. Folini, B. Van der Zwaan, and M. Wild. 2019. "Estimation of Losses in Solar Energy Production from Air Pollution in China Since 1960 Using Surface Radiation Data." *Nature Energy* 4 (8): 657–63.

Tabatabaiee, S. A., and A. Rahman. 2011. "The Effect of Urban Rail Transit on Decreasing Energy Consumption and Air Pollution in Ahvaz City." *Advanced Materials Research* 255: 2802–05.

Tan-Soo, J.-S., X.-B. Zhang, P. Qin, and L. Xie. 2019. "Using Electricity Prices to Curb Industrial Pollution." *Journal of Environmental Management* 248: 109252.

Tehran Times. 2019. "25% of Cars Failed Inspection Tests in Tehran." *Tehran Times*, July 28.

Timmerberg, S., A. Sanna, M. Kaltschmitt, and M. Finkbeiner. 2019 "Renewable Electricity Targets in Selected MENA Countries–Assessment of Available Resources, Generation Costs and GHG Emissions." *Energy Reports* 5: 1470–87.

Tripathi, B. 2019. "Gujarat's Particulate Emissions Trading Has Cut Pollution, Lifted Profits." IndiaSpend, October 30.

UITP (International Association of Public Transport). 2019. "MENA Transport Report 2019." Centre for Transport Excellence, UITP, Brussels.

UN DESA (United Nations Department of Social and Economic Affairs). 2018. *World Urbanization Prospects: The 2018 Revision.* New York: UN.

UNEP (United Nations Environment Programme). 2015. "Algeria Air Quality Policies." Policy matrix of country-level policies, UNEP, Nairobi, Kenya.

UNEP (United Nations Environment Programme). 2017. "Pushing to #BeatAirPollution on World Environment Day." Media Advisory, June 5.

UNEP (United Nations Environment Programme). 2019. "Middle East & North Africa: Actions Taken by Governments to Improve Air Quality." Report, UNEP, Nairobi, Kenya.

UNEP (United Nations Environment Programme). 2020. "Global Sulphur Levels." Interactive map, https://www.unep.org/global-sulphur-levels.

UNEP (United Nations Environment Programme). 2021. "World Lead Map, June 2021." Partnership for Clean Fuels and Vehicles, UNEP, Nairobi, Kenya. https://www.unep.org/explore-topics/transport/what-we-do/partnership-clean-fuels-and-vehicles/lead-campaign.

Venter, Z. S., K. Aunan, S. Chowdhury, and J. Lelieveld. 2020. "COVID-19 Lockdowns Cause Global Air Pollution Declines." *Proceedings of the National Academy of Sciences* 117 (32): 18984–90.

Verma, A. 2020. "Kahramaa Closes Financing for World's 2nd Cheapest Solar Power Project." Saur Energy International, July 24.

Viard, B. V., and S. Fu. 2015. "The Effect of Beijing's Driving Restrictions on Pollution and Economic Activity." *Journal of Public Economics* 125: 98–115.

von Schneidemesser, E., K. Steinmar, E. C. Weatherhead, B. Bonn, H. Gerwig, and J. Quedenau. 2019. "Air Pollution at Human Scales in an Urban Environment: Impact of Local Environment and Vehicles on Particle Number Concentrations." *Science of the Total Environment* 88: 691–700.

Wagner, F., M. Amann, G. Kiesewetter, Z. Klimont, W. Schöpp, and M. P. Heger. 2020. "The Natural Capital Index Report on Fine Particulate Matter ($PM_{2.5}$) Pollution." Unpublished report, International Institute for Applied Systems Analysis (IIASA), Laxenburg, Austria.

Waked, A., and C. Afif. 2012. "Emissions of Air Pollutants from Road Transport in Lebanon and Other Countries in the Middle East Region." *Atmospheric Environment* 61: 446–52.

Wang, Y., S. Song, S. Qiu, L. Lu, Y. Ma, X. Li and Y. Hu. 2014. "Study on International Practices for Low Emission Zone and Congestion Charging." Working paper, World Resources Institute, Washington, DC.

Wei, H. 2019. "Impacts of China's National Vehicle Fuel Standards and Subway Development on Air Pollution." *Journal of Cleaner Production* 241: 118399.

Westcott, T. 2020. "Running on Empty: Oil-Rich Libya Hit by Extreme Fuel Shortages." Middle East Eye, March 24.

Wheida, A., A. Nasser, M. El Nazer, A. Borbon, G. A. A. El Ata, M. A. Wahab, and S. C. Alfaro. 2018. "Tackling the Mortality from Long-Term Exposure to Outdoor Air Pollution in Megacities: Lessons from the Greater Cairo Case Study." *Environmental Research* 160: 223–31.

WHO (World Health Organization). 2019. "Health Consequences of Air Pollution on Populations." Departmental news item, November 15, WHO, Geneva.

WHO (World Health Organization). 2021. *WHO Global Air Quality Guidelines: Particulate Matter ($PM_{2.5}$ and PM_{10}), Ozone, Nitrogen Dioxide, Sulfur Dioxide and Carbon Monoxide.* Geneva: WHO.

Wiedinmyer, C., R. J. Yokelson, and B. K. Gullett. 2014. "Global Emissions of Trace Gases, Particulate Matter, and Hazardous Air Pollutants from Open Burning of Domestic Waste." *Environmental Science & Technology* 48 (16): 9523–30.

World Bank. 2013. "The Arab Republic of Egypt – For Better or for Worse: Air Pollution in Greater Cairo. A Sector Note." Report No. 73074-EG, World Bank, Washington, DC.

World Bank. 2014. "Plan Maroc Vert: Agriculture in a Changing Climate." Video about the Green Morocco Plan, World Bank, Washington, DC.

World Bank. 2015. "Implementation Completion and Results Report on a Loan of US$ 20.00 million to the Government of Egypt for a Second Pollution Abatement Project." Report No. ICR00003387, World Bank, Washington, DC.

World Bank. 2016. "Delivering Energy Efficiency in the Middle East and North Africa: Achieving Energy Efficiency Potential in the Industry, Services and Residential Sectors." Energy Sector Management Assistance Program (ESMAP) study, World Bank, Washington, DC.

World Bank. 2018. "The Bogota Metro Is Moving Forward with World Bank Support." Press release, August 2.

World Bank. 2019a. "Project Appraisal Document on a Proposed Loan in the Amount of US$93 Million to the Republic of Peru for a Lima Metropolitano North Extension Project." Report No. PCBASIC0185498, World Bank, Washington, DC.

World Bank. 2019b. "Sand and Dust Storms in the Middle East and North Africa (MENA) Region: Sources, Costs, and Solutions." Environmental study, Report No. 33036, World Bank, Washington, DC.

World Bank. 2020a. *Trading Together: Reviving Middle East and North Africa Regional Integration in the Post-Covid Era.* MENA Economic Update, October 2020. Washington, DC: World Bank.

World Bank. 2020b. "World Bank Provides Additional Funding for Morocco's Urban Transport Sector." Press release, November 3.

World Bank. 2021a. "Morocco Urban Transport Project (P4R)." Details page, Project No. P149653, World Bank, Washington, DC.

World Bank. 2021b. *State and Trends of Carbon Pricing 2021.* Washington, DC: World Bank.

World Bank and IHME (Institute for Health Metrics and Evaluation). 2016. "The Cost of Air Pollution: Strengthening the Economic Case for Action." Joint report, of the IHME, University of Washington, Seattle; and World Bank, Washington, DC.

World Bank and IMF (International Monetary Fund). 2021. "From COVID-19 Crisis Response to Resilient Recovery: Saving Lives and Livelihoods while Supporting Green, Resilient and Inclusive Development (GRID)." Document No. DC2021-0004 for the April 9, 2021, Meeting of the Development Committee (Joint Ministerial Committee of the Boards of Governors of the Bank and the Fund on the Transfer of Real Resources to Developing Countries), Washington, DC. https://www.devcommittee.org/sites/dc/files/download/Documents/2021-03/DC2021-0004%20Green%20Resilient%20final.pdf.

Wu, X., R. C. Nethery, B. M. Sabath, D. Braun, and F. Dominici. 2020. "Exposure to Air Pollution and COVID-19 Mortality in the United States." *Science Advances* 6 (45): 1–6.

Xue, T., T. Guan, G. Geng, Q. Zhang, Y. Zhao, and T. Zhu. 2021. "Estimation of Pregnancy Losses Attributable to Exposure to Ambient Fine Particles in South Asia: An Epidemiological Case-control Study." *The Lancet Planetary Health* 5 (1): e15–e24.

Yang, X., P. Jiang, and Y. Pan. 2020. "Does China's Carbon Emission Trading Policy Have an Employment Double Dividend and a Porter Effect?" *Energy Policy* 142: 111492.

Yang, B.-Y., Z. M. Qian, M. G. Vaughn, S. W. Howard, J. P. Pemberton, H. Ma, D.-H. Chen, et al. 2018. "Overweight Modifies the Association between Long-Term Ambient Air Pollution and Prehypertension in Chinese Adults: The 33 Communities Chinese Health Study." *Environmental Health* 17 (1): 1–15.

Zafar, S. 2020a. "Solid Waste Management in the Middle East: Major Challenges." EcoMENA, June 27.

Zafar, S. 2020b. "Renewables in MENA: An Overview." *Blogging Junction*, July 28.

Zaptia, S. 2020a. "Fuel Subsidy Reform Proposal Presented to Serraj Government." *Libya Herald*, March 3.

Zaptia, S. 2020b. "Libya's Economic Reform Salon Proposes Reforms for the Country's Fuel Subsidies." *Libya Herald*, July 27.

Zhang, X., X. Chen, and X. Zhang. 2018. "The Impact of Exposure to Air Pollution on Cognitive Performance." *Proceedings of the National Academy of Sciences* 115 (37): 9193–97.

Zhang, Q., X. Jiang, D. Tong, S. J. Davis, H. Zhao, G. Geng, T. Feng, et al. 2017. "Transboundary Health Impacts of Transported Global Air Pollution and International Trade." *Nature* 543: 705–09.

Zhang, W., J. Wang, B. Zhang, J. Bi, and H. Jiang. 2015. "Can China Comply with Its 12th Five-Year Plan on Industrial Emissions Control: A Structural Decomposition Analysis." *Environmental Science & Technology* 49 (8): 4816–24.

Zhou, L., X. Chen, and X. Tian. 2018. "The Impact of Fine Particulate Matter (PM2.5) on China's Agricultural Production from 2001 to 2010." *Journal of Cleaner Production* 178: 133–41.

Zhu, Y., J. Xie, F. Huang, and L. Cao. 2020. "Association between Short-Term Exposure to Air Pollution and COVID-19 Infection: Evidence from China." *Science of the Total Environment* 727: 138704.

ZinCo GmbH. 2020. "Project Report: Al Shaheed Park, Kuwait." Project report, ZinCo GmbH, Nürtingen, Germany. https://zinco-greenroof.com/sites/default/files/2020-04/ZinCo_Kuwait_Al_Shaheed_Park.pdf.

CHAPTER 4

Blue Seas: Freeing the Seas from Plastics

OVERVIEW

Several of the Middle East and North Africa's economies are among the world's top polluters regarding plastics leaking into their seas—threatening marine and coastal ecosystems and resulting in large public health and economic costs. In this chapter, the next section reviews the current state of plastic pollution in the region, focusing on the Mediterranean Sea given its importance to the region's local economies, whether urban or non-urban. It discusses the high amount of plastic debris already in these marine ecosystems.

Although the rate of plastic waste generation per resident of the region is often not much greater than in comparable countries, the amount of plastic debris per resident ending up in the seas of the Middle East and North Africa is the highest in the world. Thus, the region's population growth and changing consumption patterns present a challenge for the near future—especially regarding the treatment of waste—and under-mines the efforts to adopt green, resilient, and inclusive development (GRID), from which the region would benefit greatly (World Bank and IMF 2021). An alarmingly high level of waste is inadequately managed, particularly in the Maghreb and the Mashreq subregions, and often out-sourced to the informal sector with its precarious working conditions for the poor. Throughout the region, recycling rates are low, hampering the sustainable treatment of plastic waste in a circular fashion (ensuring it is reused and recycled instead of simply discarded).

The chapter then discusses the severe adverse environmental, public health, and economic impacts of the flow and accumulation of plastics in the Middle East and North Africa's seas. These impacts necessitate swift and concrete action by the region's governments to switch to a GRID growth path in this area.

The chapter's discussion of policy options makes special note of the scant knowledge about the sources of the plastics flowing into the seas and the causes for its leakage there. To efficiently reduce the volume of this flow requires knowledge about where it is generated, who generates it, and which products end up as debris in the seas. Information about the hot spots of marine-plastic debris is sparse, and even less is known about the specific causes. Gathering information and identifying the sources of plastic waste and the reasons why it ends up in the seas is hence key for effective policy making, requiring more-detailed studies for proper identification of these sources and the causes of plastic leakage.

Policy options to tackle plastic pollution in the Middle East and North Africa's seas revolve around the concept of a circular economy and adequate solid waste management (SWM). These options are discussed along the lines of the "3 R's": **Reduce** plastic consumption and production. **Reuse** plastic products. **Recycle** the plastic waste or dispose of it appropriately if neither reuse nor recycling are an option. Priority recommendations in this respect cover the following key areas:

- *Solid waste management.* Improving the weak SWM systems in many of the region's economies is a critical step to deal appropriately with plastic waste. Doing so has cross-benefits for the issue of air pollution because inadequately managed waste is often burned in an uncontrolled manner (as discussed in chapter 3).

- *Recycling.* Making recycling markets financially more sustainable is essential to increase the share of plastics that does not leave the value chain after a single use.

- *Public awareness.* Accompanying the preceding two actions with information dissemination and public awareness programs is crucial to inform the public of the adverse effects of plastic pollution, change individual behavior, and increase demand for further policies.

- *Plastic alternatives.* Close cooperation with and support for the private sector is critical to support the development of suitable plastic alternatives.

- *Pricing.* The business environment for plastic alternatives must be enhanced by tackling the price discrepancies between those alternatives and products derived from virgin plastics. These differences often stem from artificially low prices of plastic products, resulting from large subsidies for fossil fuels that are feedstock and important energy input in plastic production. Appropriate pricing would have cross-sector benefits regarding air pollution and can be supported by progressively phasing out single-use plastics (SUPs).

The chapter concludes with detailed descriptions of a broad set of policy measures to stem the plastic tide, hence putting countries on a GRID path to tackle the issue of marine-plastic pollution and related problems. It details the deficiencies of SWM throughout the region, the reasons for unsustainable consumption and production patterns, and ways to tackle those issues. It also discusses the importance of increasing public awareness about the negative consequences of marine-plastic pollution as well as ways to garner support for cleanups of plastic waste, both in the seas and on the beaches. Pondering and acting on these policy recommendations will be crucial if the Middle East and North Africa is to get a grip on the flow of plastics in its seas and address this issue consistent with a GRID framework.

THE STATE OF PLASTIC POLLUTION IN THE SEAS

Plastic-polluted seas are a consequence of both the growth in plastic use over recent decades and of poor SWM. Global plastic production has increased exponentially over the past 50 years, from 2 million tons in 1950 to more than 454 million tons in 2018 (Alessi and Di Carlo 2018). The prevalent and increasing use of plastics mainly has to do with its low cost (compared with alternatives) and its favorable physical characteristics—which, for example, increase the shelf life of perishable products and reduce transportation costs.

Unfortunately, much of the plastic that is produced ends up in the oceans, and quantities of buoyant macroplastics on the ocean surface and along coastlines could quadruple by the mid-twenty-first century, reaching an estimated 4.5 million tons (Lebreton, Egger, and Slat 2019).[1] Given these staggering numbers and the many adverse effects of marine-plastic pollution (discussed in the next section), it has been put squarely on the international agenda as a top global priority (box 4.1). These international dialogues will also set the context for the Middle East and North Africa's regional and national efforts to combat the continued flow of plastics into its seas.

Global Comparisons

The Middle East and North Africa's residents, on average, contribute the most plastic waste per person to their seas of any region in the world.

BOX 4.1

Marine-Plastic Pollution within the International Policy Agenda

The issue of marine-plastic pollution has rightly caught the attention of international policy makers and researchers in recent years. In the Sustainable Development Goals (SDGs) of the 2030 Agenda for Sustainable Development, SDG Target 14.1 refers directly to this issue: "By 2025, prevent and significantly reduce marine pollution of all kinds, in particular from land-based activities, including marine debris and nutrient pollution."[a]

Several international environmental organizations—including, among many others, the International Union for Conservation of Nature (IUCN), Greenpeace, and the Ellen MacArthur Foundation—have made informing the world about the extent, threats, and combating of marine plastics one of their top priorities. Similarly, supranational organizations such as the United Nations (UN) and the European Commission have emphasized the importance of stanching the flow of plastics into the world's oceans and seas. One prominent example, the 2021 introduction of the European Union (EU) Directive on Single-Use Plastics (SUPs),[b] tackles 10 SUP items and fishing gear that account for a large share of the waste that litters European beaches and pollutes the surrounding seas. (For more

about the EU's action plan to reduce marine litter, see box 4.9.)

The international attention to the issue of marine plastics also puts the Middle East and North Africa's role into focus, especially regarding plastics flowing into the Mediterranean Sea—one of the global hot spots in this respect—and the region's economies among the major contributors to this trend. The focus and efforts of international actors can give policy makers in the Middle East and North Africa a blueprint for how to best address the plastic tide in their seas, as this report also intends to do.

Moreover, the global momentum around this issue opens up possibilities to forge alliances and enter into cooperative initiatives and agreements that can benefit the region by supporting knowledge and technology transfers. These also include regional efforts across the Middle East and North Africa to induce learning processes and exploit knowledge spillovers tailored to the regional context. As this chapter discusses, these developments can benefit the region's economies not only from an environmental perspective but also from an economic one through increased job creation in green sectors such as recycling or the development of plastic alternatives.

a. For more information about SDG 14 on Life below Water, including its specific targets and indicators, see the United Nations' SDG 14 Knowledge Base page: https://sdgs.un.org/goals/goal14.
b. Directive 2019/904, of the European Parliament and of the Council of 5 June 2019 on the Reduction of the Impact of Certain Plastic Products on the Environment, 2019 O.J. (L 155/1).

Both of the region's drivers of increased marine-plastic pollution—increased plastic use due to large population increases and rising incomes as well as poor SWM—account for the large amounts of plastic waste flowing into the Middle East and North Africa's seas.

The average resident in the Middle East and North Africa contributes more than 6 kilograms (kg) of the plastic waste that flows into the marine spaces each year (Jambeck et al. 2015), as shown in figure 4.1. Only Sub-Saharan Africa and East Asia and Pacific come close to that level, with more than 5 kg of marine-plastic debris per capita per year. However, one should note that the heterogeneity of the region's economies regarding income, SWM adequacy, and production and consumption patterns also implies substantial heterogeneity in the amounts they individually contribute.

On aggregate, Algeria, the Arab Republic of Egypt, and Morocco are among the top 20 countries contributing the most plastic waste into oceans worldwide. In a global comparison, Egypt ranked 7th among the top 20 polluting countries in 2010, accounting for 3 percent of the discharge of plastic waste into the world's oceans and seas; Algeria was 13th, with 1.6 percent; and Morocco was 18th, with

FIGURE 4.1

Annual Per Capita Volume of Plastic Waste Entering the Sea, by World Region, 2010

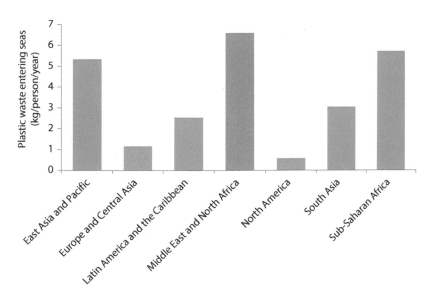

Source: Based on Jambeck et al. 2015.
Note: "North America" includes Canada and the United States. kg = kilograms.

1 percent (Jambeck et al. 2015), as shown in figure 4.2.[2] Sixteen of the top 20 countries are middle-income countries, where broader waste management is not catching up with economic development. On average, these countries mismanage 68 percent of their waste—that is, it ends up either as litter or inadequately disposed of in open dumps or uncontrolled landfills, where it is not fully contained.

Intraregional Comparisons

For most of the Middle East and North Africa economies, the amount of plastic debris entering the seas is set to double by 2025 relative to 2010 levels (figure 4.3). With rising incomes and continued population growth, waste generation and the proportion of plastics in the waste cycle are also set to rise in many of the region's economies.

Most studies suggest that the amount of plastic waste generated will increase faster than the governments' ability to plan waste management systems and develop infrastructure unless there is a shift in production and consumption patterns, improvements in SWM, and increases in the rates of reusing and recycling plastic products (Jambeck et al. 2015).

FIGURE 4.2

Top 20 Marine-Plastic Polluting Countries, 2010

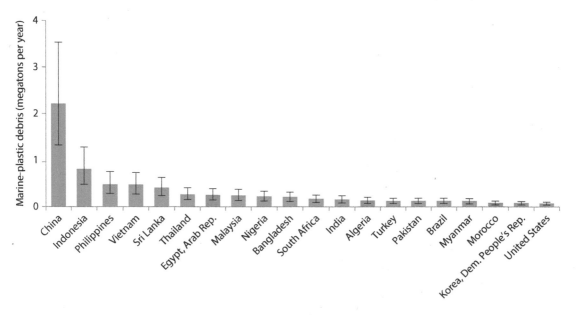

Source: Based on Jambeck et al. 2015.
Note: Orange bars designate Middle East and North Africa countries. Error bars indicate confidence intervals.

FIGURE 4.3

Volume of Plastic Debris Entering the Seas from the Middle East and North Africa, by Economy, 2010 and 2025

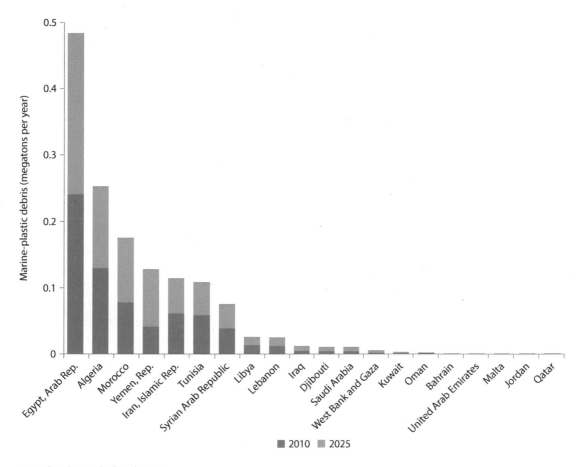

Source: Based on Jambeck et al. 2015.
Note: Figures for 2025 are projections.

Mediterranean-Polluting Countries

The Mediterranean Sea is today one of the most plastic-polluted seas in the world. This semi-enclosed area, surrounded by three continents and with intense human activity, works as a trap for plastics. For this reason, the Mediterranean is considered one of the six greatest accumulation zones for marine litter, together with the five "plastic islands" floating in the Pacific, Atlantic, and Indian Oceans (Cózar et al. 2014). The Mediterranean holds 1 percent of the world's waters but contains 7 percent of the world's marine-debris microplastics and is hence recognized as a global hot spot for targeted action (Alessi and Di Carlo 2018).

Every year, the Mediterranean receives 150,000–500,000 tons of macro-plastics and 70,000–130,000 tons of microplastics.[3]

To put this into perspective, based on this estimate and recent fish-stock accounting, more plastic flows into the Mediterranean Sea yearly than the combined annual volume of the two most commonly caught fish species (Boucher and Billard 2020).[4] Recent research shows that micro-plastics are present in the atmosphere, and synthetic textiles are the main sources of airborne microplastics, which disperse widely throughout the environment because of atmospheric conditions and human activities. During the summer months, tourists along the Mediterranean Sea gen-erate an additional 40 percent of waste that ends up in marine spaces. In total, an estimated 1,178,000 tons of plastics have accumulated in the Mediterranean (Boucher and Billard 2020).

Middle East and North Africa economies are major contributors to this continuing accumulation. Plastic waste in the Mediterranean comes mainly from four countries (of which two are in the Middle East and North Africa): Egypt (32.8 percent), Turkey (16.4 percent), Italy (10.7 percent), and Algeria (5.9 percent) (Boucher and Billard 2020). Other countries in the region (such as Lebanon, Libya, Morocco, and Tunisia) are also substantial contributors (figure 4.4). Marine-plastics hot spots tend to appear near the mouths of major rivers (for exam-ple, the Nile), which transport plastic waste from regions not directly bordering the Mediterranean Sea. They also tend to appear close to larger cities or other urban areas (Alessi and Di Carlo 2018; Boucher and Billard 2020).

Marine-Plastic Pollution of Other Seas in the Region

Relative to the Mediterranean, less is known about the pollution levels in the other seas of the region, but recent studies show increasing levels of microplastics in the Gulfs and the Red Sea. Microplastics are documented in abundance within these marine environments. Their presence in marine sediments poses a legitimate environmental concern for toxicity and food chain transfer via marine organisms.

Although the seas in the Regional Organization for the Protection of the Marine Environment (ROPME) Sea Area (RSA) are less well studied, recent research shows varying microplastics concentrations along the northern and southern coasts of the RSA but extremely high concentrations in biota along the coast of the Islamic Republic of Iran (Uddin, Fowler, and Saeed 2020).[5] The predominant fibers found are polyethylene (PE), nylon, and PET (polyethylene terephthalate), which are commonly used in plastics bags, SUPs, bottles, discarded fishing gear, and urban and industrial outflows from washing synthetic clothes

FIGURE 4.4

Average Annual Plastic-Waste Contribution of Countries Bordering the Mediterranean Sea

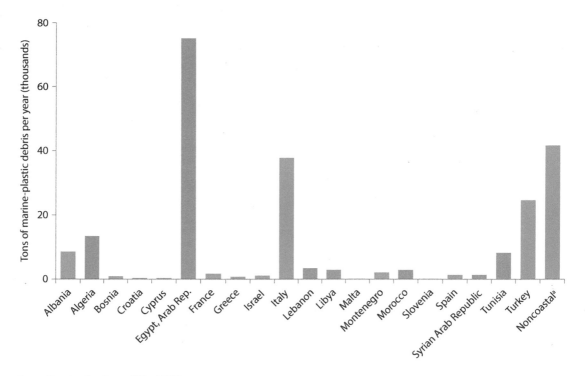

Source: Based on Boucher and Billard 2020.
Note: Orange bars designate Middle East and North Africa countries. Data are from various years.
a. "Noncoastal" countries denotes those that contribute to plastic pollution in the Mediterranean mainly through river flows, not from having a Mediterranean coastline. These countries are Bulgaria, Burundi, Ethiopia, Kenya, Kosovo, North Macedonia, Rwanda, Serbia, South Sudan, Sudan, Switzerland, Tanzania, and Uganda.

(Naji, Esmaili, and Khan 2017; Uddin, Fowler, and Saeed 2020). In the Red Sea, one in every six fish has ingested small pieces of plastic, implying that microplastic pollution has reached commercial and non-commercial fish species (Baalkhuyur et al. 2018; Martin et al. 2018).

Plastic Waste Generation in the Region

Middle East and North Africa residents generate more plastic waste per capita than residents in Asia but less than other regions. Residents in high-income Gulf Cooperation Council (GCC) countries generate around 0.2 kg per person per day, while residents in the low- and middle-income countries in the Maghreb and the Mashreq subregions generate around 0.09 and 0.07 kg per person per day, respectively (figure 4.5). These generation rates are higher than the average plastic

FIGURE 4.5

Average Daily Plastic Waste Generation Per Capita, World Regions and Middle East and North Africa Subregions, 2016

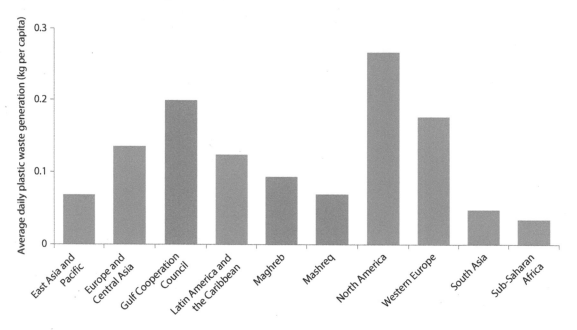

Source: Based on Law et al. 2020.
Note: Orange bars designate three groups of economies within the Middle East and North Africa. The Gulf Cooperation Council includes Bahrain, Kuwait, Oman, Qatar, Saudi Arabia, and the United Arab Emirates. The Maghreb subregion includes Algeria, Libya, Malta, Morocco, and Tunisia. The Mashreq subregion includes Djibouti, the Arab Republic of Egypt, the Islamic Republic of Iran, Iraq, Jordan, Lebanon, the Syrian Arab Republic, West Bank and Gaza, and the Republic of Yemen. "North America" includes Canada and the United States. kg = kilograms.

waste generation of 0.07 and 0.05 kg per person per day in East Asia and South Asia, respectively. North Americans generate the most plastic waste per capita in the world—around 0.27 kg per person per day.

Within the Middle East and North Africa, GCC countries such as Kuwait and the United Arab Emirates are by far the largest plastic-waste producers, each generating more 0.3 kg per person per day. At the other end of the spectrum, the residents of Morocco, the Syrian Arab Republic, and the Republic of Yemen generate the least plastic waste per capita per day, each averaging less than 0.05 kg per person per day (Law et al. 2020).

Some high-income GCC countries (Kuwait, Oman, and the United Arab Emirates) generate more plastic waste than similar high-income countries (countries above the green line in figure 4.6). However, Bahrain, Qatar, and Saudi Arabia produce less plastic waste than other income-comparable countries. As for the middle-income economies,

FIGURE 4.6

Correlation between GDP Per Capita and Rate of Plastic Waste Generation, 2016

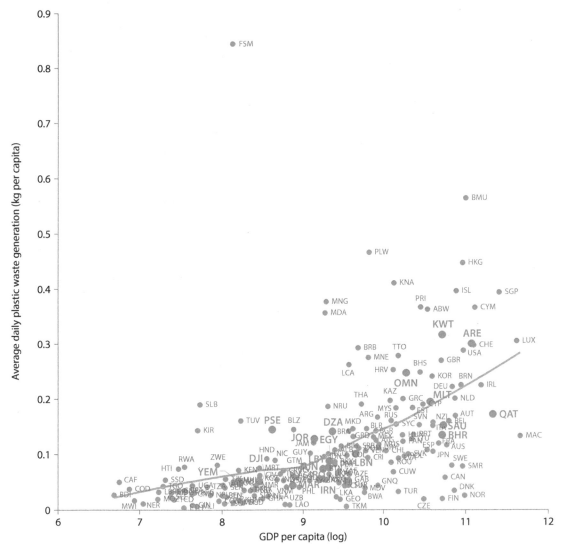

Source: Based on Law et al. 2020.

Note: Enlarged dots and codes in orange designate Middle East and North Africa economies. Smaller blue dots and codes designate economies of other global regions. Data for the Syrian Arab Republic are unavailable. Economies are labeled using ISO alpha-3 codes, as listed on the
Abbreviations page in the front matter of this report. Residents in the economies above the green line are generating more plastic waste than
what their average income level would suggest. GDP = gross domestic product; kg = kilograms.

Algeria, Jordan, and West Bank and Gaza produce more plastic waste
than would be expected for their income levels, while the rest of the
Maghreb and Mashreq countries are either on par with international
comparators or slightly below (Law et al. 2020).

The Waste Management Deficit

The lack of appropriate waste management is a major challenge for marine plastics in the Middle East and North Africa. The reasons for inadequate waste management are the lack of treatment and bad disposal management (as further discussed in the subsection on policy options to "Stop the Leakage"). Whereas the number of well-managed landfills has increased in countries such as Egypt, Morocco, Qatar, and Tunisia over the past few years and the Gulf countries have implemented a number of waste-to-energy initiatives, open dumping is still prevalent in the region.

High Rates of Poor Waste Management

Fifty-three percent of the region's total solid waste is inadequately managed, and around 12 percent of that solid waste is plastics (Kaza et al. 2018). In the Maghreb and Mashreq subregions, as much as 60 percent is inadequately managed (Jambeck et al. 2015), meaning that waste is mainly disposed of in open dump sites or uncontrolled landfills and often burned, either accidently or on purpose (figure 4.7). As a result of these practices, high volumes of plastics flow into the seas, leading to significant environmental degradation of countries' coasts. This is especially the case in the Mediterranean Sea, where poor waste management is identified as the major reason for plastic leakage (Boucher and Billard 2020). Furthermore, poor waste management contributes to air pollution, because waste in open dumps is prone to self-ignition and uncontrolled burning for longer stretches of time. (For some of the effects on air pollution, see chapter 3.)

Compared with countries of similar income, waste management practices in the Middle East and North Africa are more often inadequate. This is especially prevalent in the Mashreq subregion, where, for example, Egypt (67 percent), Iraq (63 percent), Morocco (66 percent), and Tunisia (60 percent) significantly mismanage their waste.

When comparing the shares of inadequately managed waste with the income of countries, it is worrying that the region's economies are doing worse than those with similar incomes (figure 4.8). Especially Egypt, Iraq, and the Islamic Republic of Iran are faring worse than comparable countries (represented by the green line in figure 4.8). GCC countries seem to be managing their waste relatively well relative to those with similar income levels, except for Bahrain (10 percent) and Saudi Arabia (8 percent).

FIGURE 4.7

Share of Mismanaged Waste, World Regions and Middle East and North Africa Subregions, 2010

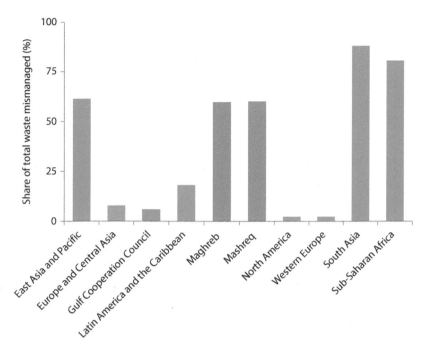

Source: Based on Jambeck et al. 2015.

Note: "Mismanaged" waste refers to waste that is disposed of either as litter or in open dumps or uncontrolled landfills, where it is not fully contained. Orange bars designate three groups of economies within the Middle East and North Africa. The Gulf Cooperation Council includes Bahrain, Kuwait, Oman, Qatar, Saudi Arabia, and the United Arab Emirates. The Maghreb subregion includes Algeria, Libya, Malta, Morocco, and Tunisia. The Mashreq subregion includes Djibouti, the Arab Republic of Egypt, the Islamic Republic of Iran, Iraq, Jordan, Lebanon, the Syrian Arab Republic, West Bank and Gaza, and the Republic of Yemen. "North America" includes Canada and the United States.

Low Rates of Recycling

The overall recycling level in the Middle East and North Africa (about 9 percent of municipal solid waste) is lower than the global average of 13.5 percent (orange line in figure 4.9) (Kaza et al. 2018). The GCC countries recycle at higher rates than the rest of the region. Qatar recycles more than 30 percent of its waste, slightly exceeding the global average for high-income countries (29 percent). However, other GCC countries recycle much less of their waste: the United Arab Emirates recycles 24 percent; Kuwait, 19 percent; Oman, 15 percent; Saudi Arabia, 12 percent; and Bahrain, close to 0 percent. Thus, on average, the GCC countries recycle only around 15 percent of their waste, well below the global average for high-income countries (green line in figure 4.9).

FIGURE 4.8

Correlation between GDP Per Capita and Share of Mismanaged Waste, 2010

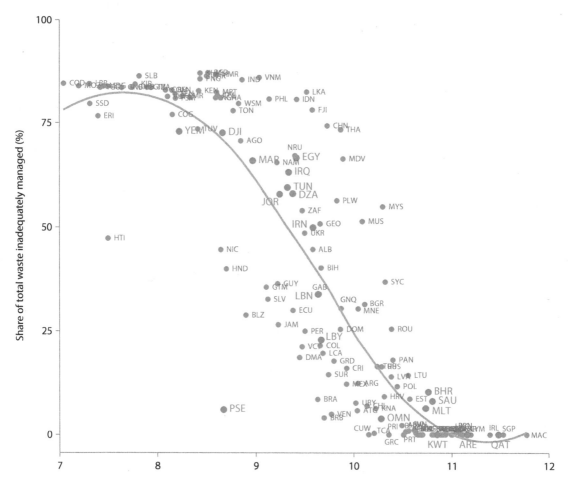

Source: Based on Jambeck et al. 2015 and the World Development Indicators database.
Note: Economies are labeled using ISO alpha-3 codes, as listed on the Abbreviations page in the front matter of this report. Enlarged dots and codes in orange designate Middle East and North Africa economies. Data for the Syrian Arab Republic are unavailable. Smaller blue dots and codes designate economies of other global regions. The green line indicates the average share of mismanaged waste for a given income. "Mismanaged" waste refers to waste that is disposed of either as litter or in open dumps or uncontrolled landfills, where it is not fully contained. GDP = gross domestic product.

The informal recycling sector is quite active across the region, especially in the Maghreb and the Mashreq. For example, an estimated 96,000 informal waste pickers are active in Cairo and account for 10 percent of the waste collected in the city (Kaza et al. 2018). Compared to average recycling rates of lower-middle-income countries globally (6 percent; gray line in figure 4.9), the Mashreq is lagging (4.7 percent), while the Maghreb is on par (6.3 percent). These rather low levels of recycling imply that most plastic is leaving the value chain after a single use, especially when considering that plastic recycling rates are generally much lower than global recycling rates (Shen and Worrell 2014).

FIGURE 4.9

Share of Municipal Solid Waste Recycled in the Middle East and North Africa, by Region, Subregion, and Economy, 2020

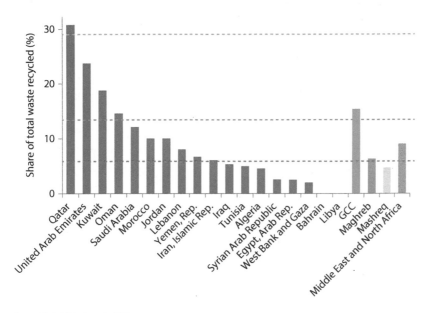

Source: Verisk Maplecroft 2020.
Note: The orange line denotes the world average recycling rate, the green line the average for high-income countries, and the gray line the average for lower-middle-income countries (grouped by World Bank income classification). The gray bar designates the Middle East and North Africa region as a whole. Light blue, orange, and yellow bars designate three groups of economies within the Middle East and North Africa, respectively, as follows: The Gulf Cooperation Council (GCC) includes Bahrain, Kuwait, Oman, Qatar, Saudi Arabia, and the United Arab Emirates. The Maghreb subregion includes Algeria, Libya, Malta, Morocco, and Tunisia. The Mashreq subregion includes Djibouti, the Arab Republic of Egypt, the Islamic Republic of Iran, Iraq, Jordan, Lebanon, the Syrian Arab Republic, West Bank and Gaza, and the Republic of Yemen. Data for Djibouti and Malta are unavailable.

Impacts of the COVID-19 Pandemic

The outbreak of the COVID-19 pandemic increased the use of plastics, hence increasing environmental pollution. The pandemic has massively increased the consumption of some plastic products including personal protective equipment and other single-use medical equipment along with packaging for food delivery services. In the Middle East and North Africa, countries like Egypt, Jordan, and Morocco mobilized industries to meet local and international demand for medical masks (Hamdallah 2020; Makki, Lamb, and Moukaddem 2020). Pharmacy customers in Lebanon were reportedly buying an average of seven masks per week after the Ministry of Health made it mandatory to wear one when leaving home (Houssari 2020). In the Gulf countries, restaurants are providing dine-in customers with up to three sets of disposable plates, cups, and cutlery for a single meal (Malek 2020).

Currently, most of the region lacks adequate infrastructure for the collection, treatment, and disposal of hazardous and medical waste, and most of the waste gets mixed with municipal solid waste, thus ending up in open dumps. Pictures of masks and gloves littering the streets, rivers, and beaches have made visible the recent impact of the increased use of plastics.

The COVID-19 global health crisis is putting additional pressure on already weakened waste management practices. Moreover, as lockdowns took effect to slow the spread of the virus, the global demand for petroleum collapsed. As a result, oil prices plummeted, making the manufacture of virgin plastics from fossil fuels much less expensive than recycling. This cost incentive, along with the lifestyle change, has complicated the challenge of overcoming plastic pollution (Adyel 2020). Although the rise in demand is expected to be temporary, behavioral barriers and misperceptions might make these "new" norms particularly sticky and hinder society's ability to transition to more sustainable practices and reduce the consumption of plastic products. As events continue to evolve and research progresses, it is expected that this new context will exert even more pressure on degraded environmental systems.

THE ENVIRONMENTAL, PUBLIC HEALTH, AND ECONOMIC IMPACTS OF PLASTIC-POLLUTED SEAS

Decades of weak SWM infrastructure and lack of proper regulation have taken a toll, and the ecological health of the Middle East and North Africa's seas is in decline. The Mediterranean Sea is among the most plastic-polluted seas in the world, with plastic concentrations comparable to the five large accumulation zones of the five subtropical gyres (Cózar et al. 2014).[6] In fact, only the garbage patch (a large area of captured marine debris) at the inner accumulation zone of the South Atlantic Gyre has a higher average concentration of plastics. When compared with five major oceanic zones,[7] the Mediterranean outpaces their average plastic-concentration rates by far (Cózar et al. 2014). Large amounts of both macroplastics (items 5 millimeters or larger) and microplastics (items smaller than 5 millimeters) are entering the Mediterranean every year (Boucher and Billard 2020).

Regarding microplastics, the Mediterranean has been found to be the most polluted in terms of particles, with estimates ranging between 21 percent and 54 percent of all global microplastic particles (van Sebille et al. 2015).[8] Such levels of plastic pollution in the Mediterranean have many negative effects. Among others, they reduce the productivity of certain blue economy sectors, result in losses of safe habitats for a host of marine and coastal species, and are detrimental to human health. This section discusses these effects, in turn, on ecosystems and biodiversity, public health, and the region's economy.

The Toll on Local Ecosystems and Biodiversity

The Mediterranean is one of the world's most highly valued seas, with a vast set of coastal and marine ecosystems that make it one of the world's biggest marine and coastal biodiversity hot spots. Sixty percent of its flora is unique to the region, and even though it represents less than 1 percent of the world's ocean surface, it is home to 28 percent of the world's endemic species (Vié, Hilton-Taylor, and Stuart 2009).

Plastic waste has severe impacts on Mediterranean ecosystems and local biodiversity. The increase in plastic pollution is reducing biodiversity, for example by entangling wildlife, which leads to high mortality rates. Of all entangled wildlife, 35 percent are birds, and hence most affected, followed by fish (27 percent), invertebrates (20 percent), and mammals (13 percent) (UNEP 2015). Recent studies show that certain consumer plastic items—plastic bags, packaging, and sheets; fishing nets and monofilament line; and balloons and other latex products— are disproportionally lethal to marine megafauna (Roman et al. 2020). Smaller items, although abundant, are seldom implicated in mortality. However, understanding which items disproportionately result in mortality and determining whether these items originate from land or sea provides an opportunity to prioritize policies that could help reduce debris-related mortality of threatened marine megafauna.

Sea-based debris (including fishing nets and ropes, fishing hooks, lines, and tackle) contribute less pollution than land-based sources, but they are an important source of animal mortality. In some geographical locations, especially in the Great Pacific Garbage Patch, fishing debris amounts to almost half of all plastic debris by weight (Lebreton et al. 2018). As for the Mediterranean, studies in Morocco confirm that marine debris from fishing contributes around 3–5 percent of total plastic debris (Nachite et al. 2019).

The impact on megafauna of ingesting sea-based debris may be underestimated because of spatial biases in data collection. Because lethality estimates are driven by the relative frequency of presence and absence of ingested items, spatial factors are ultimately less important than the physiological impact of different debris items within the animal's gut (López-Martínez et al. 2021; Roman et al. 2020).[9] Studies confirm that sea-based debris are important causes of mortality across all megafauna groups (López-Martínez et al. 2021; Roman et al. 2020; UNEP 2015).

Microplastics are the most abundant debris reported floating in the marine environment. Quantities of microplastics in marine spaces are increasing exponentially, mostly resulting from the surface-weathering degradation of plastic debris and other sources such as tire abrasion, production pellets, textiles, and personal care products carried from wastewater and surface runoff into the soil, rivers, lakes, and ultimately the oceans (Lebreton et al. 2017; Pew Charitable Trusts and SYSTEMIQ 2020).

The smaller the microplastics, the wider the range of marine organisms able to ingest or interact with them. Microplastics can also absorb and concentrate hydrophobic pollutants present in seawater at very low concentrations, increasing the adverse effects of ingestion by animals. At present, over 660 species—ranging from seabirds, fish, and mollusks to the zooplankton at the bottom of marine and food chains—are known to be affected by plastic debris (Lebreton and Andrady 2019).

Filling knowledge gaps will allow a better understanding of the tipping points and environmental thresholds for marine-plastic pollution and improve policy design specifically to address this issue and its consequences for human health.

The Consequences for Public Health

We are only beginning to understand the negative health consequences of plastics in the seas for human health; research on the impacts is at its infancy. A 2019 study by the University of Newcastle, Australia, found that an average person could be ingesting as much as 5 grams of plastics every week (Senathirajah and Palanisami 2019).[10] Other studies estimate that children and adults might ingest from dozens to 100,000 microplastic specks each day (Nor et al. 2021). Through the different ways that waste is mismanaged (for example, ending up in marine spaces but also often burned and thereby entering the air), the exposure routes have been expanded from the food chain to contaminated food and drinks and, more recently, to inhalation (Zhang et al. 2020).

In sum, plastics have been found in the food we eat (fish that have eaten plastics), the water we drink (microplastics in the drinking water), and the air we breathe (airborne plastic particles from uncontrolled waste burning).

Exposure Routes

Microplastics have recently been detected in the atmosphere of urban, suburban, and even remote areas far from source regions of microplastics, suggesting long-distance atmospheric transport of microplastics. In addition, emerging evidence suggests the presence of microplastics in human stool and colectomy specimens, confirming its presence in the human colon through ingestion (D'Angelo and Meccariello 2021; Ibrahim et al. 2021).

Microplastics are commonly found in marine-related produce such as seafood and table salt. In humans, most of the microplastic ingestion from seafood is likely from species eaten in their entirety, such as mussels, oysters, shrimps, crabs, and some small fish. For example, microplastics have been found in the digestive tract of many commercial species, such as Atlantic mackerel (*Scombrus scombrus*), herring (*Clupea harengus*), and plaice (*Pleuronectes plastessa*) as well as the digestive tract within the shell and in the muscle tissue of wild tiger prawns (*Penaeus semiculcatus*) and brown shrimp (*Crangon crangon*). Microplastics have been found in all samples of mussels purchased from UK supermarkets (CIEL 2019). Regarding seaweed, at high exposure, microplastic particles could stick to the surface of edible species (such as *Fucus vesiculosus*), although washing reduced the number of particles by 94.5 percent.

Microplastic particles have been found in commercial table salt derived from sea, lake, and rock salt, which suggests the high-level contamination of the environmental background (CIEL 2019). Consumers who drink three cups of coffee in disposable paper cups are ingesting about 75,000 microplastic particles from the thin layer of plastic inside the cup (D'Angelo and Meccariello 2021).

Morbidity and Mortality Risks

Evidence regarding microplastic toxicity and epidemiology is emerging. From a human health perspective, the effects of inhaled or ingested microplastics depend on factors such as size, chemical composition, and shape. The absorbed particles can affect the body through chemical toxicity, and the smallest particles can be taken in by cells, potentially being transferred to human body tissues and causing inflammations comparable to the impacts of particulate matter. The interaction between microplastics and other gut contents, including proteins, lipids, and carbohydrates, appears to be highly complex (CIEL 2019; Dalberg Advisors 2019a; Lim 2021; Pew Charitable Trusts and SYSTEMIQ 2020) but potentially dangerous. Initial results showed that the accumulation of microplastics in the human body could lead to inflammation, tissue damage, cell death, or carcinogenesis (Wright and Kelly 2017).

In addition, there is growing evidence that plastics may be contributing significantly to exposure to complex mixtures of chemical contaminants (such as chemicals either intentionally added during the production process, originating from ultraviolet [UV] radiation, coming from the waste-recycling process, or absorbed from environmental pollution) that cause endocrine disruptions from inhalation, ingestion, or both inhalation and ingestion of microplastics (Gallo et al. 2018).

Finally, some of the most recent research has examined potential reproductive effects. Microplastics accumulate in placentas during pregnancies and are a potential threat to male fertility. Several microplastic fragments were detected in placenta samples in a recent study collected from pregnant women. Possible entry points include the bloodstream but also respiratory and gastric organs (Ragusa et al. 2021). Ingested microplastics can also bioaccumulate in mammalian tissue, affecting rodents' semen quality, as a consequence of inflammation and oxidative stress damage. Furthermore, the morphological features of microplastics can make them an ideal vehicle for additional environmental pollutants (D'Angelo and Meccariello 2021). That microplastic exposure affects sperm quality in animals highlights possible reproductive risks for humans as well, a topic where further research is needed.

The Costs to the Blue Economy

The Mediterranean's "blue economy" is among the most valued in the world, and its coastal and marine areas represent one of the Middle East and North Africa region's most important economic assets. The Mediterranean Sea's vast coastal and marine ecosystems deliver important economic and environmental benefits. The "shared wealth fund"—that is, the value of the Mediterranean coastal and marine assets dependent on functional ecosystems—was calculated to total about US$5.6 trillion, comprising marine fisheries (US$39 billion), sea grass (US$716.9 billion), productive coastline (US$4.65 trillion), and carbon absorption (US$173.5 billion), and generating economic output of about US$450 billion per year (Randone, Di Carlo, and Constantini 2017). The tourism sector accounts for 92 percent of the total value, followed by ecosystem services enabled by the ocean (6 percent), and fisheries and aquaculture (2 percent). Before the COVID-19 pandemic, the Mediterranean region attracted one-third of all global tourism, and several Middle East and North Africa countries rely on this sector for much of their income.

Moreover, the Mediterranean Sea brings innumerable other benefits not included in this valuation, including ecosystem services such as coastal protection and climate regulation. The Mediterranean also provides strong interdependencies with other critical sectors for the region's economy such as transportation, clean energy, and cultural tourism.

Marine-plastic pollution causes heavy economic losses to this economic wealth. Losses from plastic pollution are estimated at €641 million per year for the Mediterranean, including up to €268 million in tourism,

€235 million in the maritime industry, and €138 million in fisheries (Dalberg Advisors 2019d), as follows:

- *Tourism-related costs* are linked to the expenses incurred by coastal towns to clean up beaches from additional waste generated during the tourist season. Marine litter also affects the aesthetic value (which is difficult to quantify) and attractiveness of the beaches and shorelines for recreational purposes, such as diving, snorkeling, and recreational fishing (during which plastics can affect catch amounts and damage gear) (UNEP 2016).

- *The maritime industry* reliant on propeller boats (in marine transportation and fishing, for example) is particularly vulnerable to collisions with large plastic objects that become entangled with propeller blades and clog water intakes for engine cooling systems. Costs are calculated in vessel downtime, delays, and additional maintenance costs.

- *Port facilities* are at risk of damage from plastic pollution, including the clogging of waterways, which creates delays and causes additional cleanup costs.

- *For the fishing sector*, the largest costs come from vehicle damage and maintenance caused by plastic debris, collision with plastics, and delays when fishing nets fill up with plastics rather than fish (Dalberg Advisors 2019d).

In addition to the significant costs of plastic waste to the tourism, fishing, and shipping industries, the plastics-producing industry also consumes large amounts of fossil fuels, directly as feedstock and indirectly in the form of produced electricity, leading to high climate and air pollutant emissions. Over 90 percent of plastics produced are derived from virgin fossil feedstocks (EMAF 2016). Feedstock prices are the most influential factor in determining regional production advantages because these prices are a major part of the overall cost structure for petrochemicals (IEA 2018) and are a major determinant of overall costs for plastic production.

In the Middle East and North Africa, more precisely in the Middle East, feedstock prices account for around one-fourth to one-third of total costs (for products using ethane as feedstock), while comparable feedstock accounts for about 50 percent of total costs in Europe and the United States. The Middle East and North Africa's low prices for fossil fuels and (in some countries) heavy subsidies to the fossil fuel sector drive down overall costs for petrochemicals and plastic production (as the Policy Review section discusses in more detail). Research indicates that 8 percent of the world's oil production is used to make plastics, half of which is used as material feedstock and half as fuel for the production process (BP 2015; Hopewell, Dvorak, and Kosior 2009; Plastics Europe 2015). This is equivalent to the oil consumption of the global aviation sector (EMAF 2016). If the current strong growth of plastics continues as business as usual, the plastics sector will account for 20 percent of total oil consumption by 2050.[11]

POLICY REVIEW: HOW TO GET CLEAR, BLUE, PLASTIC-FREE SEAS

This section discusses some of the main steps toward "blueing" the Middle East and North Africa region's seas and making them free of marine plastics while also considering the differences between economies.

First, the "Priority Recommendations" section addresses how plastic pollution of the region's seas exemplifies the damages that a traditional linear economy causes to environments and their inhabitants and economies. This part of the report recommends transforming the life cycle of plastics through a circular-economy approach that will preserve environmental resources, improve residents' well-being, and advance the region's economies. The recommendations presented here can be implemented while the region's economies make the necessary invest-ments to obtain country-specific data and evidence to combat plastic pollution—most importantly, identifying sources of marine-plastic pollution by location and economic sector. Although a wealth of infor-mation about plastic pollution exists at the global level, including for the Mediterranean, local source information on marine plastics is critically missing. Such information is key for designing effective policy measures to achieve a circular-economy approach that will synergistically benefit environments, residents' health, and their economies.

Second, a section on "The Sources of Marine Plastics" discusses the limited regional information currently available on sectors, types of plastics, hot spot locations, and leakage points. It also briefly discusses the global literature on the main sources of leakage, then summarizes the recent work of some Middle East and North Africa economies to identify hot spots.

Third, the "Comprehensive Policy Options" section reviews the full range of policies to stem the region's plastic-pollution wave. It discusses the principles of a circular economy and presents examples of how both private sector and public sector actors can tackle some of these issues to improve the circularity of plastics while transforming the region's economy and waste management.

Priority Recommendations: Actions to Stem the Plastic Tide

Reducing marine-plastic pollution requires (a) comprehensive solutions along the plastic life cycle that are (b) linked to a circular-economy approach and (c) tailored to each country's specific context and needs. The building blocks of this integrated approach are a set of components commonly known as the 3 R's: reduce, reuse, recycle.

The 3 R's of a Circular Economy

Reduce. New ways to *reduce* the use (and thereby the manufacture) of plastic might include, for example, incentivizing the use (and thereby the manufacture) of reusable rather than SUP food-service-related materials such as plastic bags, cups, dishes, knives, forks, and spoons. These new ways must take into account that it is currently much easier and cheaper for consumers and manufacturers to see such products as single-use throwaways. Governmental entities must create appropriate incentives and corresponding disincentives to change such throwaway-oriented consumer and producer behaviors to limit the current—and future— harms of plastic pollution and achieve the environmental, public health, and economic benefits of a circular-economy approach.

Reuse. New approaches are needed to foster the *reuse* of plastic products by changing consumer and producer behaviors. As with *reducing the use* of plastics, *increasing the reuse* of plastics requires government incentives and disincentives to foster circular-economy approaches that benefit the entire spectrum of individuals, communities, private enterprises, and government programs. To that end, it is important to redirect the financial resources that stakeholders now pay for dealing with the consequences of plastic pollution, enabling them instead to benefit from a circular-economy approach to the use of plastics.

Recycle. New methods for *recycling* plastic products can help ensure that nonreplaceable natural resources currently used in the manufacture of virgin plastics can be saved for more essential uses. The use of natural resources for the extensive manufacture of virgin plastics is *not* an *essential* use, because recycled plastics can serve that same purpose. Furthermore, recycling reduces the stream of waste that imposes

- Direct costs (for operating and expanding dump sites);

- Indirect social costs (for health care to treat air-pollution-related diseases, since dump sites in the Middle East and North Africa are often openly burning (both intentionally and by accident), which produces airborne toxic particles);

- Costs incurred by fishers and port operators from macroplastics damage;

- Demonstrated health impacts on humans from ingesting the microplastics in contaminated fish; and

- Forgone economic benefits (when tourists avoid plastic-polluted beaches and other coastal areas).

All of those costs would be dramatically reduced if not eliminated by a circular-economy approach to the use of plastic.

Priority Steps toward Achieving a Circular Economy

In parallel with the 3 R's of a circular-economy approach to plastic pollution of the region's seas, governmental entities must continue and enhance cleanup methods to restore ecosystems affected by high levels of plastic pollution—including the beach areas vital to Middle East and North Africa economies' income from international and intranational tourists. Essential to such restoration is the upgrading of SWM policies, practices, and facilities.

Although pollution prevention is preferable to pollution remediation, the reality is that it will take time for the region's economies to shift to a circular-economy approach regarding the use of plastics. Moreover, the leakage of plastics and other pollutants from poorly maintained dump sites poses substantial current threats to residents' health, to local and national economies, and to environmental resources. Dump-site leakage of pollutants contributes to residents' morbidity and mortality, damages local and national economies, and spoils—in some cases, irretrievably—the region's natural resources. Piecemeal approaches will at best yield fragmentary and partial solutions. The only comprehensive solution to plastic pollution of the Middle East and North Africa's seas and coastal areas is a circular-economy approach to the use of plastics.

For a 3 R's approach of reduce, reuse, and recycle—along with appropriate SWM (including cleanups) of plastics not amenable to reuse or recycling—to work, government entities (in cooperation with academia, nongovernmental organizations [NGOs], and the private sector) must raise public awareness about the negative impacts of plastic pollution on their health, their family members' health, and on the economy. Meanwhile, governments must support those residents and enterprises that experience financial challenges in the transition to a circular-economy approach, while also continuing to advance the circular economy's comprehensive overall benefits. The priority policy recommendations below are divided into five main categories; however, all of them are interdependent. Table 4.1 summarizes these priority recommendations.

TABLE 4.1

Priority Policy Options to Tackle Marine-Plastic Pollution in the Middle East and North Africa, by Subregion

Main objective	Measure	Subregion[a]	Timeline for implementation	Financial cost	Effectiveness
Stop plastic leakage by improving SWM	Convert open dumps to engineered landfills	Maghreb, Mashreq	medium	high	high
	Improve municipal financing for SWM projects	Maghreb, Mashreq	medium	high	medium
	Upgrade vehicle fleet for collection services	Maghreb, Mashreq	medium	high	medium
Decrease production and consumption of plastic items	Increase input prices for producers by removing fossil fuel subsidies	Mashreq, GCC	short	medium	high
	Introduce taxes, charges, or both for plastic products	Maghreb, Mashreq, GCC	short	medium	high
	Increase quality and quantity of recyclable materials and plastic reprocessing capacity	Maghreb, Mashreq, GCC	medium	high	high
Ensure reliable, high-quality supply of recycling materials	Mandate minimum recycled contents	Maghreb, Mashreq, GCC	medium	high	high
	Create pathways to formalization for informal recyclers	Maghreb, Mashreq	short	medium	medium
	Support enterprise-development programs for recyclers' associations and cooperatives	Maghreb, Mashreq	short	low	high
Generate long-term demand for recycled plastics	Develop fiscal, market, and regulatory instruments	Maghreb, Mashreq, GCC	medium	low	high
	Introduce EPR systems and DR schemes	Maghreb, Mashreq, GCC	long	medium	high
	Execute social marketing, campaigns, and educational programs	Maghreb, Mashreq, GCC	short	low	medium
Enable new manufacturing processes and technologies, innovation, green jobs, and development of alternative green materials	Add subsidies for R&D, and implement blended finance schemes for companies developing new alternative products and recyclable materials	GCC	medium	medium	high
	Align and standardize recycling of packaging	Maghreb, Mashreq, GCC	medium	medium	high
	Progressively phase out SUPs (for example, food containers, straws)	GCC	long	high	high

Source: Based on World Bank data.

Note: DR = deposit-refund; EPR = extended producer responsibility; R&D = research and development; SUPs = single-use plastics; SWM = solid waste management.

a. The Gulf Cooperation Council (GCC) includes Bahrain, Kuwait, Oman, Qatar, Saudi Arabia, and the United Arab Emirates. The Maghreb subregion includes Algeria, Libya, Malta, Morocco, and Tunisia. The Mashreq subregion includes Djibouti, the Arab Republic of Egypt, the Islamic Republic of Iran, Iraq, Jordan, Lebanon, the Syrian Arab Republic, West Bank and Gaza, and the Republic of Yemen.

No. 1: Improve SWM to Minimize Plastic Leakage

To move toward a circular economy, improve SWM to minimize plastic leakage. This also provides important co-benefits for the urban environment (such as reducing the costs of cleaning up such leakage) and public health (reducing morbidity and mortality related to air pollution from open burning at dump sites).

Although SWM is generally a local responsibility, national governments can help cities improve their capacity to organize and regulate the collection and disposal of urban waste. Consequently, adequate financing schemes must be put in place that are tailored to local solutions. These improvements include not only the phaseouts of harmful disposal practices such as open dumps or uncontrolled landfills but also upgrading of collection and processing equipment to reduce leakage.

The COVID-19 crisis increased the pressure to develop adequate medical and other hazardous waste management systems, which are largely missing in the region; therefore, this circumstance should be addressed, because the COVID-19 situation remains unstable. One first step would be to sort and segregate waste at the source, separating organic from inorganic waste where possible.

No. 2: Make Recycling Markets Work

Well-functioning recycling markets require a reliable supply of materials to *be* recycled. The quality and quantity of recycled material and plastics in the region is low, and processing capacity is limited. Setting mandatory minimum amounts of recycled content in manufactured items would guarantee a domestic demand and encourage investments in the plastic-recycling industry. To this end, guidelines and standards for recycling across the value chain must be developed, together with uniform labeling systems that indicate recyclability to inform and guide consumers and recyclers. There is a potential US$6 billion recycling industry in GCC countries alone if 40 percent of the waste were recycled (Menachery 2020).

Creating an enabling environment to support a transparent local market and long-term demand for recycling is a must to implement a functional circular-economy approach to the manufacture and use of plastics. In most Middle East and North Africa economies, the collection of recyclable material relies on the informal sector. Appropriate government policies can create more jobs and improve the working conditions of informal waste pickers by building capacity for delivering higher-quality products through training and through small and microenterprise development programs. In addition, government policies can provide the informal industry with a pathway to formalization, licensing, and compliance to ensure a transparent market and level playing field with formal recyclers who comply with environmental, health, and safety standards and other requirements.

As markets develop, recovery and recycling facilities can provide economies of scale, reducing the costs of recycling. To make recycling markets functional and successful, there is a need to generate long-term demand by developing national standards combined with support to local governments to build capacity for ensuring local-level separation, recycling, and technology upgrades.

No. 3: Raise Public Awareness about Improving SWM

Programs to improve SWM must be accompanied by information and public-awareness campaigns. Because improved SWM implies actions and costs that include the collaboration of residents and consumers, residents must be made aware of the advantages of improved SWM—particularly its potential to prevent the damage that plastic littering causes not only to public health, drainage channels, rivers, shorelines, and sea life but also to local and national economies.

Campaigns can build support for behavioral change to increase the use of recyclable products and separate waste at the source. Most current recycling programs in the Middle East and North Africa fail because they lack local residents' participation and because they mismanage waste at the source (World Bank 2021d, 2021f). Information and logistical support to enable separation (for example, through well-designed communication campaigns, provision of clearly marked containers, and convenient disposal sites) are critical actions to increase recycling levels and the quality of collected materials.

Educational programs in schools are also important to raise awareness of environmental responsibility and build understanding of the principles of circular economies. Similarly, volunteer beach and river cleanup programs have proven to raise awareness of environmental values among residents—while those programs and their participants perform the valuable task of removing plastic litter and other litter from beaches and thereby also prevent a sizable portion of such litter from ending up in the region's seas.

No. 4: Collaborate with the Private Sector toward Phasing Out SUPs

Collaboration with the private sector supports the development and manufacturing of alternatives to SUPs, aligns recycling standards, and helps to progressively phase out SUPs. As a result, reducing overall plastic use has significant potential to create new markets and local jobs in a postpandemic context. Commitments by governmental entities to reduce overall plastic use can stimulate innovation, develop new technologies, and provide collaboration opportunities for dynamic businesses in a changing world—while preserving the natural resources upon which all nations depend.

The new circular economy requires new forms of government-business collaboration to standardize types of packaging with increased recycled content. This requires a new set of regulations and blended financing schemes to help leapfrog technologies and develop recycling markets. Governments are important enablers—indeed, they are the only organizations with the authority, ability to establish incentives and disincentives, and other resources—to promote this approach. And only they can enact and implement the legal and regulatory frameworks needed to break down barriers to investments by using economic and financial instruments such as those described in the "Comprehensive Policy Options" section below.

SUPs are the most problematic plastics because they account for the great majority of items found to be polluting the seas and beaches. On Moroccan beaches, five SUP items (cigarette butts, bottle caps, food wrappers, beverage bottles, and shopping bags) accounted for 59 percent of the marine debris (World Bank 2021d). Recent European bans on SUPs show how to set progressive measures over the next few years with an initial focus on products that have non-SUP alternatives readily available while supporting the development of additional new alternatives.

In this regard, both the private and public sectors have critical roles to play—the private sector in driving innovative solutions to the plastic crisis and the public sector in creating an enabling environment to support the phaseout of SUPs, the development of alternatives, and the setting of ambitious goals. Middle East and North Africa governments and companies can work together to gradually replace SUPs and increase the availability and affordability of greener alternatives. These programs can be part of a stimulus plan with blended finance schemes for sustainable recovery in a postpandemic world.

No. 5: Level the Playing Field for Green Alternatives

Reducing fossil fuel subsidies can help level the playing field for greener alternatives. The current alternatives to plastics (for example, wooden spoons, paper cups, bioplastic food wrapping, starch-based bowls, and so forth) are significantly less available and more expensive than the plastic versions. In the Middle East and North Africa, this price gap is at least a factor of four and in most cases much larger.

The main feedstock of plastic products (oil, gas, and other fossil fuels) is heavily subsidized for domestic plastic producers, as is energy derived from these sources—the energy that plastic producers use in their manufacturing and delivery activities. Hence, as chapter 3 argued regarding air pollution, fossil fuel subsidy reforms are also called for to address plastic pollution in the Middle East and North Africa's seas. Such reforms must be meticulously planned and should engage the full spectrum of stakeholders. However, they are critically necessary to decrease the production and consumption of plastics and hence the amount of plastics that flows into the region's seas.

The Need for New, Collaborative Frameworks

The new set of solutions will require new frameworks of collaboration. Municipalities, ministries, corporations, and society play important roles in tackling plastic pollution. In the Middle East and North Africa, low collaboration between the public and private sectors, on the one hand, and high collaboration between companies, on the other hand, limit the wide implementation of circular-economy models. The Thailand Public-Private Partnership for Plastic and Waste Management, the Malaysia Sustainable Plastic Alliance, and the Philippine Alliance for Recycling and Materials Sustainability are examples of public-private collaboration platforms that merit examination and emulation (World Bank 2021a, 2021b, 2021c).

To summarize the central finding of this part of the report: Any and all efforts to stem the flow of plastic pollution into the region's seas merit consideration. However, the only way to comprehensively minimize this flow and its inevitable damages to the region's environment, natural resources, residents' health, and economies is through the implementation of a circular-economy approach to the use of plastics.

One Must Measure What One Would Manage: The Sources of Marine Plastics

Effective policy making for reducing marine plastics requires an understanding of the causes and sources of plastic pollution. Policy makers need to know (a) the *sources of the plastics* (including the cities, localities, or river basins that are the major contributors); (b) the *contributing sectors* (tourism, fisheries, manufacturing, retail, and so forth); (c) the *estimated quantities*; and (d) the *waste-disposal practices* that contribute to the problem. Other key information includes the composition of plastics that enter the seas (such as bags, cups, packaging, and fishnets, among others) to design counteracting policies accordingly.

Sources and Sectors

The sources of plastic pollution in the Middle East and North Africa are not assessed thoroughly regarding which cities or sectors and types of plastics contribute most to the problem. Evidence suggests that the vast majority of plastics that enter the seas are in the form of macroplastics and SUPs. Fragmentation of macroplastics into microplastics makes this category a large contributor to the high levels of microplastics in the Mediterranean Sea.

About 80 percent of marine-plastic litter comes from land-based sources and 20 percent from marine sources (such as fishing, aquaculture, and maritime transport) (Li et al. 2016). As for economic sectors, tourism, recreational marine activities, and shipping discharge the most plastics directly into the marine environment and along shorelines (UNEP 2015). In addition, waste management and disposal practices are important sources of leakage: municipal uncollected waste contributed 61 percent of total leakage in 2016 and waste mismanagement, 39 percent (Pew Charitable Trusts and SYSTEMIQ 2020).

Macroplastics make up the biggest portion (around 94 percent) of all plastic items entering the Mediterranean every year (Boucher and Billard 2020). Of all plastic products, SUPs and packaging material contribute the most to plastic leakage into the ocean. Global estimates of land-based sources report that about 80 percent of this plastic leakage comes from single-use packaging material. Of that, 51 percent consists of flexible monomaterials (for example, plastic and flow wrap, light grocery bags, food containers); 29 percent from multilayers and multimaterial films (for example, condiment and shampoo single-portion items, chips, and sweets packets); and 20 percent from rigid materials (for example, bottles, pens, toys, combs, toothbrushes, durable goods, and buckets) (Pew Charitable Trusts and SYSTEMIQ 2020). SUPs are by far the biggest contributors to marine-plastic leakage, mainly because of their low recyclability, low value, and limited reusability.

The level of microplastics in the oceans is expected to increase, although the sources that contribute to the leakage in the Middle East and North Africa are not well understood, and more research is needed. Microplastics account for about 6 percent of the current total flow of plastics into the Mediterranean. Four sources of microplastics account for almost all of it: dust from tire abrasion (53 percent), textiles (33 percent), microbeads in personal care products such as cosmetics (12 percent), and production pellets (2 percent) released into the environment as microsized particles (usually less than 5 millimeters) (Boucher and Billard 2020). The current mass of microplastics in the oceans and on beaches is expected to double by 2050 from 2020 levels as current material left in the environment slowly degrades into smaller pieces (Lebreton, Egger, and Slat 2019). Globally, 15–31 percent of microplastics come from a primary source, with the rest resulting from degraded macroplastics (Boucher and Friot 2017).

Maghreb Models for Emulation

At the subregional level, some Maghreb countries are making notable progress in identifying the sources of plastic leakage and hot spots that require urgent action. For example, Morocco, in response to its concerns about marine pollution and building on previous initiatives, is taking initial steps toward formulating a national strategy for a "plastic-free coastline" (littoral sans plastique, or LISP) dedicated to the reduction of marine pollution and plastic leakage and to the promotion of circular-economy models and blue-economy models in coastal regions.

In 2019, with the support of the World Bank, Morocco undertook a comprehensive analysis to identify hot spots and anticipate main areas of intervention and challenges. (Box 4.2 presents more details about the study methodology.) Through a participatory approach in 17 coastal towns involving all stakeholders at the local level and different methods of research (collecting primary data, using mapping tools, and reviewing the literature to identify plastic-leakage hot spots), the analysis showed the following (World Bank 2021d):

- The main sources of plastic leakage in those areas were high levels of household and plastic waste in highly populated coastal regions; waste mismanagement (in some towns, as much as 60 percent); and high quantities of plastic waste generated by tourism, agriculture, and marine activities.

BOX 4.2

Identifying the Hot Spots of Marine-Plastic Debris along Morocco's Coasts

Identifying hot spots of marine-plastic leakage is a core element of policy action because it answers questions such as "where to act" (in which location or which industrial or economic sector); "what is leaking" (which polymer, application, or both polymer and application are most commonly found); and "why is there leaking" (which cultural behaviors and context features, such as a lack of garbage bins, are driving the leakage). Locating hot spots is a technical task that is key to identifying the sources of plastic pollution and key to formulating relevant interventions and policy instruments.

The government of Morocco—with the help of the World Bank—has been identifying the main waste-leakage points or hot spot areas related to marine-plastic pollution. The study is based on guidance developed by the United Nations Environment Programme (UNEP), International Union for Conservation of Nature (IUCN), and the public-private Life Cycle Initiative to identify plastic-leakage

(continued)

BOX 4.2

Identifying the Hot Spots of Marine-Plastic Debris along Morocco's Coasts (*Continued*)

hot spots, assess their impacts along the plastic value chain, and prioritize actions. This methodology is based on global studies that identify different land-based activities as the main sources of marine pollution delivered to the sea by various transport factors.

The methodology includes two major components:

- *Quantifying marine waste:* The goal is to identify and quantify the most harmful plastic items (or group of items) and activities that are sources of marine debris (for instance, fishing, tourism, and recreational activities as well as SUPs, bottles, plastic bags, and so forth).

- *Identifying pressure points:* To determine the coastal hot spots of plastic leakage, a set of indicators and criteria are applied to evaluate different categories of pressures (such as coastal populations, waste management practices, economic activities, and environmental status).

The results of these studies helped to identify 17 coastal cities within the two Moroccan seafronts (5 Mediterranean cities and 12 Atlantic cities), based on a set of indicators relating to (a) population size; (b) rate of mismanaged waste; (c) amount of plastic waste not collected; (d) amount of plastic

waste in open dump sites; (e) amount of plastic waste delivered to the sea; (f) amount of marine waste; (g) percentage of plastics in total marine waste; and (h) percentage of medical, sanitary, or both medical and sanitary waste in total marine waste.

For each of these categories, several indicators were used and rated on a scale of 1 to 4, where higher values indicate worse performance. The maximum score for a given site is 124, while the minimum is 31, with higher values signaling that these sites are faring worse and should be treated as hot spots. The results of this analysis showed that

- Hot Spot Priority (A) is the city of Casablanca, which had the highest possible score (124), mainly because of population pressure and waste disposal in the Mediouna open dump site; and

- Hot Spots Priority (B) are the cities of Kenitra (with a score of 105 out of 124 possible points), Tangier (94), and Tetouan (83) because of their high populations and the delivery of waste to landfills.

Other cities such as Nador, Rabat-Salé, Mohammadia, El Jadida, Safi, Agadir or Sidi Ifni are classified as relatively sensitive areas with scores ranging from 58 to 82.

Source: World Bank 2021d.

- Plastic debris represents between 40 and 90 percent of total waste in the seabed.

- Recreational activities generate most of the plastic waste (58 percent), followed by port activities (30 percent) and agriculture (12 percent).

- In addition, 98 percent of the plastic waste stemming from recreational activities consists of SUPs (such as bottles, food packaging, plastic cups, and food containers).

Similarly, Tunisia is carrying out initial studies to develop a policy framework for a coastal strategy free of plastic pollution. As part of the collaboration between the Tunisian government and the World Bank for the development of a circular and blue economy strategy, a series of studies addressing marine-plastic debris have been carried out with the aim of developing an integrated strategy and series of policy measures to reduce plastic pollution. These studies revealed the following (World Bank 2021f):

- The concentration of Tunisia's population along the coastlines (72 percent of the total population in 13 coastal governorates) is a major driver of waste generation and plastic pollution.

- The portion of beach waste that is plastic ranges from 48 percent at Sfax Beach to 78 percent in places like the Kerkennah Islands (where there is high fishing activity and no sustainable waste management regulations for this type of activity).

- Products such as bottle caps, food packaging, plastic bags, and other plastic fragments measuring up to 2.5 centimeters were the top five products found in the fieldwork.

The same study found that no specific system has been put in place by Tunisia's tourist municipalities for integrated waste management, especially for hotels and restaurants. A detailed value-chain analysis of the plastic and packaging sector sheds light on the current sector bottlenecks as well as the preliminary impacts of COVID-19 on the surge of SUPs (World Bank 2021f).

In conclusion, plastic pollution is the result of failures across the entire plastic life cycle, including production, consumption, waste management, and secondary markets for recycled material. Identifying the sources of plastic pollution is a complex issue that involves multiple sources and actors. Addressing plastic pollution for blue and clean seas will require all stakeholders across the value chain to join forces and intervene at various levels to obtain results in the short run. Efforts to properly identify the hot spots, sources, and causes of marine-plastic pollution in the Middle East and North Africa should be initiated and supported at all levels.

These efforts are a necessary prerequisite to formulate appropriate, location-specific policy responses to rein in the plastic tide that is entering the region's seas.

Comprehensive Policy Options for Reducing Marine-Plastic Pollution

Reducing marine-plastic pollution in the Middle East and North Africa's seas should be part of a comprehensive circular-economy approach. This is crucial to protect the natural resources upon which the region's residents (like the residents of all countries) depend, while reducing pollution's damages to human health and to countries' economies. Effectively reducing the region's marine-plastic pollution is one component of a five-component approach, here called the "3 R's + 2":

Reduce, Reuse, Recycle

+ *Appropriate SWM* (including cleanups) of plastics that cannot be reused or recycled

+ *Awareness-raising* of all stakeholders.

The latter "stakeholders" must include the general population to foster understanding of, and support for, the benefits of a circular-economy approach.

Crucially, reducing marine-plastic pollution requires the improvement of SWM systems to stop leakage. This is an important complement to the other policies, helping producers rethink how plastics are produced and how new consumption models for reuse, recycling, and cleanup methods are needed to restore currently degraded ecosystems. This section of the report discusses a wider set of solutions for addressing these issues, highlights best practices in the region and other countries, and begins with a look at the circular-economy approach.

The Circular-Economy Approach to Plastics

The circular economy proposes a set of principles to (a) synergistically improve—at the local, country, and global levels—the environment, human health, and economic development by (b) minimizing the wasting of resources and the pollution of the environment through a comprehensive approach of reducing, reusing, and recycling materials. In other words, the circular economy is an economic system in which materials constantly flow around a closed-loop system rather than being used once and then discarded (figure 4.10).

In the case of plastics, circular concept means retaining the value of plastics in the economy without harmful leakage into the natural environment.

FIGURE 4.10

A Circular Economy for Plastics

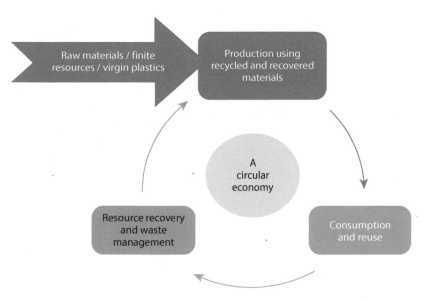

Source: Adapted from Defra and EA 2018. Permission for adaptation granted under the terms of UK Open Government Licence v3.0.

Unlike the traditional "*take-make-dispose*" linear economy, the circular economy proposes a fundamental rethinking of the product life cycle—improving recycling, promoting reuse, creating a market for recycled materials, and redesigning products with solutions that keep products' end of life in mind. The circular-economy approach includes as a core principle the collaborative engagement of all relevant stakeholders along the value chain (EMAF 2017; Geissdoerfer et al. 2017).

Implementing this framework, although simple conceptually, requires that various challenges be addressed in the way plastics are consumed, produced, and discarded. In today's economy, plastic packaging represents 40 percent of the total production of plastic products and contributes 80 percent of plastic leakage into the ocean (Pew Charitable Trusts and SYSTEMIQ 2020). Plastic packaging is typically single-use, ubiquitous, and extremely difficult to recycle. The global negative impacts of plastic packaging alone are estimated at US$40 billion annually and expected to increase significantly if production continues to increase under a business-as-usual scenario (EMAF 2016).

To overcome these challenges would bring great benefits. Increasing the life-span of products and materials through circular-economy principles dramatically reduces the production of those items, the consumption of raw materials, and the emissions of greenhouse gases from unnecessary production, consumption, and waste disposal processes. At least as importantly, a circular-economy approach protects and enhances human health and advances economic development while protecting the environment—sea, land, and air—locally and globally.

Spheres of Implementation

This section of the report lays out a range of solutions for implementing the principles of a circular economy to free the Middle East and North Africa's seas from plastic pollution—in general, by improving SWM, incentivizing recycling markets, reducing plastic consumption, redesigning production processes, and reimagining business models to stimulate reuse. For this new approach to manufacturing, consumption, and disposal of plastics to work, new policies and initiatives are required in several spheres of activity:

- *Improving SWM.* Plastic, depending on how it is handled, can pose a significant threat to climate and environment when it reaches the waste phase of its life cycle. Because plastics continue to pollute long after their use, disposal mechanisms must be improved, as for other types of waste.

- *Incentivizing recycling markets.* By incentivizing the development and organization of recycling markets to provide clean, high-quality materials for product manufacturing, private sector participation can be encouraged through better transparency, technology, and information. Increasing plastic recycling to mitigate environmental challenges can unlock new economic growth opportunities for plastic-producing economies in the Middle East and North Africa.

- *Reducing plastic consumption.* Initiatives to ban or reduce nonessential plastics such as SUPs are intended to reduce unnecessary and excessive use of materials through changes in products, processes, and behaviors in consumption patterns.

- *Redesigning production processes and material recirculation.* New policies, technologies, and processes are necessary to ensure that products are designed and managed throughout their life cycles for reuse and continuous recycling instead of for discarding and disposal. These processes include setting and reinforcing standards to regulate waste—such as extended producer responsibility (EPR) schemes—and improving the design and end-of-life handling of products.

- *Reimagining new business models.* To stimulate *reuse*, current strategies for material recirculation face systemic challenges. By themselves, pledges to increase recycling rates, even dramatically, are unlikely to effectively address the local, national, or global environmental, human health, economic—or climate—impacts of growing plastic production. Accompanying, or providing the context for, such pledges must be governmentally enforced incentives and corresponding disincentives applied all along the chain from plastic producers to distributors to end users.

Potential Rewards from a Circular Economy

The rise in population in recent decades, the increased exploitation of resources, and generalized open-dumping practices are the main drivers of waste management problems in the Middle East and North Africa. However, the treatment-capacity deficiency can be addressed in a way that creates new revenue streams, fully using waste as a resource. In this way, among others, the circular-economy approach presents an opportunity—and a cost-effective one—to rethink some of the region's current waste policies. For example, GCC countries could save US$138 billion by 2030 through an integral circular-economy approach, according to a report for the 2019 World Government Summit (Anouti et al. 2019).

Another important expected outcome from the transition to a circular economy is its net beneficial effect on job creation. Between 2012 and 2018, the number of jobs linked to the circular economy in the European Union (EU) grew by 5 percent to around 4 million (EC 2020). With adequate policies and regulatory frameworks to support the development of, and transition to, new industries, governments can ensure that the circular economy creates more and better jobs, including jobs for vulnerable groups.

Notably, the circular-economy approach creates more jobs than waste management systems that primarily burn or bury waste. Recycling can create over 50 times as many jobs as landfills and incinerators, repair systems can create 200 times as many, and remanufacturing 30 times as many (Ribeiro-Broomhead and Tangri 2021). The remanufacturing sector is highly promising, because materials like baled paper and aluminum are used as feedstock for the manufacture of consumer goods. Although these calculations can vary depending on the material in question and how individual facilitates operate, the results show that recycling, repair, and reuse can create thousands of new jobs across cities of different incomes and contexts. The same study estimates that the job growth in the high-recovery-rate scenario is particularly dramatic in cities with low current recycling rates. Cities with lower collection rates could see even greater jobs gains as municipal waste services are expanded. Anywhere from 10 to 60 jobs can be created in composting, recycling, and remanufacturing for every job lost in disposal methods (Ribeiro-Broomhead and Tangri 2021).

A recent report from the United Nations Development Programme (UNDP) about the potential of circular economy in Indonesia shows important opportunities not only for the plastics sector but also for five economic sectors that represent one-third of the country's gross domestic product and employed more than 43 million people in 2019: food and beverages, textiles, construction, wholesale and retail trade, and electrical and electronic equipment (Bappenas and UNDP 2021). A circular-economy approach would reduce waste in each sector by 18–52 percent, reduce carbon dioxide (CO_2) emissions by 126 million tons and water use by 6.3 billion cubic meters, and create 4.4 million net cumulative jobs by 2030 (Bappenas and UNDP 2021). By creating new opportunities, making supply chains more resilient, and providing business opportunities—especially for small and medium enterprises (SMEs)—a circular economy can be a key economic component in the region's economic recovery plans.

For the Middle East and North Africa region, adopting a circular-economy approach to the design, manufacture, use, and recycling of plastics (reduce, recycle, reuse) can help decouple plastics from fossil fuel usage, promote economic recovery in a post–COVID-19 context, and support a transition to a more sustainable development pathway. Morocco, Tunisia, and the city of Dubai in the United Arab Emirates are already taking steps to optimize resources and minimize waste by engaging key industry actors, understanding the sources of pollution, and identifying the locales and sectors that contribute most to the problem (as noted earlier in the "Maghreb Models for Emulation" subsection and below in the "Policies to Recycle Plastic Waste" subsection).

Experience shows that there is no single solution for managing plastic pollution of the oceans and that governments, in conjunction with the private sector and NGOs, will have to use an integrated approach with new ways to deliver the important benefits of today's plastics. Among these efforts, the SwitchMed project aims to promote the circular economy in countries along the Mediterranean basin, supporting industry trainings, awareness programs, and legislative procedures in several of the region's economies, including Algeria, Egypt, Jordan, Lebanon, Morocco, Tunisia, and West Bank and Gaza.[12]

Policies to Manage Plastic Pollution Using a Circular-Economy Approach

Economies differ in their levels of plastic pollution, sources infrastructure, and approaches to managing waste. Designing effective interventions will need to take into account each economy's policy contexts, governance structures, economic resources, and stakeholder participation levels and will need to develop cost-benefit analyses for the options being considered.

Based on the Middle East and North Africa's current plastic-pollution situation and drawing on sources in the literature, measures to improve the circularity of plastics fall into five broad types of instruments (figure 4.11):

FIGURE 4.11

Circular Economy Solutions around Consumption, Production, and Management of Plastic Waste, by Instrument Type

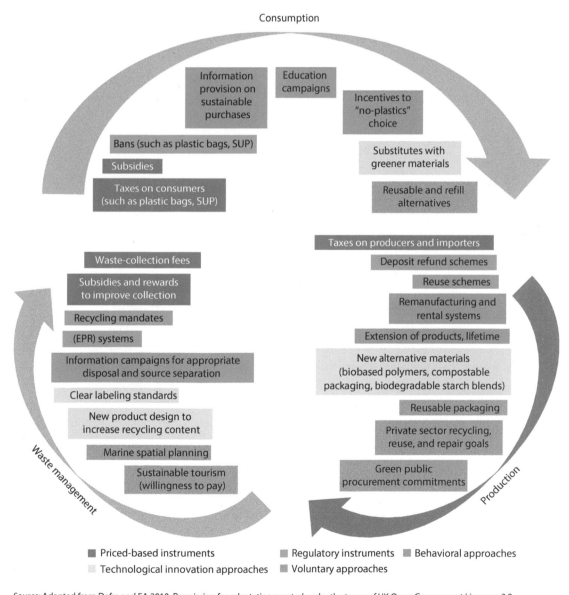

Source: Adapted from Defra and EA 2018. Permission for adaptation granted under the terms of UK Open Government Licence v3.0.
Note: EPR = extended producer responsibility; SUP = single-use plastic.

- *Price-based instruments*, which can discourage the use of plastic goods or inputs by changing their relative price and help to collect the necessary revenues for waste management and subsidize mechanisms to improve collection;

- *Regulatory instruments*, which directly determine waste levels by banning certain products, encouraging recycling, and addressing barriers to promote higher sustainability standards in product manufacturing processes;

- *Behavioral approaches*, which use people's social preferences to influence behavior (possibly encouraged by NGOs and the private sector) in favor of lower plastic use and improved waste management practices;

- *Technological innovation instruments*, which can change the amount and types of plastics produced and influence the recyclability and disposal of products—particularly if private sector investment in research and development (R&D) in this area is incentivized by policies that encourage changes in consumer preferences, spur innovation, and promote and guide investment flows; and

- *Voluntary approaches*, which consumer groups and firms can adopt in response to government policies—or in the absence of such policies, in anticipation of upcoming government regulations—either to support the optimal use of limited resources or for some of the other aforementioned motivations.

Policies to Stop the Leakage: Improve SWM Systems

In the Middle East and North Africa, inadequate SWM is the main culprit for the large volume of plastics entering the seas. In the Maghreb and the Mashreq subregions, around 60 percent of waste is mismanaged—that is, burned, disposed of as litter, or left in uncontrolled dump sites (figure 4.7). These deficiencies in management lead to leakage of plastic waste into the environment, with substantial amounts ending up in marine spaces. Along with the 3 R's (reduce, reuse, and recycle), ways to address poor SWM are among the most important measures to lower the flow of plastics into the region's seas.

Growth in Waste Generation

Total waste generation in the region varies considerably from one economy to another and is expected to double by 2050 from 129 million tons to 255 million tons. This figure will likely increase in the short run as the population continues to urbanize and as incomes rise (Kaza et al. 2018). Many of the largest waste generators are high-income countries, mainly

FIGURE 4.12

Average Daily Waste Generation Per Capita, Globally and in the Middle East and North Africa, by Economy and Subregion, 2016

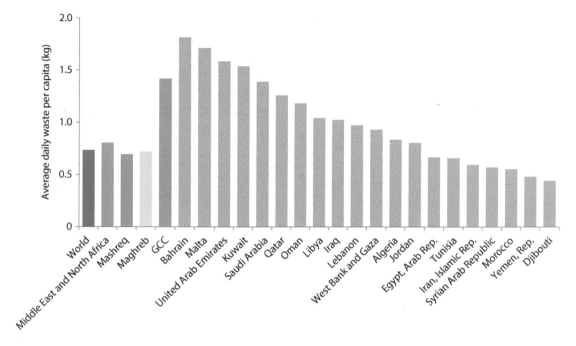

Source: Based on Kaza et al. 2018.
Note: Orange, yellow, and gray bars designate three groups of economies within the Middle East and North Africa, respectively, as follows: The Gulf Cooperation Council (GCC) includes Bahrain, Kuwait, Oman, Qatar, Saudi Arabia, and the United Arab Emirates. The Maghreb subregion includes Algeria, Libya, Malta, Morocco, and Tunisia. The Mashreq subregion includes Djibouti, the Arab Republic of Egypt, the Islamic Republic of Iran, Iraq, Jordan, Lebanon, the Syrian Arab Republic, West Bank and Gaza, and the Republic of Yemen. kg = kilograms.

those in the GCC subregion (figure 4.12). Residents in these countries are generating substantially more waste than the global average, while the Maghreb and Mashreq subregions, as well as the Middle East and North Africa overall, are roughly on par.

On average, plastics constitute around 12 percent of the region's total waste, a share expected to grow with rising incomes (Kaza et al. 2018)—implying that poor SWM overall has adverse repercussions on the flow of plastics into the region's seas. Furthermore, waste generation in the region's cities, especially in the city-states of the GCC, is significantly higher (1.38 kg per person per day) than the region's average (0.81 kg per person per day) (Kaza et al. 2018).

Inefficiencies, Disparities in Urban and Rural Waste Collection

In some of the region's economies, waste collection coverage is relatively comprehensive. In some of those for which data are available, waste collection in the urban areas reached 100 percent or nearly 100 percent in 2016 (Kaza et al. 2018). Notable exceptions are Egypt (57.5 percent), Tunisia (80 percent), and Morocco (85 percent) (figure 4.13, panel a).

However, rural coverage is generally lower than in the urban areas of most economies. On this score, Egypt and Tunisia again report the lowest levels, with rural coverage of 15 percent and 5 percent, respectively (figure 4.13, panel b). It should also be noted that waste collection statistics are not available for some of the poorest countries in the region, including Djibouti and the Republic of Yemen. Waste collection services in these countries, as well as in those rocked by recent conflicts, can be expected to be rather low.

Even though formal waste collection rates in the Middle East and North Africa seem to be relatively high, operational inefficiencies and outdated equipment result in leakage and littering. Collection is predominantly done door-to-door by trucks, and equipment in the waste treatment process is often outdated, bringing about operational inefficiencies. For instance, trucks used for collection in Cairo are typically old and do not have sufficient capacity for the waste they are supposed

FIGURE 4.13

Shares of Urban and Rural Populations Covered by Waste Collection Services in Selected Middle East and North Africa Economies, 2016

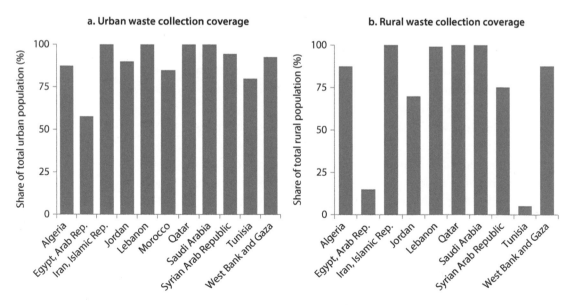

Source: Based on Kaza et al. 2018.
Note: Only countries and economies for which data were available are shown. Figures for West Bank and Gaza refer to the share of households covered by waste-collection services, as opposed to the share of total population covered (as presented for the other locations).

to collect (Mostafa 2020). This leads them to overflow and litter their routes with trash during transportation. Furthermore, lack of coordination about vehicle routes and collection frequency makes it difficult to strike a balance between fulfilling waste collection requirements and the number of visits by waste collection vehicles (Mostafa 2020).

Mismanaged Waste Disposal

Much of the waste collected in the Middle East and North Africa ends up in open dumps. Several of the region's economies have such high rates of inadequately managed waste not because it isn't collected but because of its treatment or disposal. In 2016, the share of waste disposed of in open dump sites in the Middle East and North Africa was 53 percent (Kaza et al. 2018), but this share varied widely across countries (see lighter blue bars in figure 4.14).

Furthermore, even in the region's high-income countries, most landfills are not engineered landfills and effectively operate as dumps

FIGURE 4.14

Share of Waste Going into Open Dumps or Unspecified Landfills in the Middle East and North Africa, by Economy, 2016

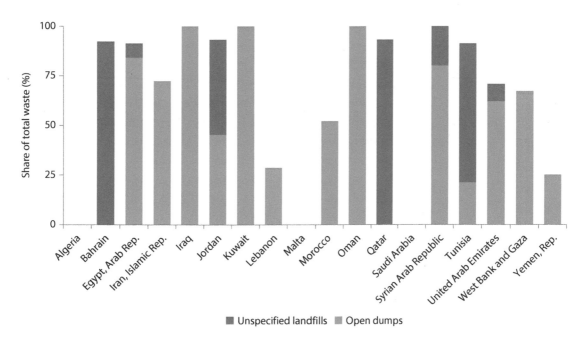

Source: Based on Kaza et al. 2018.
Note: "Unspecified" landfills are ambiguous about their ability to prevent waste leakage. Kaza et al. (2018) report treatment separately for controlled, sanitary, and unspecified landfills. Significant amounts of waste disposed of in "unspecified" facilities are mismanaged, especially in low- and middle-income countries (Law et al. 2020). Data for Djibouti and Libya are not available.

(Kaza et al. 2018). For instance, GCC countries such as Bahrain and Qatar exhibit comparatively high rates of waste disposal in "unspecified" or unengineered landfills (see dark blue bars in figure 4.14). The category of "unspecified" landfills is ambiguous about their potential for proper waste management; it is assumed that waste disposed of in such facilities is partly mismanaged, especially in low- and middle-income countries (Law et al. 2020).

However, use of controlled landfills has been growing in recent years. For example, Morocco has increased disposal into controlled landfills from 10 percent in 2008 to 53 percent in 2016 (Kaza et al. 2018). Data on disposal into controlled and sanitary landfills for other economies are as follows: Algeria, 91 percent; Saudi Arabia, 85 percent; Lebanon, 48 percent; and West Bank and Gaza, 33 percent.

Disposing of collected waste in dump sites and uncontrolled landfills undermines achievements stemming from proper waste collection. Such disposal, indicating poor operational practices, makes it easier for plastic waste to flow into waterways and reach marine areas. For example, even though collection rates in the Philippines are high (around 84 percent nationwide), around 17 percent of the collected plastic waste still ends up in the marine ecosystem (Engle, Stuchtey, and Vanthournout 2016). This ocean-leakage rate is even higher for uncollected waste, of which around 31 percent ends up in the sea.

The False Economy of Mismanaged Waste

Open dumping is prevalent in many areas because it is much cheaper than the alternatives. Setting up open dumps requires virtually no starting capital, while the costs of establishing landfills are higher but still comparatively lower than other alternatives (Kaza et al. 2018). The direct operating costs arising from open dumps for low-income and lower-middle-income countries are about four to five times cheaper than the costs of controlled landfilling (table 4.2).

The comparatively high costs of recycling—including its typically higher collection and transfer costs (Boskovic et al. 2016)—contribute to low rates of adoption in the region despite recycling's many environmental advantages. Strained budgets may restrict the ability to pursue ambitious recycling programs right away; however, the leakage of plastic waste could still be reduced by ensuring that waste is disposed of in properly controlled landfills and comparable disposal sites.

However, the cost of openly dumped waste exceeds the cost of proper disposal many times over if social and environmental costs are considered. Even though open dumping is about four to five times cheaper than proper landfilling, after considering the negative externalities, open dumping is actually more expensive.

TABLE 4.2

Costs of Waste Treatment, by Type and Country Income Level, 2016

US$ per ton

Waste management type	Country income group			
	Low income	Lower-middle income	Upper-middle income	High income
Collection and transfer	20–50	30–75	50–100	90–200
Controlled landfill to sanitary landfill	10–20	15–40	20–65	40–100
Open dumping	2–8	3–10	n.a.	n.a.
Recycling	0–25	5–30	5–50	30–80
Composting	5–30	10–40	20–75	35–90

Source: Kaza et al. 2018.

Note: Sample is global. Country income groups are according to World Bank classifications. n.a. = not applicable.

Incorporating social and environmental costs is a complex task and depends on many location-specific characteristics. The leakage of solid waste such as plastics from open dumps into the environment, including into marine spaces, is one important side of the coin in this respect. Another important issue concerns leachate—liquid that passes through untreated waste and ends up in surface water and groundwater, containing large amounts of harmful dissolved and suspended waste matter, leading to health as well as environmental damages and associated costs. Furthermore, waste at uncontrolled open dumps is often burned, either intentionally to reduce waste or spontaneously, increasing air pollution associated with severe health effects (Cogut 2016). Internalizing these external costs of waste (including plastics) and its improper disposal renders controlled-disposal alternatives superior from an economic perspective in the Middle East and North Africa (Loukil and Rouached 2012).

Put simply, leaving waste uncollected and hence completely unmanaged carries costs that typically exceed the costs of properly managing it severalfold—often 5-fold to 10-fold (UNEP 2015). A ton of uncollected waste carries an average loss of US$375, based on the economic losses in tourism, fisheries, and health care.[13] Comparing this to the costs of proper waste collection and treatment puts the high costs of unmanaged waste in the spotlight—and highlights that the most effective solution to the plastic pollution of the Middle East and North Africa's seas, beaches, and coastal areas is a circular-economy approach: it minimizes the amounts and sources of pollutants while simultaneously increasing benefits to residents' well-being, enterprises' opportunities to develop and serve new market segments, and furthering the development of local and national economies.

Impediments to Adequate SWM

A major impediment to SWM performance in the Middle East and North Africa is the highly centralized waste management in many of the region's economies, which hampers the capabilities of local, decentralized service providers. Even though municipalities are the main executive bodies in waste management, they often have (a) little autonomy to deal with local waste management needs, and (b) insufficient financial support for those tasks (Mahjoub, Jemai, and Haddaoui 2020). For example, these authors find, high centralization and the prevalent tendency for a top-down approach is a major hindrance to efficient waste management in Tunisia.

In Lebanon, around 80 percent of rural households surveyed were willing to support local, decentralized waste management perceived to be more efficient in providing waste management services. On average, residents' willingness to pay was US$48 per year, representing a 30 percent increase over current council taxes (Al Ahad et al. 2020). However, this study also identified a lack of financial and technical resources as well as an insufficient amount of well-suited land as major obstacles to introducing such a decentralized scheme.

Municipalities often have insufficient funds for SWM. Especially among low- and middle-income economies in the Middle East and North Africa there is lack of financing as well as technical capacities to set up an SWM system and other issues are perceived as more pressing (Mostafa 2020). The costs for the collection of waste are often borne by municipalities, whose cost recovery for waste collection, transfer, treatment and disposal, which requires coordination and planning by the local government to establish a fee-collection system, is so limited that they have outdated equipment. For example, recovered costs account for only 15 percent of the total SWM costs in the Greater Tunis metropolitan area (Mahjoub, Jemai, and Haddaoui 2020). Typically, municipalities finance these costs through local tax collection, as central governments are often outsourcing the responsibility for waste management to local governments without providing adequate funding; however, these taxes are often not exclusively allocated to waste management, and diversion to other causes is common. In Egypt, waste management fees had been incorporated into electricity bills, but this practice was recently discontinued because of the lack of allocations to waste management (Aly 2020; *Egypt Today* 2020).

Green Bonds and Other Ways Forward

Issuing green bonds. Although limited financing for SWM systems is an obstacle for many Middle East and North Africa governments, innovative approaches could be a way forward. In other parts of the world, issuing green bonds to finance waste management systems has been an

established practice, and municipalities in emerging countries such as
India are increasingly adopting them as well. Egypt is the first country in
the Middle East and North Africa to move in this direction by recently
issuing sovereign green bonds to increase affordable financing for environmental projects, including pollution management (box 4.3).

BOX 4.3

Green Bond Financing for SWM Systems

Since 2007, when the European Investment
Bank issued the first green bond, the green
bonds market has become a booming sector
for financing green infrastructure projects,
growing at an annual rate of 94 percent. In
the United States, issuing bonds to finance
the establishment and operation of waste
management systems has a long history,
with the establishment of tax-exempt private
activity bonds (PABs) in 1968. These PABs
can be issued by private companies to raise
capital for a variety of infrastructure projects, including solid waste disposal facilities
and projects like airport or port construction (Ruth 2017). Recent issuances for solid
waste management (SWM) in the United
States included the September 2020 place-
ment of bonds worth US$40 million for
Castella Waste Systems Inc., which provides
SWM services in the northeastern United
States, and the issuance of bonds worth
US$150 million for Waste Management
Inc. by the California Municipal Finance
Authority (CMFA 2020; CTBH 2020).

Internationally, including in low- and
middle-income countries, the issuance
of bonds by municipalities or related
corporations for infrastructure investments is
a more recent phenomenon, but the issuance
of green bonds is growing. For example, the
Ahmedabad Municipal Corporation in the
Indian state of Gujarat raised over US$26
million in January 2019 for "green projects"
by issuing municipal bonds. It intends to
use the funds for several projects, including
waste management investments, and this
was the fifth time the municipality raised
funds by issuing such bonds. In the 2018/19
fiscal year, eight local governments in India
floated municipal bonds worth more than
US$400 million (Verma 2020).

In September 2020, the Arab Republic
of Egypt became the first country in the
Middle East and North Africa to issue
green bonds as part of a set of projects
for renewable energy and pollution
management. The issuance floated US$750
million worth of bonds, which were five
times oversubscribed. The projects financed
are part of a five-year plan to satisfy growing
demand from investors and to increase
affordable financing in these sectors in an
innovative way (Barbuscia and Ramnarayan
2020). Egypt has a portfolio of eligible green
projects worth US$1.9 billion, of which
almost 40 percent is aimed at pollution
reduction and control. Financial experts
expect green bonds in Egypt to become
widely popular following a global trend;
such bonds will have a positive impact on
Egypt's budget and increase the volume of
foreign investment (Abu Zaid 2020).

Investing in disposal infrastructure. However they are financed, investments in formal waste disposal and establishment of safe disposal sites are critical to address plastic pollution and prevent leakage. Formal disposal is very low in the Middle East and North Africa region. Increasing the share of formal waste disposal to 50 percent by 2040, largely by replacing dump sites with managed landfills, can reduce vast amounts of plastic leakage into the ocean (Pew Charitable Trusts and SYSTEMIQ 2020). Improving the efficiency and convenience of disposal, scaling up waste separation at the source, and improving the logistics and economic viability of whole waste management system are critical to ensure the environmental and economic benefits of clean urban and rural places.

Upgrading capacity and equipment. Upgrading existing equipment in the collection process and investing in more efficient pickup services are also important. Given the obsolescence of large parts of the region's equipment and collection fleet, investments in these areas are imperative, especially given the region's growing population and increasing waste generation. The adoption of proper capacity plans, collection frequency, and vehicle-routing plans for existing fleets may reduce leakage while simultaneously reducing total vehicle mileage, traffic congestion, and hence carbon emissions caused by trucks picking up waste (Mostafa 2020).

Studies show that direct dumping of postcollected waste could be reduced by 80 percent by combining technological innovation and stronger regulatory control. For instance, the movement of waste collection vehicles could be monitored through telemetry, which enables cost-effective vehicle tracking. This technology has already been applied in some cities in low- and middle-income countries (Pew Charitable Trusts and SYSTEMIQ 2020).

Integrating SWM systems. Moving toward an integrated solid waste management system that integrates the "3 R's" is key to decreasing plastic leakage into the seas. A strategy focusing only on collection and disposal, with an emphasis on safe and controlled disposal methods, is likely insufficient by itself to solve marine-plastic pollution. As cities and countries continue to grow, plastic waste will worsen as it grows faster than the ability to expand waste infrastructure.

The impacts of fast growth of waste due to the increased consumption of disposable products and medical waste can be seen in the waste generation associated with COVID-19. Some countries lack a special infrastructure for hazardous and medical waste, which ends up mixed with household waste that is burned or deposited in landfills, increasing the risk of environmental contamination. Reducing the amount of waste in the system and replacing plastics with other materials is essential to reduce marine pollution and decrease plastics' impacts on human health.

Employing new business models to drive up formal collection rates—including new models for waste aggregation and decentralization and management of waste—can bring many benefits in new jobs and more investment.

By combining technology, an adequate business environment, and incentives, waste management can become a source of income generation. Heading toward such a system may be a steep climb for many economies in the Middle East and North Africa, but it would increase their resilience to future challenges associated with growing populations and increased waste generation. Improving the circularity of waste in general, and of plastics in particular, requires a rethinking of current practices; however, a circular-economy approach will yield a broad spectrum of health and economic benefits at the local and national levels, as well as, of course, advancing the blueing of the region's skies and seas. The alternatives to a circular-economy approach—a business-as-usual or piecemeal approach—will only lead to additional pollution, additional morbidity and mortality, and increased constraints on local and national economic development.

Policies to Reduce Production and Consumption of Plastics

Solving the plastic trap requires reducing consumption and production levels. It requires rethinking to cut production and consumption in ways that will nudge plastic-related industries to rethink business models to still thrive economically and boost new market opportunities. Many large private sector companies, including ones in the Middle East and North Africa, are embracing the opportunities for a circular economy and adapting to consumer demands by making important commitments to increase recyclable materials, improve capacities, and eliminate the stream of plastics to landfills and open dumps that pollute rivers and oceans. The following subsections discuss pathways for such rethinking.

The Low Price of Plastics Relative to Green Alternatives

In the Middle East and North Africa, prices for SUPs are significantly lower than greener alternatives, according to a desk assessment of prices on the websites of major supermarkets and online retailers in selected countries (figure 4.15).[14] The prices discussed here for SUP products and their comparable green alternatives refer to a single unit of each good obtainable in supermarkets and the like.[15]

In the countries under consideration, plastic cups cost one-third, or less, what comparable biodegradable cups cost. Similarly, for single-use alternative cutlery (such as spoons, forks, and knives), prices are on average four times higher, and they can be substantially higher than that in

FIGURE 4.15

Price Comparison of Selected SUP Items and Green Alternatives in the Middle East and North Africa, 2020

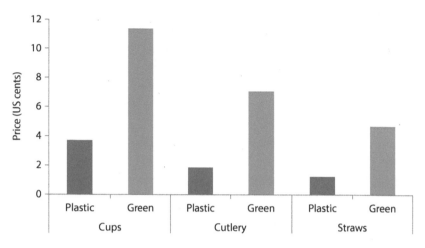

Source: Desk assessments of prices shown on major supermarkets and online retail websites.
Note: The figure shows end-consumer prices per unit in US cents, converted using exchange rates as of December 2, 2020. Plastic products refer to single-use plastic (SUP) items. Green products refer to items made of biodegradable, environmentally friendly resources. Countries consulted were those for which comparable pricing data were available: the Arab Republic of Egypt, Jordan, Lebanon, Morocco, and all of the Gulf Cooperation Council countries (Bahrain, Kuwait, Oman, Qatar, Saudi Arabia, and the United Arab Emirates). Prices were standardized to represent a single unit of each good; however, where batch sizes varied, some variation was unavoidable, which could distort per unit prices. Purchases from wholesale distributors by commercial clients, such as restaurants or catering services, likely involve different prices (and price discrepancies) and are not reflected in this analysis.

some countries, such as Bahrain and the United Arab Emirates. Roughly the same price comparisons hold for plastic straws and their alternatives made of paper or other biodegradable materials, again with intercountry variations.

A major reason for the large discrepancies between the prices of plastic products and their greener alternatives is the heavy subsidization of input prices for petrochemical companies. The Middle East and North Africa's oil-exporting countries use the petrochemical industry to diversify their economies, and it is an important part of those economies, with a large part of demand coming from the packaging industry (box 4.4). The main inputs for the petrochemical industry—energy and raw materials—are heavily subsidized in the Middle East and North Africa (Fattouh and El-Katiri 2013). Such subsidies favor plastic-producing industries in several ways.

First, these subsidies favor plastics by fixing prices on fossil fuels, which are the main inputs for petrochemicals. These implicit subsidies depress the price of plastic feedstock to artificially low levels (Skovgaard and van Asselt 2019; Tobin 2012).

BOX 4.4

A Snapshot of the Petrochemical Industry in the Middle East and North Africa

The petrochemical industry is an important economic sector in Middle East and North Africa economies, especially in the Middle East. To shield themselves from the volatility of oil prices and diversify their industry away from their heavy dependence on oil and gas exports, several of the region's economies have made large investments in the petrochemical industry, with plastic production being the dominant part (GPCA 2014). The sector has seen immense growth in recent decades and continues to attract major investments, with almost US$100 billion of planned projects for 2020–24. The Arab Republic of Egypt leads the list of committed investments in petrochemicals, followed by the Islamic Republic of Iran and Saudi Arabia (Benali and Al-Ashmawy 2020).

In the Gulf Cooperation Council (GCC) countries, the polymer industry alone supported over 150,000 jobs in 2016, with 40,000 employees directly employed in the sector (GPCA 2016). Polymer production in the GCC grew by a compounded annual growth rate of 11 percent between 2006 and 2016 and is expected to grow in the future by around 5 percent (GPCA 2016, 2019). Within the GCC, Saudi Arabia is by far the largest producer of plastic resins, capturing around two-thirds of the GCC's overall production (APICORP 2016). Globally, the petrochemical industry is expected to account for most of the growth in oil demand by the end of the current decade (Benali and Al-Ashmawy 2020).

In the GCC, the consumer packaging industry accounts for a major part of the demand for polymers. Around 44 percent of polymers produced in the GCC were used in the packaging industry (GPCA 2016). The low costs of plastics and plastics' high preservation abilities make plastics widely used in the packaging industry. Environmentally less harmful packaging materials, such as biodegradable and bio-based plastics, could serve as alternatives in this industry segment. However, considering only their prices on a weight basis, fossil-based plastics again emerge as a clear winner, costing between one-third and one-tenth of their biodegradable alternatives (van den Oever et al. 2017).

These alternatives' significant additional expenses relative to customary plastics are a major detriment to their broader adoption in the Middle East and North Africa, based on economic incentives alone (Market Data Forecast 2021). Nevertheless, several of the region's economies, especially in the GCC—following increased public awareness about the problem of single-use plastics (SUPs)—have banned businesses' use and production of common plastic bags, mandating their replacement with bags made of these alternative materials. Such moves can also enhance a country's reputation for environmental friendliness. And an enhanced "green reputation" acts as an additional incentive for purchases by environmentally aware consumers.

Foreign consumers' increased demand for environmentally friendly options will affect plastic-producing companies in the short term. The advent of more and more initiatives to phase out SUPs in some major export markets of the plastic industry

(continued)

BOX 4.4

A Snapshot of the Petrochemical Industry in the Middle East and North Africa (*Continued*)

(including China, India, Italy, and Turkey) could have severe repercussions on the demand for the GCC's plastics, and exports may be reduced substantially (Al Sarihi 2019). This will in turn hamper future sales revenues and put pressure on the sector's profitability. To meet these domestic and international shifts in demand, plastic producers and the packaging industry have been active in developing and producing greener alternatives to their traditional products and making their products more efficient in terms of their life cycles and ultimate disposition (GPCA 2018, 2019).

With momentum gaining for the EU's circular-economy legislation,[a] the need for adaptation by GCC's petrochemical sector—for which Europe represents a major export market—has

already been acknowledged by the Gulf Petrochemicals & Chemicals Association (GPCA) as necessitating innovation in product development and consideration of products' life cycle and ultimate disposition (Al-Sadoun 2019). For high-income countries such as those in the GCC, fostering further innovation in these sectors is crucial for reducing plastic waste, which could also be driven by removing subsidies and by pressing petrochemical companies to invest more in research and development (R&D) for alternative products that are less harmful to the environment. Here again, the principles and practices of a circular-economy approach would beneficially serve multiple purposes for producers, consumers, and their local and national economies.

a. "First Circular Economy Action Plan," European Commission: https://ec.europa.eu/environment /topics/circular-economy/first-circular-economy-action-plan_en.

Second, these subsidies also favor plastics because the petrochemical industry and plastic production are highly electricity-intensive (and simultaneously energy-inefficient) industries (Schlüter and Rosano 2016). The production of electricity uses fossil fuels to a large degree; consequently, the electricity industry is a main beneficiary of subsidized fossil fuel prices, and those cost benefits transfer to plastic producers (El-Katiri and Fattouh 2017).

Hence, subsidies for fossil fuels favor plastic production by reducing the prices of its inputs and lowering their production costs (Moerenhout and Irschlinger 2020). In 2015, around three-fourths of plastic consumption in the Middle East was of items produced in that region, which indicates that low input prices are passed on to end products and hence end users (EUROMAP 2016).[16]

In many Middle East and North Africa economies, large subsidies benefit the petrochemical sector by keeping its input prices fixed at artificially low levels. A prime example for the practice of input subsidization is the Saudi Arabian petrochemical giant Saudi Basic Industries Corporation (SABIC), one of the world's largest producers of plastic raw materials. Despite some reforms, SABIC has still been able to buy its main inputs at large discounts from Saudi Aramco in recent years, paying a fixed price of US$1.75 per million Btu (British thermal unit) for ethane, US$1.25 per million Btu for natural gas, and propane for only 80 percent of its market value.[17] Saudi Aramco in turn receives large subsidies from the Saudi government for supplying these and other products such as gasoline at lower prices to the domestic market—subsidies that totalled over US$40 billion in 2018. In 2020, Aramco acquired a 70 percent stake in SABIC, and the payment of equalization fees to Aramco for supplying cheaper feedstock to SABIC came under scrutiny (Bakr 2020). Ethane represents the main input for petrochemicals and hence plastic production in the Middle East and North Africa, while naphtha is the main input for Asian plastic producers. Artificially depressing input prices for plastic production in this way contributes heavily to the low prices of virgin plastic and in turn to the discrepancies between plastic products and their greener alternatives. Additionally, the high subsidization of petrochemicals and the plastic production put strains on public budgets, which are already under pressure.

The region's low prices for fossil fuels make plastic feedstock prices and their contribution to overall plastic costs the lowest worldwide. Despite the reforms implemented in major producing countries such as Saudi Arabia, key input prices for petrochemicals, like those for ethane, remain significantly below international benchmarks. Feedstock costs are the most influential factor for determining regional production advantages (IEA 2018). In the Middle East and North Africa, the ethane-based petrochemicals (from which plastic products are made) account for around one-fourth to one-third of total costs, while comparable feedstock accounts for about one-half of total costs in Europe and the United States. For naphtha, another major feedstock for plastic production next to ethane, price discrepancies are not as pronounced. These feedstock prices drive the overall production costs of petrochemicals, and hence plastic production, and give petrochemicals in the Middle East and North Africa a significant cost advantage. This can also be seen in the price of ethylene (a main input for PET production) (Rubeis et al. 2016). The Middle East (which accounts for the lion's share of petrochemicals and plastic production in the region) has a distinct price advantage, especially for ethane-based ethylene.

Plastics' low input prices are an impediment to the broader adoption of green alternatives. These low prices, by artificially driving down the prices of plastics in the Middle East and North Africa, undermine the potential emergence of greener alternatives to plastics. By granting the petrochemical industry such low input prices, at least partly through subsidies for feedstock, the region's petrochemical sector can produce plastics at low costs and pass them on to the prices of plastic products. Plastic's low prices explain, at least partly, its widespread adoption and contribute to consumers' unwillingness to switch to environmentally less harmful alternatives.

Measures to reduce plastic consumption should include pricing as an important consideration. However, it is also important to recognize that such measures could, similarly to fossil fuel price increases, lead to affordability issues for low-income households. It is therefore necessary to accompany the pricing reforms discussed below with support measures for these households as well as communications and awareness campaigns to minimize public discontent.

Taxes on Consumers and Producers to Reduce Plastic Consumption

Taxes on plastic material and on certain uses of plastics (such as bags and other forms of SUPs) can help reduce their unsustainable consumption. Well-designed taxes can lead to the use of alternatives that are more durable, sustainable, or both—for instance, redesigned plastic options that are more readily recyclable or compostable, or more durable plastic or nonplastic alternatives manufactured from wood, metal, or glass. The most widely used taxes for reducing plastic pollution take several forms: levies on plastic bags, taxes and charges on packaging and plastic products (such as kitchenware), and weight-based fees for plastics being part of a product (OECD 2015). Common ways of taxing consumers and producers to reduce plastic consumption are noted below.

Taxes on consumers (such as taxes on SUPs). Consumer fees on plastic bags have become a popular measure to reduce consumption. Worldwide, 30 countries charge consumers fees for plastic bags at the national level, and 27 countries tax the manufacture and production of plastic bags (UNEP 2018b). The fees vary by country, often based on the thickness and material content of the plastic bags regulated. In the Middle East and North Africa, Jordan and Tunisia tax the manufacture, production, and import of plastic bags; more recently, the United Arab Emirates imposed a consumer fee on plastic bags (UNEP 2018b).

Many cases worldwide show success in implementing plastic fees. Ireland introduced a tax per bag in 2002, resulting in a gradual 90 percent reduction in the use of plastic bags (Convery, McDonnell, and Ferreira 2007). Thus, marine-plastic litter, which had represented 5 percent of the national composition of total marine litter before the adoption of the levy, fell by around 22 percent by 2004. The success factors were as follows:

- Setting the tax, following a willingness-to-pay survey for plastic bags, that was six times higher than the average maximum willingness to pay;

- Extensive consultation with all stakeholders (the public and retail industry); and

- Accompanying information campaigns that explained the policy objectives and tax-revenue destinations, paving the way for widespread awareness and buy-in.

In Scotland in 2014, a mandatory charge was introduced on all types of retail bags, and that charge contributed to reducing carrier bag use by about 80 percent across the main retail chains in its first year of application (McElearney and Warmington 2015). In Portugal, a plastic-bag tax was implemented in 2015 and consequently reduced consumption by 74 percent (Martinho, Balaia, and Pires 2017).

The effectiveness of taxes on plastics varies according to policy context and enforcement levels. Because most of these policies are quite recent, it is too early to draw robust conclusions about the environmental impact that taxes and levies have had. In 50 percent of cases, information is lacking partly because some countries have adopted them only recently and partially because monitoring is inadequate. In countries that do have data, about 30 percent have registered drastic drops in the consumption of plastic bags within the first year.

The remaining 20 percent of countries have reported little or no change, mainly for two reasons: a lack of enforcement and a lack of affordable alternatives. The latter has led to cases of smuggling and the rise of black markets for plastic bags or the use of thicker plastic bags that are not covered by the bans (UNEP 2018b). For example, the South African government implemented a tax on plastic bags, but the strategy failed because the levy was too low and customers ended up paying the tax, creating a steady increase in demand for plastic bags (Lam, Ramanathan, and Carbery 2018).

Although plastic shopping bags account for only a small proportion of the plastics used in packaging, research shows they are a major source of the plastic pollution that ends up in the oceans, causing high rates of animal deaths and microplastic pollution. Imposing fees on plastic items can both disincentivize their use and provide revenue for governments or funds for environmental purposes. Morocco, for example, distributes the proceeds to environmental funds (Powell 2018). In Ireland, the tax on plastic bags not only reduced their use by about 90 percent but also raised tax revenues of around €12 million to €14 million that were earmarked for an environment fund (Convery, McDonnell, and Ferrerira 2007). Recently, the United Arab Emirates announced plans to make the country free of SUPs by gradually introducing fees on them, eventually ending in a ban (box 4.5).

For effective taxation, a clear target and good communication are important principles. A clear target should define *what* message the price signals are intended to send and *who* the intended recipient is for that message. If the objective is to reduce the amount of plastic that ends up in the environment, then it is necessary to understand *what* behavior change *by whom* will be necessary to accomplish that. For example, is the primary goal to change how products are made, or is it to alter price signals that influence what consumers buy and use? Good communication to relevant stakeholders is also essential. The political reality of taxation is that it creates winners and losers, and new taxes are often not popular—although plastic-bag charges have received broad support in most countries. Clear communication about *what* the purpose of the tax

---

BOX 4.5

Eliminating SUPs in the United Arab Emirates

After a study identified the most common products that cause the largest amount of marine waste, the United Arab Emirates announced a plan to become single-use plastic (SUP)-free. The country uses 11 billion plastic bags annually, equivalent to 1,184 plastic bags per person per year compared with a global average of 307 plastic bags per person per year (Ahmad 2020). In 2020, the Abu Dhabi emirate (of the United Arab Emirates) also announced its intention to declare Abu Dhabi free of SUPs by the end of 2021.[a] The 16 prioritized items such as plastic bags, cutlery, cups, plastic bottles, and abandoned fishing gear account for around 70 percent of total marine litter. Such items will be subject to increasingly stricter regulations.

The policy instruments include fees on items having clear alternatives available and a ban on their free distribution to the end consumer by, for example, fast-food outlets and caterers. With the proceeds from these fees, an environment fund is being established to finance environmental activities and campaigns. These regulations and initiatives will be put into action gradually, with the goal of a 100 percent reduction in the use of SUP bags and a significant reduction in the use of other SUP items. Further goals include a 50 percent collection rate of plastic bottles by the end of 2021 and a ban on SUPs in the operations of government entities in Abu Dhabi.[b]

a. The regulations outlined target the consumption of SUP items in Abu Dhabi. The policy framework explicitly states, however, that it will not affect the export of such items.
b. Emirate of Abu Dhabi. 2020.

is, *whom* it is being levied on, and *why* are essential characteristics of good tax design (Powell 2018).

Producer taxes (on manufacturing, exports, and imports). Taxes directed at producers aim to discourage the production of plastics, especially SUPs. A range of different product taxes could be used for environmental policy purposes. For example, 29 countries globally have enacted some type of tax, mostly on SUPs, either as a special environmental tax or fee or in the form of higher excise taxes (UNEP 2018b). These taxes aim to reduce SUPs as a category of waste, to better manage plastic waste, or to increase the rate of postconsumer recycling. In the Middle East and North Africa, Morocco and Tunisia have introduced taxes on only the manufacturing and import of plastic products (Framework Law on the Environment and Sustainable Development, Morocco; and Ecotaxes Finance Law, Tunisia (boxes 4.6 and 4.7, respectively). Some European

---

BOX 4.6

Morocco: Implementing an Ecotax on Plastic Production

In recent decades, Morocco passed legislation and implemented policies in its Framework Law on the Environment and Sustainable Development to position itself as a rising climate leader and drive the improvement of waste collection and treatment systems in line with its environmental and climate goals. However, the lack of appropriate institutional structures, particularly at the regional level, and limited funding have been barriers to successful implementation. Skills and expertise are insufficient in many cities, clear responsibilities still need to be defined, and monitoring and control systems are still inadequate.

In 2008, Morocco launched a national municipal solid waste management (SWM) program (the National Program of Management of Household Waste [PNDM]) with the goal of achieving a collection rate of 85 percent by 2016 and 100 percent by 2030. The program aimed to create controlled landfills to serve all urban centers by 2020 and rehabilitate and shut down all illegal dump sites by 2020. It also set a target of achieving a recovery or recycling rate of 20 percent by 2022.

In 2012, Morocco's Department of the Environment prepared a business plan for development of the plastic-recycling sector. It estimated funding needs of DH 165 million annually. To ensure this financing, a 1.5 percent ad valorem plastic ecotax was introduced in 2013 under the Framework Act 99/12 on the sale, factory output, and importation of plastics and articles falling under chapter 39 of the Harmonized System (HS). In 2016, this rate was reduced to 1 percent. The tax revenues contributed to subsidizing recycling stations, financing them, or both; developing the plastic-recycling sector; and integrating the informal sector.

Despite these much-needed policies, the results achieved have been very modest. Collection rates and deposits in controlled

(continued)

BOX 4.6

Morocco: Implementing an Ecotax on Plastic Production (*Continued*)

landfills improved from 10 percent in 2008 to 32 percent in 2015. Fourteen controlled landfills were completed, but as of 2020 an estimated 300 uncontrolled dumps still exist around the country that should be closed and rehabilitated. In 2014, the methods for redeploying the ecotax were the subject of a governance study, in close consultation with the stakeholders concerned, which showed that the activities supported to date were limited to those relating to setting up sorting centers at controlled landfills, plastic-bag collection campaigns, and awareness campaigns. The lack of capacities

at the local level for implementing the initiatives was also a constraint. Overall, systematic results on the effectiveness of the ecotax are scarce and there has been little documentation.

For the future, this tax will probably require more reviews, because a more recent study showed that the plastic sector is expected to continue to increase in Morocco, reaching 15 million tons, with a turnover of around DH 32 billion by 2030. The automotive, aeronautics, and electric and electronics sectors will continue to be the main customers of this industry.

Source: World Bank 2021d.

BOX 4.7

Tunisia: The ECOLEF Program to Increase Recycling

The ECOLEF program was the first program to manage consumer packaging in Africa and the Middle East and North Africa. Tunisia's Ministry of Environment launched ECOLEF as a public-private partnership in 1997. The ECOLEF program developed a national system for recovery and recycling postconsumer packaging, primarily focused on plastic waste. It is administered by the National Agency for Waste Management (ANGed) in cooperation with private companies and governed by several decrees that specify the methods required for collecting and managing bags, plastic bottles, and other high-density polyethylene (PEHD) plastics.

The ECOLEF program has improved the market for collecting postconsumer packaging, improved recycling rate, and created thousands of recycling jobs. The system encourages individual and informal collectors to gather used plastics and metals and deliver the materials to ECOLEF collection centers. In return, waste collectors are paid based on the type and quantity of packaging collected. Those who participate in the ANGed program and sell their materials to an ECOLEF collection center receive a subsidized price instead of the local market prices.

Since its launch, the system has enabled the regulation of the sector and

(continued)

BOX 4.7

Tunisia: The ECOLEF Program to Increase Recycling (*Continued*)

facilitated creation of specialized companies— currently comprising 180 microcollection companies contracted by ANGed, 80 micro-recycling companies, and 45 collection and storage points and centers. It has also created an estimated 18,000 jobs and collected more than 150,000 tons of plastic-packaging waste. For the recycling of plastics, polyethylene terephthalate (PET)—generally used to make bottles—is collected, cleaned, ground up, and then exported to other countries. Other plastics like PEHD waste (such as plugs and rigid boxes) are collected, cleaned, crushed, and processed into raw materials.

Because of the international context at the time the program was created, it also had a significant impact on increasing the exports of PET waste. Those exports increased by 96 percent between 2001 and 2013, averaging US$3 million in value. However, since 2014, and following the China plastic-import ban in 2017, Tunisia's plastic-waste exports have been declining significantly (figure B4.7.1).

After the Tunisian revolution (or Jasmine Revolution) in 2011, waste management in urban and rural areas deteriorated. This had a significant negative impact and disrupted the main phases of the cycle from waste collection to recycling. The ECOLEF system has faced several difficulties, resulting in decreased quantities collected and reduced numbers of active recyclers in the system.

FIGURE B4.7.1

PET Waste Exports in Tunisia, 2000–18

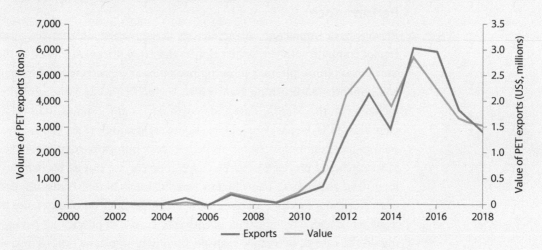

Source: United Nations Comtrade database.
Note: PET = polyethylene terephthalate.

Source: World Bank 2021f.

countries have decreased taxes on "clean" goods to lower their final costs and incentivize their use (for example, compostable or bioplastic bags, coffee cups, and cutlery) (OECD 2015).

In the case of taxes, as throughout this analysis of plastic pollution of the Middle East and North Africa's seas, the principles and practices of a circular-economy approach provide the best—and, indeed, the only fully workable—means to address this problem, because this approach (a) reduces pollution by *reducing the use and manufacture of the pollutant* (in this case, plastics); (b) reduces the *impacts* of *that pollution* on residents' physical and economic well-being; and (c) simultaneously *incentivizes the invention and development* of new products and processes, which *fosters employment and economic growth*.

Taxes can be used to reduce waste management costs by decreasing the amount of waste packaging and end-of-life products being discarded. They can also discourage the sale of certain products (such as plastic bottles and food containers) that involve no recyclable packaging content and have high end-of-life waste management costs as well as to encourage consumers and producers to switch to alternatives. However, other instruments can be considered as "producer taxes" such as deposit-refund schemes (DRSs) and extended producer responsibility (EPR) schemes that target aspects of waste management and recycling rates that cannot be easily addressed through taxes levied at the point of product sale (OECD 2015).

Policies to Address the Issue of Plastics' Superior Performance

Plastics have important characteristics that greener alternatives often cannot compete with yet. In addition to their low prices (discussed in the previous section), plastic packaging preserves and protects food longer, consequently diminishing food waste—which is a key issue globally, including in the Middle East and North Africa and especially in GCC countries.[18] Plastic packaging is also lightweight and flexible. Alternatives such as glass, aluminum, paper, and other more innovative materials (such as starch-based products) usually weigh more or are not as resilient and long-lived as plastics. Their higher weights incur higher transportation costs, both in a strictly economic sense and in an environmental sense by using more fuel, which further incentivizes the use of plastics for packaging. In the case of beverages, transporting them in plastic bottles instead of glass bottles uses around 40 percent less fuel (Pew Charitable Trusts and SYSTEMIQ 2020).

Some of plastics' positive properties must be emulated by their innovative alternatives to make alternatives more competitive. Alternatives that are environmentally less harmful face an uphill battle emulating these properties, while at the same time having to be price competitive. Achieving this requires concerted efforts supporting R&D on green alternatives and the diffusion of successful technologies.

Some plastics (such as SUPs) are more problematic than others; therefore, it is important to identify which products might require greater attention than others. Plastic film and multilayer packaging have low rates of recycling and high rates of leakage into the environment, particularly in low- and middle-income countries (Pew Charitable Trusts and SYSTEMIQ 2020), which merits concerted attention regarding alternatives. It is important to differentiate between options and examine on a case-by-case basis whether substitutes should be subsidized.

Promote R&D and Technological Diffusion of Alternatives

Materials (paper, coated paper, and similar items that can be composted) are available to replace most of the packaging and flexible plastics, particularly in low-income and lower-middle-income countries. Recyclable paper is widespread globally now. For example, 85 percent of paper and cardboard is recycled, and this material is undergoing rapid innovation, leading to better barrier properties for food conservation and cost/weight performance (Pew Charitable Trusts and SYSTEMIQ 2020).[19] Coating is important to increase barrier properties, particularly for food applications. Materials that can be composted are available, and new material formats that are under development (including nonplastic and compostable plastic materials) should meet relevant local composting standards.[20]

Because local conditions and the trade-offs of using the substitute materials play such important roles, key stakeholders in the Middle East and North Africa (local authorities, the private sector, and manufacturers) must work together to evaluate any proposed substitutions by conducting a full-cycle analysis to move forward. Studies by neutral bodies according to recognized standards make it easier to delineate the path toward decreasing plastic use while avoiding unintended consequences. Variables such as transportation; costs of food-production changes (switching to other materials involves new production and end-of-life disposal costs); new waste streams (coatings for certain papers or compostable materials might require specific recyclable technologies); and health impacts (food safety and chemical release of certain recyclable paper can lead to certain health risks) are some issues needing in-depth study when analyzing substitute materials (Pew Charitable Trusts and SYSTEMIQ 2020).

The private sector in GCC countries is already developing a renewable feedstock alternative that, with a conducive policy environment, could be scaled up in the region. In Saudi Arabia, SABIC has developed a polycarbonate (PC) based on certified renewable feedstock.

This solution reduces CO_2 emissions by up to 50 percent and fossil depletion by up to 35 percent during production, compared with fossil-based PC production. This is a product of collaboration with the European value chain to develop more environmentally friendly products given the important role of the circular-economy agenda in this region. Spanish Petroleum Company (Cepsa), a Madrid-based multinational oil and gas company, has been a strategic value chain partner in this project, supporting SABIC through the production of renewables inputs. In addition, SABIC has formed downstream collaborations with other multinationals such as Unilever, Tupperware Brands, Vinventions, and Walki Group to develop new circular polymers from mixed plastic waste.

Win-win partnerships have proven strategic potential to grow the alternatives-to-plastics sector, and government policies should encourage these circular-economy approaches to reducing, reusing, recycling—and replacing and reformulating—plastics. SABIC's new material is currently produced in Geleen, the Netherlands, and is not yet available globally. This certified PC resin may be used for applications in all market segments, such as automotive, consumer, electronics, electrical, construction, and health care (SABIC 2019).

High-income countries in the Middle East and North Africa could promote innovation and awareness of green alternatives through subsidies for companies conducting R&D on alternatives to plastics. Such subsidies could come in the form of tax breaks, lower import duties for these products, or simply increased public investment in companies whose R&D focuses on alternatives to plastics or on products with increased recyclable content in packaging.

Such subsidies for green alternatives could incentivize producers and retailers to develop and offer a broader palette of more sustainable products while also stimulating demand for greener products by consumers through lower prices and higher awareness. Often, businesses trying to invent or promote alternatives to plastics have difficulty borrowing on capital markets and face higher costs of capital (Barrowclough and Birkbeck 2020). Thus, introducing new and easier ways for such companies to attract investments and financing could play an important role in supporting viable alternatives to plastics (Barrowclough and Kozul-Wright 2018). With the market for green products in the Middle East and North Africa still in its infancy, supporting this market is crucial to advance the use of such alternatives in the region. Some of these companies are already setting up and expanding their businesses in the region (box 4.8).

Reducing the volume of plastics being used and manufactured as well as replacing plastics require significant innovation,

---

BOX 4.8

Emerging Alternatives to SUPs in the Middle East and North Africa

Two companies with their regional head-quarters in the United Arab Emirates are at the forefront of providing green alternatives to single-use plastic (SUP) items in the Middle East and North Africa. MyEarth is driving innovation with its starch-based, fully compostable product lines.[a] These include biodegradable bags, straws, and food boxes. Its products try to bridge the gap between versatility and affordability and offer eco-friendly, economically viable alternatives to plastic items. MyEarth has established partnerships with several companies and is planning to expand within the region and beyond.

A second company, Avani Middle East, established in 2014 as a social service enterprise in Bali, launched its operations in the United Arab Emirates in 2017 as one of the country's first companies with a strong commitment to reducing plastics in everyday lives.[b] Its most famous product, the cassava bag, is made of a cheap and common root vegetable and dissolves within a period of months, both at land and at sea. Other products include sugarcane-fiber-based houseware, cornstarch straws, and wooden cutlery. Several companies have switched to Avani Middle East's products; for example, the global retailer Virgin Megastores introduced the cassava bag in all its stores in the United Arab Emirates.[c]

These examples showcase the potential for green alternatives to plastic items in the region and highlight the importance of a level playing field to expand the use of green alternatives.

a. See the MyEarth website: https://www.myearth.ae/.
b. See the Avani website: https://www.avanime.eco/.
c. https://avanime.eco/blog/our-news-1/post/virgin-mega-store-to-roll-out-avani-s-bio-cassava-bag-1.

---

incentives, and well-funded research and data development about waste management. Today, plastic manufacturing is classified as a medium-intensity R&D industry, and the waste management sector is classified as low-intensity R&D (Pew Charitable Trusts and SYSTEMIQ 2020). If these industries transition (while keeping production costs low) toward more competitive market dynamics wherein consumer demand shifts toward more environmentally friendly products, then rapid innovation is mandatory to keep up with these shifts. Developing smart policies and regulatory frameworks, alternative business models, new materials, and more effective collection and recycling systems is critical to reduce plastic leakage into the oceans and make the most of changing market demands. Circular-economy approaches provide the best—indeed the most effective and efficient—means to address these circumstances.

Bans on SUPs

It may take time for the prices and the performance of alternatives to plastics to become competitive with plastics. However, plastic-polluted seas are a pressing problem now that must be addressed as soon as possible. Hence, banning SUP items may be a necessary step to stem the plastic tide in the short term.

In the Middle East and North Africa, only Saudi Arabia and the United Arab Emirates have national bans on SUPs. Saudi Arabia bans the manufacture, advertisement, sale, import, and use of polypropylene (PP) and polyethylene (PE) plastics intended for one-time use, including personal care products; plastic bags intended for one-time use; and disposable food-service products such as spoons, plates, and cups.[21] The United Arab Emirates bans the manufacture and import of nonbiodegradable semirigid plastic packaging for food, magazines, consumer durables, garbage bags, shrink-wrap, pallet wrap, and other disposables.[22]

However, many of the region's economies have some sort of SUP-bag ban in place, in line with the global trend. Globally, most countries have adopted some form of legislation banning plastic bags—127 countries by 2018 (UNEP 2018b). These regulations span a range of interventions to reduce the manufacture, distribution, use, and trade of plastic bags, but the most common is the ban on free retail distribution, which 91 countries have adopted.

In the Middle East and North Africa, Jordan, Saudi Arabia, the United Arab Emirates, and the Republic of Yemen have adopted the approach of regulating market entry, manufacture or production, importation, and retail distribution of SUP bags. Other economies in the region have opted for partial bans or restrictions, mostly in the form of thickness requirements—for example, Tunisia bans bags thinner than 40 microns; the Republic of Yemen, 60; Jordan, 200; and Saudi Arabia, 250, while most European countries accept 50 microns—or they have imposed production-volume limits. Morocco banned the production, sale, and use of nonbiodegradable plastic bags in 2015. Although the use of virgin-plastic bags dropped, illegal production continues, and data on the growth in use of biodegradable bags are not available (Dalberg Advisors 2019b).

Egypt has put in place several retail initiatives to boost awareness, but there is no overarching framework and insufficient technical capacity to replace the use of plastic bags. Consumption is not controlled, and it averaged 124 bags per capita in 2015, equivalent to 12 billion per year (CEDARE 2020). A National Initiative for Reduction of Plastic Bags Consumption was launched in 2017, which led to the decree issued by the Governorates of the Red Sea and South Sinai in 2019 to restrict the single use of plastic bags. Table 4.3 summarizes the plastic-bag regulations in Middle East and North Africa countries.

TABLE 4.3

Plastic-Bag Regulations in Middle East and North Africa Countries, 2018

Country	Restrictions on plastic bags
Algeria	Import restrictions on plastic bags
Bahrain	Regulates disposal only at national level (solid waste/litter regulation)
Djibouti	Ban on nonbiodegradable plastic bags
Iraq	No law found
Jordan	Ban on plastic bags with thickness of 200 microns or less
Kuwait	No law found
Lebanon	Ban on local production, importation, marketing, and use of plastic-packaging bags
Morocco	Prohibition of the manufacture, import, export, marketing, and use of plastic bags
Oman	Regulates disposal only at national level (solid waste/litter regulation)
Qatar	Regulates disposal only at national level (solid waste/litter regulation)
Saudi Arabia	Disposable plastic products made of polypropylene and polyethylene with film thickness equal to or less than 250 microns that are generally used for packaging, such as carrier bags, wraps, and similar applications must be of the oxo-biodegradable type and bear the prescribed logo
Syrian Arab Republic	No law found
Tunisia	Ban on production, import, marketing, possession, and distribution of bags with a thickness of less than 40 microns or bags of low volumes with a capacity of less than 30 liters except for authorized biodegradable bags
United Arab Emirates	Manufacturers and suppliers of plastic bags must meet prescribed standards for oxo-degradable bags and distribute only complying products
Yemen, Rep.	Ban on manufacture of plastic bags below 60 microns and import of plastic bags below 70 microns

Source: Adapted from UNEP 2018b.

Globally, certain other restrictions address the materials used to manufacture plastic bags, with the goal of phasing out nonbiodegradable plastic bags or to incentivize the production, import, or use of bags that are biodegradable, compostable, or both. For example, two countries require a certain type of recycled material: Austria requires plastic bags to have a certain amount (by weight) of materials that can be recycled. Italy bans nonbiodegradable bags and requires that bags intended to carry food products consist of at least 30 percent recycled plastics. The EU has also taken some important steps toward reducing the amount of littered plastic waste on its beaches and in the surrounding seas (box 4.9).

A ban on plastic bags might include exemptions for specific uses. The exemptions can relate to certain activities and certain products. At the global level, 25 countries expressly provide exemptions in their bans, including for the handling and transport of perishable and fresh food items, carrying small retail items, scientific or medical research use, and waste storage and disposal. In the Middle East and North Africa, Saudi Arabia has exempted primary packaging for fresh, perishable, and other loose food as well as pharmaceutical products (UNEP 2018b).

BOX 4.9

The EU Plan to Reduce SUP

In Europe, around 25.8 million tons of plastic waste are generated annually, and less than 30 percent is collected or recycled (EC 2018). The impacts of plastic litter (especially single-use and disposable items) are growing each year as more plastic litter accumulates on European beaches, in oceans, and in the overall environment.

The first "European Strategy for Plastics in a Circular Economy" was adopted in January 2018, followed by the June 2019 approval of Directive (EU) 2019/904 to prevent and reduce the impact of single-use plastic (SUP) products on the environment (particularly the marine environment) and on human health as well as to promote the transition to a circular economy with innovative, sustainable business models. The SUP directive was initially proposed to tackle the SUP items most frequently found on beaches as well as lost and abandoned fishing gear. Through this directive, member states are expected to take the necessary measures to achieve an ambitious, sustained reduction in the consumption of SUPs, in line with the European Union's (EU) waste policy and leading to a substantial reduction in consumption trends. Those measures must contain measurable quantitative goals for a set of most-used products (described below), according to a regional study.

By July 3, 2021, member states submitted descriptions of the measures they have adopted, notified the Commission, and made the descriptions publicly available. These measures included national consumption-reduction targets, reusable alternatives to SUP products available at the point of sale to the final consumer, and economic instruments such as extra charges for SUP products at the point of sale to the final consumer. The measures may vary depending on products' environmental impact over their life cycle, including when they become waste.

Among the main takeaways of the 2019 directive are the following:

- An EU-wide ban on SUP straws, plates, cutlery, beverage stirrers, balloon sticks, oxo-degradable plastics, and expanded polystyrene food containers, beverage containers, and beverage cups by mid-2021

- Extended producer responsibility (EPR) schemes covering the costs of collection, transport, and treatment; cleanup of litter and awareness-raising measures regarding food containers, packets, and wrappers, cups for beverages, beverage containers with a capacity of up to 3 liters, lightweight plastic carrier bags, and fishing gear by December 31, 2024, and for packets and wrappers by January 5, 2023

- EPR schemes for the costs of cleanup of litter, awareness-raising measures, and data gathering and reporting for balloons and wet wipes by December 31, 2024, and for tobacco products by January 5, 2023

- A significant, sustained reduction in the consumption of food containers and

(continued)

BOX 4.9

The EU Plan to Reduce SUP (*Continued*)

cups for beverages by 2026, with the possibility for EU countries to adopt national consumption-reduction targets, the promotion of reusable alternatives, the implementation of economic instruments (such as deposit-refund schemes [DRSs]), or market restrictions (completely or for only certain applications) on food containers and cups.

- A requirement for all beverage bottles with a capacity of up to 3 liters to

  o Have tethered caps by 2024; and

  o Incorporate 25 percent recycled plastic content by 2025 (polyethylene terephthalate [PET] bottles) and 30 percent by 2030 for all types of beverage bottles.

Source: European Commission, https://ec.europa.eu/environment/strategy/plastics-strategy_en.

Policies to Reuse Plastic Products

In the linear-economy model, consumers are often encouraged to replace a damaged item with a new one, and the Middle East and North Africa currently lacks the infrastructure to support a circular model. In a circular-economy model, the goal is to extend product life by encouraging reuse by other consumers by reselling, remanufacturing, facilitating repair, or renting. This practice preserves the stock of natural resources and minimizes the impact of dealing with waste.

Most plastic waste accumulating in the Middle East and North Africa stems from the use of plastics as packaging materials. New business models must be developed to create a reuse ecosystem that works for consumers and producers. Replacing just 20 percent of the region's SUP packaging with reusable alternatives would create a market worth at least US$10 billion (EMAF 2020). Reuse-business models can provide many benefits for the Middle East and North Africa, including cost reduction for companies, increased brand loyalty, and a better user experience. The region presently lacks the legislative environment and technological infrastructure to effectively support the reuse of plastic products. Ways to support plastics' reuse are discussed below.

Extended Producer Responsibility (EPR) Schemes

EPR schemes have been evolving for several decades. The goal of these frameworks is to ensure that those who place products on the market (that is, producers and importers) are responsible financially, logistically, or both, for their products when they become waste at the end of their life cycle. In an EPR system, the cost for the final recycling or disposal of materials is borne by the producer of that item. EPR programs implemented in individual countries differ widely, including the industries and products covered; the policy contexts in which they have been introduced; the nature of the responsibilities placed on producers; and the social, economic, and cultural contexts in which programs operate.

Efforts in the Middle East and North Africa. There is scope for rolling out sustainable waste management systems adapted to local contexts and based on shared responsibility in the Middle East and North Africa. Despite high rates of waste mismanagement, there has been progress mainly in cities. Most of the region's economies have established ministries or specialized authorities for waste with qualified personnel—for example, the National Waste Agency (NDA) in Algeria, the Waste Management Regulatory Authority (WMRA) in Egypt, the Oman Environmental Services Holding Company (be'ah), or ANGed in Tunisia. Cities have improved their infrastructure and developed waste collection capacities, and decision makers are more aware of the concept of separated-waste collection. In some countries, such as Egypt, Morocco, and Tunisia, resident-engagement initiatives and financial investments are under way. Many high-income countries in the GCC are finding ways to increase sustainable disposal through waste-to-energy projects, new regulations and programs, and private sector initiatives with new circular business models to take advantage of the circular economy and recovery after COVID-19.

As for EPR programs, some of the region's economies have identified the opportunity to implement them, but more capacity is needed. Morocco, for instance, stipulates integration of the principle of EPR under the 2013 Framework Law 99-12 under the National Charter of the Environment and Sustainable Development (CNEDD). Its Law 28-00, which lays out the fundamental rules and principles for the management and disposal of waste, sets up a system of accountability based on the "polluter pays" principle. And a bill amending and supplementing this law to strengthen the integration of the EPR principle is being finalized. However, this process has been delayed, and currently there is no functioning mechanism for working with the private sector on this matter (World Bank 2021d).

More recently, Jordan announced that it will begin implementing the EPR system on a voluntary basis in major waste-generating companies. The government is working with producers to establish an association to identify the types of waste on the market and draw up an action plan for the recovery of recyclable material, with the goal of creating job opportunities as well (WAM 2020).

A Greek EPR model. Setting up local waste management structures and strengthening local expertise to ensure sustainable operation is key for program success. In Greece, a producer responsibility organization

(PRO)—the Hellenic Recovery Recycling Corporation (HERRCO)—to implement a collective system for packaging waste has been key to setting up and sustaining such operations. The main activity of HERRCO, which covers 95 percent of the nation's territory, has been developing and operating a network of blue bins for packaging waste, in close cooperation with municipalities.

In 2003, HERRCO introduced its blue-bin recycling system for the collection of mixed recyclable packaging waste such as paper, cardboard, metal, glass, and plastics. Producers are obliged to pay fees of approximately €66 per ton for plastic packaging put on the market. Between 2011 and 2015, the percentage of Greece's population covered by the blue-bin system increased from 75 percent to 92 percent (HERRCO 2015).

Specific awareness-raising and education about how to recycle at the household level is important because contamination in the packaging presents a significant challenge to the separate collection of recyclable plastic waste. However, only 6 percent of all plastic waste is placed in blue bins, and an estimated 50 percent of the bins' content is contaminated (Dalberg Advisors 2019e). Therefore, there is great potential for increasing the quantity of recycled materials. The informal sector in Greece plays a role in obtaining recycling materials and thereby creates a secondary market. In recent years, the growing number of immigrants in Greece has resulted in the increased removal of high-value materials from recycling bins, although this typically consists of paper, cardboard, and metals.

The Greek system has been facing challenges because of the lack of mandatory participation on the part of the private sector, lack of enforcement and consumer awareness, and low fees. Not all producers, importers, and online retailers are registered in the system. HERRCO members account for only 10 percent of the plastics produced and mainly include large multinational companies, with a large proportion of SMEs not fulfilling obligations (Dalberg Advisors 2019e). This is reflected in the amount of material collected. In 2015, HERRCO reported that around 356,000 tons of recyclables were collected, and 202,000 tons were recycled. The difference between collected and recycled quantities indicates a loss rate of around 43 percent, highlighting that the system could improve its overall performance (Elliott et al. 2020).

Recommendations to improve system effectiveness are in line with the need to increase fees charged to producers (currently some of the lowest in Europe) to cover the full costs of end-of-life management, including litter (in line with EU policy). In addition, producers of the least recyclable forms of packaging should be taxed at rates that reflect the "polluter pays" principle and incentivize packaging-design changes that incorporate recycling, recycled content, and reuse (Elliott et al. 2020).

EPR schemes for certain types of plastic products have been successful in Greece. For example, ECOELASTIKA, a Greek initiative for collecting, transporting, and recovering end-of-life tires, collects around 95 percent of end-of-life tires in Greece (ECOELASTIKA 2014). Members of this nonprofit organization are companies that import tires and vehicles. ECOELASTIKA processes used tires, including retreading and used-tire trading as well as the production of rubber crumb.

Plastic pollution in the Middle East and North Africa's agriculture sector is an increasing problem that can be solved by working with local stakeholders. Agri-plastics is a sector with increased interest in sustainable waste management, and other countries' best practices can be helpful to MENA. For example, the Hellenic Plant Protection Association (ESYF) offers a separate management system for agri-plastics, particularly for empty plastic packaging from crop-protection products such as pesticides. The ESYF represents 20–25 companies involved in the crop-protection industry that produce, standardize, and distribute most of plant-protection products in the Greek market.[23]

Determining correct fee levels and enforcement are key aspects of program success. EPR systems could significantly increase the amount of funding available for optimizing existing waste collection and management systems in line with the needs of municipalities. Fees charged to producers should reflect the costs required to deliver a well-functioning waste collection, transport, and treatment system (Elliot et al. 2020). Consumer-awareness programs and incentives for correct separation are key elements of program implementation and contribute to long-term success.

Deposit-Refund Schemes

Deposit-refund schemes (DRSs) are one of the best ways for reusing plastic waste, particularly beverage packaging. The goal of DRSs is to recover certain products by incentivizing consumers to return the product packaging or end-of-life products. A typical example is a deposit payment to the retailer when a beverage in PET is bought, with the deposit refunded when the empty container is returned.

Typically, legislation establishing a DRS mandates specific actions on the part of producers and retailers and may set up new institutions to handle the collection and processing of returned products. Many studies have shown the positive impact of DRSs on reducing the inappropriate dumping of waste and the economic benefits in the form of savings in waste management (Calabrese et al. 2021; Dinan 1993; Lavee 2010; Linderhof et al. 2019; Zhou et al. 2020). One of the most successful cases is in Germany where the adoption of a DRS resulted in 98 percent of one-way PET packaging being returned to appropriate collection sites (Zhou et al. 2020).

The implementation of DRSs varies depending on the context. Ten of 27 EU member countries have implemented DRSs, although this will change, because countries must follow the most recent Plastic Strategy and EU Directive (2019/904-6) (mentioned in box 4.9). Research shows that different DRS models produce different return rates, depending on urban densities; container types and sizes; socioeconomic, cultural, and geographic factors; and how the chosen model is implemented and operated (Calabrese et al. 2021; Zhou et al. 2020).

In 2001, a DRS went into effect for beverage containers in Israel—the only place in the region that has a DRS. In 2010, the law was amended, setting up a collection target of 77 percent for all deposit containers in the market and prohibiting manufacturers from applying for exemptions from the target. Supermarkets and shops are obligated to take up to 50 containers per customer per day (Reloop 2020). The program's manager has been the ELA Recycling Corporation, a nonprofit organization that serves all major suppliers (covering 95 percent of the market). Its goal is to promote, coordinate, and fund the selection, collection, sorting, and recycling of bottles and beverage containers. ELA is officially recognized by the Ministry of Environmental Protection as the agent through which producers are expected to meet requirements such as the recycling target.

The system has been successful and has achieved significant waste management savings. A cost-benefit analysis of the program shows that total benefits exceed total costs by slightly over 35 percent (Lavee 2010). The DRS represents significant cost savings within the municipal waste management system. By taking beverage containers out of regular waste containers, a municipality can enjoy significant savings in waste management costs (consisting mainly of waste storage and collection).

There are practically no DRSs in the Middle East and North Africa. Only the Abu Dhabi emirate is implementing such a system, albeit without a binding legal framework (as described in box 4.5).

The selection of a DRS or an EPR system is a complex task and multiple variables should be considered in the implementation, including previous policy frameworks and local structures. For example, in Spain, even if the DRS would reach 90 percent of the package-return index, the existing EPR system would obtain significantly better environmental results because of the DRS's environmental (and financial) costs in transporting recovered packages (Abejón et al. 2020). In some north European countries, both systems coexist. Issues such as the number and distribution of counting plants (the network of plants) and the distances from manual collection points to DRS packaging points are factors to consider when comparing EPR and DRS systems (Abejón et al. 2020).

In most cases, policies should specify the deposit to be charged on certain beverage containers and a minimum recycling target. A DRS combines two types of economic incentives: (a) a tax on the purchase of the container that should reflect the potential for inefficient disposal (for example, burial in landfill or littering in public spaces); and (b) a subsidy to whoever returns the container so that it can be disposed of in the environmentally preferred way (recycling). In general, the most effective systems are run by the beverage industry as a form of producer responsibility, with minimal government intervention (Hogg et al. 2010). For a DRS to succeed, it is key to work with consumers to educate them about the benefits of a DRS and how they can obtain a refund of their deposit.

Policies to Recycle Plastic Waste Properly

To fully implement a circular-economy approach and stop the leakage of plastics, recycling levels in the Middle East and North Africa should be higher. As shown earlier, waste management in the region is often inadequate, leading to leakage of plastic waste into the environment and marine spaces. Hence, improving SWM is an important precondition for recycling. It is crucial to enhance the currently short useful life-span of plastics and retain used plastics in the value chain as long as possible. However, recycling faces some challenges that must be addressed to present it as a viable alternative to traditional disposal methods and widen its adoption in the region.

In the region, the level of recycling—including for plastics—is very low except in some high-income GCC countries. The proportion of recycled waste in Jordan and Morocco is 10 percent; in Lebanon, 8 percent; in the Republic of Yemen, 7 percent; in the Islamic Republic of Iran, 6 percent; in Algeria, Iraq, and Tunisia, 5 percent each; in Egypt and Syria, 3 percent each; in West Bank and Gaza, 2 percent; and in Bahrain and Libya, 0 percent each (Verisk Maplecroft 2020). Among the GCC countries, Qatar leads the region with 31 percent of waste being recycled, although high-income countries such as the United Arab Emirates (24 percent), Kuwait (19 percent), Oman (15 percent), and Saudi Arabia (12 percent) remain below the average of countries of similar income and even lower than the world average (figure 4.9).

Challenges from an informal recycling sector. Most recycling collection is informal—conducted separately from the SWM system by the informal sector consisting of waste pickers, collectors, and wholesalers. These circumstances result in an informal economy for the collection and treatment of high-value recyclable material. In Morocco, 90 percent of recycling is essentially informal despite several government policy programs to formalize this activity. This informal recycling competes intensively with formal recyclers that bear the costs of taxation and compliance with health and safety regulations (World Bank 2021d).

The waste pickers' activity is independent, seasonal, permanent, and occasional for some—taking place in the streets, landfills, or near trading and waste-recovery platforms—and it is poorly controlled. Collectors are those with space to aggregate the material collected; collectors pay very low prices to ragpickers. Wholesalers often offer equipment to itinerant waste pickers to improve their performance. Tunisia's collection and recycling industry involves 15,000–18,000 jobs directly and indirectly, underscoring the importance of the sector as a source of jobs (World Bank 2021f).

Challenges from low-priced virgin plastic production. In the Middle East and North Africa, the current pricing of virgin plastics and the low costs of traditional disposal are the primary drivers for the low viability of plastic-recycling schemes. The price of recycled plastics versus virgin plastics, and the costs of setting up and operating a recycling scheme versus other forms of final disposal, are important barriers that have prevented the region from achieving higher recycling rates. Plastic litter is costly to recover, and both the amount of plastic litter and its associated costs will continue to increase as the region's economic growth continues.

The production of virgin plastics receives important fiscal benefits and incentives relative to the recycling industry. Subsidies to the petrochemicals sector combined with low global oil prices are driving a wedge between the prices of recycled and virgin plastics. During 2020, the price of virgin PET resins, the most widely recycled plastic, was around half of the price for recycled PET flakes (Hicks 2020; Staub 2020). As a regional example, the estimated 2019 cost in Morocco of collecting and preparing enough recycled plastics to produce 1 ton of plastic bags was DH 20,000 (around US$2,100). The same number of bags could be produced from virgin plastics for only DH 12,000 (around US$1,250) (Dalberg Advisors 2019b).

These lower production costs also translate into lower costs for the consumer, disincentivizing the broader adoption of recycled products. Through the removal of distortive fossil fuel subsidies, price discrepancies could be reduced, leveling the playing field between recycled plastic products and those derived from virgin plastics. This is much in line with the discussion in the preceding section and also has important cross-sector benefits for local air pollution (as discussed in the chapter 3 section, "Policies to Reduce Vehicle Emissions").

Challenges from lack of scale and regulations. The Middle East and North Africa's current recycling schemes lack economies of scale. Even though there are some recycling schemes operating in the region, scale and volume are needed to justify investment and infrastructure capable of extracting more value. In Morocco, the estimated potential of plastic waste is approximately 794,000 tons, of which only 198,500 tons (25 percent) is recovered and recycled (World Bank 2021d). This means that 75 percent of valuable material is lost in the current economy. The Plastic Recovery Association reported that, before the pandemic, recycling units ran at less than 50 percent of production capacity (World Bank 2021d).

The absence of regulations governing collection, sorting, and separation at source makes it even more difficult and costlier to obtain valuable material. Most recyclable material ends up with household waste and is subsequently buried in controlled or uncontrolled landfills, or partially collected by formal and informal waste pickers with low technical and sanitary capacities to produce a high-quality item.

Detailed mapping of value chains in each country and market potential could be an important gate opener to engage more with the private sector. For example, in Morocco, the resins with the highest rates of recovery are predominantly PET, at 30 percent; low-density polyethylene (LDPE), 29 percent; high-density polyethylene (HDPE), 27 percent; and polyvinyl chloride (PVC), 10 percent. Therefore, these four recyclable resins could have the largest market potential to scale up. A detailed mapping of the plastics value chain in the Middle East and North Africa for most recyclable and consumed resins—and the identification of the local industries that consume the most plastic—could provide valuable information in setting reliable regional or national recycling targets and road maps. For example, using a plastic-value approach, Malaysia, the Philippines, and Thailand evaluated the plastics-recycling industry and its role supporting a circular economy and found that less than 25 percent of available plastic is recycled, and 75 percent of valuable material is lost—the equivalent of US$6 billion per year across the three countries. This situation is especially pronounced for SUPs and represents a significant untapped business opportunity if key market barriers are addressed (box 4.10). Therefore, more attention to each locality's challenges is required for more accurate policy design.

Create Economically Viable Recycling Markets

To increase plastic-recycling rates, a reliable supply of and long-term demand for recyclable materials are needed, along with recycling design standards (figure 4.16). Major global brands are making voluntary global commitments to incorporate larger amounts of recycled plastic materials into their products and packaging. There is an increasing level of interest in augmenting the levels of recycling content in products because consumer preferences are changing in favor of more sustainable products. However, government policies lack clarity about the use of recycled resins in food-contact applications. Similarly, they lack clarity about the recycling standards for packaging products that producers, recyclers, and

BOX 4.10

Plastics Circularity and Market Potential: Examples from Malaysia, the Philippines, and Thailand

Asia is responsible for over 80 percent of the world's marine leakage, and 8 of the top 10 contributing countries are from that region—among them, Malaysia, the Philippines, and Thailand. This situation has led to an increased awareness regarding plastic management, bringing the topic of plastic pollution to the forefront of consumers' consciousness and resulting in policy development and the setting of ambitious national goals in each country:

- *In Malaysia*, the government recently launched "Malaysia's Roadmap Towards Zero Single-Use Plastics 2018–2030," while also developing a "Circular Economy Roadmap" to address plastic production, consumption, recycling, and waste management.

- *In the Philippines*, the government is currently developing new strategies. These include (a) finalizing the "National Plan of Action for the Reduction of Marine Litter" (published as part of the Philippine Development Plan (PDP) 2017–2022), which will contribute to the national target to divert 80 percent of national waste by 2022; (b) implementing the Ecological Solid Waste Management Act (Republic Act 9003), which is an integrated solid waste management (SWM) plan based on the 3 R's (reduce, reuse, recycle); and (c) implementing the "Philippine Action Plan for Sustainable Consumption and Production" (PAP4SCP).

- *In Thailand*, policy makers committed to protect the marine environment, strengthen regional cooperation, and set a national "Roadmap on Plastic Waste Management 2018–2030" as a policy framework to manage the country's plastic-waste problem.

Using a plastic value-chain approach, a series of studies were developed with the support of the World Bank to examine untapped economic opportunities to promote plastics circularity and address marine debris in Malaysia, the Philippines, and Thailand. These studies primarily assessed the market for plastics recycling in each country to engage and increase private sector participation and improve the enabling policy environment for implementing circular-economy business models. In these three countries, more than 75 percent of the material value of plastics is lost when single-use plastics (SUPs) are discarded rather than recovered and recycled. This is equivalent to US$6 billion per year across the three countries, representing an untapped business opportunity if key market barriers can be addressed.

Like the reality in Middle East and North Africa economies, most recycling in these three countries happens separately from the SWM system via the informal sector, and most suppliers of recycled resins are small and medium enterprises (SMEs) challenged by the lack of scale, management systems, technologies, and informal supply networks that are neither fully integrated

(continued)

BOX 4.10

Plastics Circularity and Market Potential: Examples from Malaysia, the Philippines, and Thailand (*Continued*)

nor bear costs similar to the costs borne by regular businesses. Based on consultations with industry stakeholders, several actions could create an enabling environment to increase investment in plastic recycling and reduce waste:

- Demand-side incentives to establish a strong market for recycled plastics (for example, recycled-content targets and green public procurement)

- Government support to reduce capital-investment risk (for example, mandating source segregation and setting up extended producer responsibility [EPR] frameworks)

- Widening existing government incentives for investments in the adoption of newer technologies and processes (for example, matching grants); increasing the supply of quality plastics (through improved recycling standards, industry targets for the collection of plastics); and sharing know-how, best-in-class innovations, technologies, and processes.

These measures would be a turning point to enable equal opportunities for, and the growth of, a resilient plastic-recycling industry with high-quality products that retain high material value and the ability to increasingly replace virgin materials.

Sources: World Bank 2021a, 2021b, 2021c.

plastic associations need to capitalize on these opportunities. Most recycling companies in the Middle East and North Africa are micro, small, and medium enterprises that are disconnected and challenged by a lack of scale, technology, and technical capacities to take advantage of these market trends.

Some initiatives in GCC countries such as waste-to-power projects are being developed, while recycling plants have started to segregate waste for reuse and to employ technologies for integrated waste management. For example, Abu Dhabi currently recycles 28 percent of nonhazardous waste and composts 6 percent. Dubai's Smart Sustainability Oasis recycling project currently segregates 18 types of household waste and aims to reduce overall generated waste, thereby diverting 75 percent of waste from landfills by 2021 (*Gulf News* 2017; WAM 2018). Sharjah's Bee'ah, a waste management company in the United Arab Emirates, has the highest landfill diversion ratio in the GCC at 76 percent and it is aiming for 100 percent (Ibrahim 2020).[24] In 2021, a Bahrain company in collaboration with an Australian firm has started treating hazardous waste and converting it into raw materials for the construction and steel industries.

FIGURE 4.16

Principles for Making Recycling Markets More Financially Sustainable

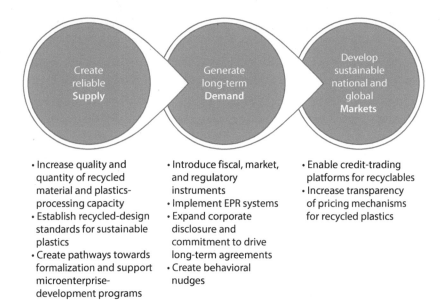

• Increase quality and
  quantity of recycled
  material and plastics-
  processing capacity
• Establish recycled-design
  standards for sustainable
  plastics
• Create pathways towards
  formalization and support
  microenterprise-
  development programs

• Introduce fiscal, market,
  and regulatory
  instruments
• Implement EPR systems
• Expand corporate
  disclosure and
  commitment to drive
  long-term agreements
• Create behavioral
  nudges

• Enable credit-trading
  platforms for recyclables
• Increase transparency
  of pricing mechanisms
  for recycled plastics

Source: Personal communication with Delphine Arri, World Bank Senior Environment Engineer, and James Michelsen, IFC Senior Industry Specialist.
Note: EPR = extended producer responsibility.

Saudi Arabia and the United Arab Emirates are developing eco-parks that will use recycled materials. Saudi Arabia is also finalizing its national strategy for the circular economy. All these are important initiatives taking shape in the region, showing the increased attention to circular-economy models.

National and sector goals and policy road maps are important to give vision to the sector, create an enabling environment, and outline necessary steps toward making the most of these opportunities. National targets for used recycled materials and plastics that consider current baselines and potential capacities are helpful to incentivize conversations with all industry stakeholders. High-income countries and regions such as the EU have set goals based on industry analysis and potential to be achieved in a relatively short period of time. For example, the EU mandated that, by 2025, at least 55 percent of all plastic packaging in the region must be recycled. As of 2017, this rate was 42 percent. The clear requirements and targets mean that there is less confusion and better enforcement. The best policies that deal with plastics are those that engage with stakeholders across the value chain and have clear targets that enable stakeholders to understand what is required of them.

Develop Inclusive Recycling Systems

Recycling is still in the development phase, and most of it lies in the hands of the informal sector. Dry recyclables such as paper and cardboard, plastics, and metals are collected from the streets, landfills, or open dump sites by individuals and are recycled or sold abroad depending on the options available in the particular country. Egypt, Morocco, and Tunisia have large, informal solid-waste and recycling sectors whose integration is key to modernize SWM (Scheinberg and Savain 2015).

Informal recyclers are increasingly recognized for creating value for their cities. They are usually responsible for most of the recyclable materials, and local authorities have begun to see these benefits. Informal recycling also secures livelihoods for many low- and semi-skilled workers and secures waste management services for poor and marginal areas and difficult-to-reach parts of the cities. The objective of several programs and initiatives in the last few years has been to integrate informal recyclers into urban waste management systems and strengthen their access to local and international value chains. The goal is to develop a modern, high-performance, inclusive urban recycling system that relies on cooperation between the informal sector and local SWM authorities (Pew Charitable Trusts and SYSTEMIQ 2018; Scheinberg and Savain 2015).

Recyclers in Morocco. Despite some progress, recycling remains largely informal and operates under precarious conditions. In Morocco, recycling is not clearly defined in the current Law 28-00. Most pretreatment is performed manually, which lowers efficiency. Recyclers sort polymers primarily depending on demand and market value: PET making up 30 percent of all recycled plastics; LDPE and HDPE making up 29 and 27 percent, respectively; and PVC making up 10 percent. Some recyclers are organized or semiorganized and provide some pretreatment of waste (like washing, grinding, and so forth) for producing materials in the recycling industry (powder, granules, flakes, and so forth).

According to the Association de Valorisation du Plastique, in 2019 the plastic-recycling sector in Morocco had 70 plastic-recycling units (co-ops and associations), processes the equivalent of 60,000 tons of waste per year (for a turnover of DH 720 million, which is around US$78 million), and provides approximately 2,000 direct jobs and 5,000 indirect jobs. Although waste pickers are part of the first link of the value chain, they remain socially excluded and exposed to health risks and abuse by intermediaries and wholesalers. Most recyclers have low education levels and are often illiterate (World Bank 2021d).

Crucially, recyclers in Morocco, like elsewhere, face the challenge of providing a material that is often more expensive than virgin plastics. Aside from the subsidized plastic production, this is due to the lack of investment and capital. In addition, this sector's place in the value chain suffers from (a) lack of equipment for washing, compressing, and crushing waste to increase the sales price; (b) lack of training and capacities to preserve workers' health and maintain work equipment and tools; (c) abuse of workers by intermediaries and wholesalers; and (d) lack of awareness by the general population regarding sorting at source, which would dramatically help recovery rates (World Bank 2021d).

Recyclers in Egypt. Similarly, the informal sector in Egypt is complex, with different groups of people recovering materials and selling. The *Zabbaleen*—a term for many in the informal recycling sector (coined from the Egyptian Arabic word for "garbage people" and hence understandably perceived as derogatory—is estimated to be a community of over 250,000 people. They provide collection, recovery, processing, and recycling services to the Greater Cairo area of approximately 20 million people in 2020. In 1986, the newly established Cairo and Giza Cleansing and Beautification Authorities (CCBA and GCBA) began licensing them (EcoConServ 2010).

Of all the actors in the current waste system, the traditional informal waste pickers recover the largest volume of materials from the city. They do so with a regularity that has granted them an established role in the nation's waste recovery and recycling. They collect an estimated 65 percent of Cairo's waste, recycling around 85 percent of that amount (EcoConServ 2010).

Other groups of vulnerable people working with recyclables are the *Sarriiha* (singular *Sarriih*) and *lae'ita*. *Sarriiha* are those who roam the streets buying, trading, and exchanging recyclable waste items, and *lae'ita* are those who scavenge and collect the waste by picking through dumps, landfills, and street bins. They roam the country in both rural and urban areas either with pushcarts or on donkey-pulled carts. Unlike the *Zabbaleen*, the *lae'ita* do not collect from households or regular commercial and institutional clients. They are not organized into nonprofit organizations, cooperatives, or trade associations and thus have limited support. Basically, they exchange mainly plastics and metal that housewives set aside for them in return for household items of utility. Usually, these groups are linked to a trader (a *mo'allem*) who owns a depot and supplies their donkey cart and the day's cash for immediate transactions.

Principles for a more inclusive recycling sector. Although formal integration is increasing its visibility and moving rapidly, the situation in the Middle East and North Africa remains complex and requires more attention. Decision makers' growing awareness of structural integration of informal workers is developing in a supportive and global environment. More interventions are targeting professionalization of recycling workers, although there is still a need to build better consensus about how and at what pace. International experiences provide several broad principles for inclusion of informal waste pickers to support the objectives of a circular economy and the recycling of waste:

- Provide pathways to formalization, licensing, and compliance to ensure market transparency for informal waste pickers

- Recognize informal recyclers' work as an occupation and call them "recycling workers"

- Facilitate enterprise development programs through cooperatives or associations for informal waste pickers and collectors

- Provide financial support for establishing small, medium, and large recycling units through suitable incentives

- Through community-based organizations, implement awareness-raising programs for waste pickers regarding health, safety, and financial matters

- Implement awareness programs for other stakeholders about informal waste pickers and their mainstreaming into the economy.

With appropriate policies, governments can create more jobs and improve the working conditions of informal waste pickers by building capacity for the delivery of higher-quality products through training and small enterprise or microenterprise development programs. In addition, the informal industry can be provided with a pathway to formalization, licensing, and compliance to ensure a transparent market and level the playing field with formal recyclers that comply with environmental, health, and safety standards and other requirements (as discussed in box 4.11, with examples from Latin America).

More generally, the region's economies could profit from increased job potential induced by higher rates of recycling. With China banning imports of lower-grade waste in early 2018, the case for developing proper waste management infrastructure in the Middle East and North Africa and developing its domestic recycling capabilities has been strengthened (Al-Sadoun 2018). In the GCC countries alone, an increase of approximately 40 percent in recycling rates would cut CO_2 emissions by 10–12 million tons per year and decrease primary energy consumption. Furthermore, the creation and advancement of the recycling industry in these countries could create 50,000 new jobs with a total market potential of around US$6 billion per year, and investors can expect operating margins of above 15 percent in various opportunities across the value chain (Menachery 2020). In Tunisia, the ECOLEF

BOX 4.11

Integration of Recyclers into Local Waste Management Systems: Examples from Latin America

In 2011, the Colombian Constitutional Court declared that "waste pickers are historically the holders of an environmental role of great importance." Meanwhile, in Peru, Law No. 27419 specified that "the State recognizes the activity of waste pickers, promotes their formalization and their integration into the waste management system." These were the instrumental laws that approved the transitional regime for the formalization of "waste pickers" into key stakeholders within their respective countries' recycling materials value chains. Both instruments gave recyclers a few years to organize themselves into cooperatives and to complete the technical, administrative, and commercial requirements to participate in the tendering process that would increase recycling rates in the capital cities of Bogotá and Lima, respectively.

Recycler organizations had to demonstrate that they had sufficient accounting systems in place, the capacity to control collecting routes with georeferencing systems, mechanisms for managing grievances, and the management capabilities to compete with other companies in transparent tendering processes for the collection of recyclables.

Local urban waste management authorities pay for the services provided based on the number of recyclable materials collected and the cost of their transport to approved weighing centers. The amount of remuneration was set according to the costs that otherwise would have resulted from the nonmanagement of recyclable waste. In addition, recycler associations could sell clean recycled materials to other companies.

Source: World Bank 2021e.

---

program has generated an estimated 18,000 jobs in the local economy while also collecting 150 kilotons of plastic waste since its inception in 1998 (Dalberg Advisors 2019c).

However, it should also be acknowledged that such a transition would not carry benefits for everybody. Transitioning toward a more circular concept of the economy is a major overhaul and leaves those people who lack the necessary skills at risk of job displacement. It is hence important to implement policies that cushion these potential adverse effects such as educational and vocational training programs, preferably at a low-threshold level to include the large share of informal workers in the Middle East and North Africa.

Improve Labeling and Recycled-Content Targets for the Most Problematic Products

Lack of knowledge about products' recyclability leads to high inefficiencies in the recycling process. The public's knowledge gaps regarding what can be recycled lead to the mixing of different waste types and higher levels of contamination, reducing recyclability levels. Having a lot of unsorted waste either necessitates sorting at the waste-treatment facility or leads to low levels of overall recyclable material. Both outcomes undermine efforts to recycle, raising inefficiencies in the process. For example, several source-sorting pilot projects in Morocco that sought to achieve recycling levels of 20 percent were tested, with mixed results (World Bank 2021d). In Tunisia, the lack of a legal framework establishing the obligation to separate waste has hindered the creation of a recovery system for plastic waste, increasing the flows of plastic waste to nature and landfills (World Bank 2021f).

The use of instruments such as ecolabels is an important tool to inform consumers about their choices and has been successful in some contexts. Recently, there has been increased use of such labels to inform consumers whether a product or service meets certain environmental performance standards. This type of certification provides a level of trust to consumers, covering a wide range of environmental impacts from production to design and disposal. The UK government, for example, has developed a program to ensure that consumers receive better information, working with key stakeholders including industry, trade associations, and standard-setting bodies to develop options for domestic ecolabels that indicate an item's suitability for recycling (Defra and EA 2018).

Clear, regulated labeling of plastic products and packaging is necessary to integrate the end consumer into the recycling process at the time of purchase. Deficiencies can be counteracted by clear labeling of these products, indicating their recyclability, and conveying that information to the consumer in an easy-to-comprehend way. Having clear, standardized labels for goods that can be recycled, as well as the degree to which they are recyclable, allows the end user to make informed choices. The issue of labeling plastic products is more and more recognized at the global level (Tsakana and Rucevska 2020), including by Middle East and North Africa economies. For example, the GPCA, which brings together the largest plastic producers in the region, also recognizes that product labels indicating the recyclability of plastics can be used to increase demand for recyclable goods.

Setting recycled-content standards for plastic products is a gradual way to decouple recycled products from virgin plastics. Setting a determined content by regulation promotes domestic demand and encourages investments in the plastic-recycling industry. Recycled-content targets must be set in consultation with the relevant industries to ensure product quality, performance, and safety. A feasibility study is recommended to evaluate and recommend specific recycled-content targets and set milestones to arrive at these targets. To successfully implement this type of measure, it is essential to assess the local waste management infrastructure and policy options as well as to conduct sensitivity analysis of prices and cost-benefit analysis for each target. These studies are prerequisites before any targets for recycled content are set. A recent World Bank study on recycling markets in the Philippines recommends that any recycled-content targets should be set at a resin-level or at an end-use-application level (World Bank 2021c). These types of measures can be effective because they stimulate the local secondary market for recycled products.

Policies to Increase Public Awareness and Social Marketing for Effective Policy Making

An effective policy framework requires social awareness and educational programming to encourage changes in consumer behavior and promote a gradual transformation toward a circular economy. Public-awareness campaigns and regulatory and economic policy instruments are all needed for lasting change.

In the Middle East and North Africa, public awareness about recycling possibilities is low. For example, in 2017, only around 38 percent of GCC residents had an informed view about the recyclability of plastics (GPCA 2017). Similarly, a survey in Jordan revealed that knowledge about proper recycling practices was low despite the high willingness of respondents to learn more about these practices (Aljaradin, Persson, and Al-Itawi 2011).

Nonetheless, consumers' concern about the widespread use of plastics in everyday items and plastics' impact on the environment is growing in the region. In 2018, a representative survey conducted by YouGov in the United Arab Emirates revealed that more than half the respondents were seriously concerned about the excessive use of plastics in their city (YouGov 2018). Close to two-thirds of survey respondents were concerned about the impact of excessive plastic waste on environmental degradation, and more than 80 percent of respondents believed that reducing the use of plastics is the best way to protect the environment. Interestingly, persons older than 40 seem to be more concerned than younger people about the indiscriminate use of plastics. In Dubai, 90 percent of respondents surveyed stated that they make efforts to reduce their own plastic consumption (Lindo 2020). In Oman, 88 percent of respondents surveyed supported a plastic-bag ban and gave numerous suggestions regarding the recycling of plastic bags (*Muscat Daily* 2018).

Knowledge about legislation is also lacking, further indicating that better outreach and communication are needed. In the same YouGov (2018) survey in the United Arab Emirates, residents were asked about government initiatives to reduce the use of plastics. Although most respondents knew about the concept of "reduce, reuse, recycle," only two out of five respondents were aware of government legislation to discourage plastic use. Middle East and North Africa governments must enhance the dissemination of information about legislation to increase public awareness of such legislation and adherence to it. Better communication regarding existing legislation, together with information on the adverse effects of plastic pollution, could help shape public opinion about lowering the use of plastics in everyday life.

People can change their behavior toward the environment with a supportive policy framework, effective public outreach, and the availability of more environmentally friendly options. People can change their behavior to be more responsible toward the environment by, for example, estimating the costs relating to the use of plastics and identifying co-benefits from reducing plastic use (Marazzi et al. 2020). A combination of approaches of social and education campaigns and decision-supporting tools can have positive, long-lasting results. An evaluation of a Europe-wide program showed that participatory events designed to facilitate dialogue on solutions brought together 1,500 stakeholders and revealed support for cross-cutting pollution prevention measures (Veiga et al. 2016). Therefore, more environmentally sustainable choices and actions can become new norms and widely accepted social practices if robustly supported by private and public sector initiatives, well-enforced policies, and evidence-based communications campaigns (Marazzi et al. 2020).

Initiatives to raise public awareness about recycling have been undertaken in Middle East and North Africa's economies in the past. For example, in a cooperation with the Agence Française de Développement (AFD), a multiyear project to raise awareness among residents of 12 municipalities was undertaken from 2014 to 2017 by the Lebanese nonprofit organization arcenciel.[25] It involved awareness campaigns to encourage waste reduction and recycling. Arcenciel also published a "Municipal Waste Management Guide" for the sorting and recycling of household waste. The Tunisian National Waste Management Agency, in cooperation with the Sweepnet network, carried out awareness-raising projects and established a dedicated communication and awareness office. And in the course of the EU-funded project SwitchMed and with support by the UNEP, Jordan has launched an awareness campaign titled "One Dead Sea Is Enough" in reference to the Dead Sea bordering the country. The project also included requirements for manufacturers to provide biodegradable plastic bags, job trainings for the industry to raise resource efficiency, and development of the Sustainable Consumption and Production Action Plan currently under implementation (SwitchMed, n.d.; UNEP 2018a).

The young need to be engaged in the fight against marine plastics, but specific approaches are needed. A project in the United Arab Emirates (and India) spearheaded by the Ervis Foundation engages with youth to bring about a generational change in plastic consumption and disposal.[26] Their program involves activities such as beach cleanups organized with the help of student ambassadors and the development of a mobile app to spread knowledge about plastic pollution and its effects as well as support ideas for change. This social enterprise is also supported by UNEP in its efforts to foster positive behavior change in the younger generation.

Policies to Reduce Marine Litter from Beach-Tourism Recreational Activities

Understanding how beach activities influence marine litter in the Mediterranean Sea region is important for policy design because the region's bathing season is long and beaches are generally crowded and close to population centers. Because tourism is a pillar of the region's economy, involving tourists in the design of policies and care for the environment should be a high priority for waste managers to preserve the region's natural beauty. Beach litter may be up to 40 percent higher in the summer and consists primarily of plastics (bottles, bags, caps, lids, and so forth); cigarette butts; aluminum (cans, pull tabs); and glass (bottles), and originates from shoreline-recreational activities (Galgani, Hanke, and Maes 2015; Munari et al. 2016).

In the Middle East and North Africa, high levels of tourism contribute to the plastic-pollution problem. In Morocco, resort-area beaches with significant hotel infrastructure and major coastal cities or "urban beaches" are significantly more polluted during the bathing season (for example Agadir, Casablanca, Rabat, Saâidia, and Tangier) (World Bank 2021d). Of all the debris found, an average of 60.5 percent was related to coastal-recreational activities (CRA) and to smoking-related activities (SMA). CRA debris consisted of food containers, cups, wrappers, trays, straws, cutlery, plastic bottles, cans, and so forth, and SMA debris consisted of cigarette butts, lighters, and tobacco packaging (Nachite et al. 2019).

Regardless of their exact percentages of marine litter, recreational activities on beaches play an important role in the generation of marine litter. Although it is key to encourage litter-prevention interventions at the source, programming to influence beachgoer behavior is important and should be implemented in addition to "end-of-pipe" interventions such as beach cleanups by municipalities and volunteers. Data show that, in the Mediterranean Sea region and other coastal areas, most of the litter washed up on the beaches was left by beachgoers (Galgani, Hanke, and Maes 2015; Munari et al. 2016; Portman et al. 2019; Thiel et al. 2013).

Plastics are the most ubiquitous beach-litter items, and therefore considering beachgoers' behaviors in the design of policies is beneficial. Plastics are the most abundant materials found in the Mediterranean and found to be the most abundant litter item on beaches in Israel, Morocco, and Tunisia (Alkalay, Pasternak, and Zask 2007; Laglbauer et al. 2014; Pasternak et al. 2017; Portman and Brennan 2017).

Efforts to increase beach-waste infrastructure for environmental behavioral change should consider beachgoers' and beach managers' needs. Considering elements of design for sustainable behavior to reduce litter can improve litter collection on beaches, highlighting the importance of interdisciplinary work to address the problem of marine litter from land-based sources.

Principles of sustainable design for public spaces can be applied to one of the most common and problematic types of litter on beaches: cigarette butts. Around 30 percent of the litter on beaches comes from smoking-related activities (cigarette butts, lighters, cigarette packs) (UNEP/MAP and Plan Bleu 2020). Therefore, specific policies to mitigate this type of impact are crucial.

Other possible policy measures such as smoking bans in public areas, including beaches and parks, can reduce marine pollution. Because tobacco-product waste became a serious problem along Thailand's coastline, the government issued the 2017 Tobacco Product Control Act, which states that the Ministry of Public Health can propose new regulations such as a ban on tobacco use on beaches. Voluntary smoking bans in areas that struggle with smoking-related litter may be appropriate.

Policies to Tackle Marine Sources of Plastic: The Fishing and Shipping Sectors

Although maritime sources of plastic waste in the Mediterranean are rather low relative to land-based sources, the fishing and shipping industries are significant contributors to the problem. Abandoned, lost, or otherwise discarded fishing gear represents a significant source of pollution with serious environmental and socioeconomic impacts. In the European seas, it is estimated that between 2,000 and 12,000 tons of fishing gear are lost each year from the active fishing fleet, with even higher levels expected in the coastal areas from aquaculture gear loss or abandonment (estimated at between 3,000 and 41,000 tons) (OSPAR 2020).

The harm caused by fishing gear as marine litter is well documented, and wildlife is most affected. Marine megafauna are known to mistakenly eat anthropogenic debris and die from consequent gastrointestinal blockages, perforations, and malnutrition as well as suffering sublethal impacts (Roman et al. 2020). Marine ingestion occurs in over 1,400 species, among which marine mammals, sea turtles, and seabirds are well represented (Claro et al. 2019).

Other impacts include harm to habitats through smothering and abrasion, potential support to the spread of invasive alien species, and transportation of additives and organic pollutants to habitats and throughout the food chain (OSPAR 2020). A main concern is that fishing gear degrades very slowly over decades, with fibers found to be shedding from the surface of trawl fragments older than 30 years, while the netting itself remains sturdy and robust (OSPAR 2020).

Independently of the gear type, the main materials used are plastics: PP, PE, and polyamide (nylon/PA6). Fishing gear can also include single and mixed materials containing metals, PVC, polystyrene, polyvinylidene difluoride, Dacron (PET and polyester), rubber, foam, and various hazardous materials (for example, lead weight and copper coatings). Overall, the supply chain for fishing gear is complex and country-specific, with local assembly undertaken locally (OSPAR 2020).

Currently, only a small proportion of fishing gear is recycled at end of life, although it would be possible to scale up if recycling markets were well developed. Fishing gear can contain multiple types of mixed polymers, which require high levels of preprocessing (sorting and dismantling) to be recycled, with high costs and time involved.

Circular-economy approaches to the design of fishing gear are still nascent and primarily driven by functionality and cost, not environmental impact and waste management. Hazardous materials are still used in fishing gear (for example, copper coating, lead).

Increasing management aimed at preventing the loss of fishing gear (or parts thereof) could give the quickest gains for now. For this, international fishing organizations recommend that it is important to undertake national mapping exercises for the life cycle of fishing gear as well as analysis of national legal frameworks related to the end-of-life fishing gear waste (OSPAR 2020).

Solutions could include the introduction of minimum quality standards for fishing gear, repair, and port disposal of damaged nets, and enforcing penalties for dumping, failure to retrieve lost items, and restricting fishing activity in locations and conditions where loss is likely (Roman et al. 2020).

Cleanups to Reduce Marine-Plastic Pollution

Cleanups are important responses to the alarming levels of plastic pollu-tion affecting the Mediterranean coasts of many Middle East and North Africa countries because they restore and preserve ecosystems and main-tain the beauty of the local beaches that attract millions of tourists every year to the region. Most macroplastics in the Mediterranean end up on coastlines before they are sucked into the sea, showing the importance of introducing cleanup solutions on land.

Decreasing the flow of plastics into the seas should be the primary objective, but it is also important to reduce the amount that is already floating in marine spaces and washed up along the shores. More than 1 million tons of total plastic has accumulated in the Mediterranean (Boucher and Billard 2020). While a steady stream of more than 200,000 tons of plastics enters the Mediterranean every year, 1,000 metric tons are washed up along its shores (Boucher and Billard 2020). By not dealing with and removing them appropriately, more and more of it ends up in these spaces permanently. Moreover, through contin-ued fragmentation of macroplastics into microplastics and ingestion by fish and other species, plastics enter the food chain, severely affecting biodiversity and human health.

Plastic concentrations in the Mediterranean Sea are highly variable. As a semi-enclosed sea, the Mediterranean has a very low flush rate as well as significant environmental variability (Kaandorp, Dijkstra, and van Sebille 2020; Portman and Brennan 2017). Often plastic waste remains in coastal areas. Adding to various sources of plastics, many surface currents end in the eastern basin because of downwelling, increasing the problem of these locations. More importantly for policy design, the highest fluxes of sinking (of more than 1 kg per square kilometer per day) occur just next to the coast, where the nonbuoyant plastics immediately sink.

Given the amount of plastic pollution already accumulated in marine spaces, introducing cleanup technologies can be part of the solution. At least 52 technologies exist for the prevention or collection of plastic pol-lutants. Of these, 14 technologies focus on leakage prevention (such as wastewater treatment for microplastics in clothing), and 38 focus on col-lecting macroplastic waste already in waterways (Schmaltz et al. 2020).

The Ocean Clean-up is the best-known technology for corraling and capturing floating plastic items in the North Pacific Gyre.[27] Other examples are the Seabin Project that collects floating debris from ports and marinas in Sydney, Australia,[28] and "Mr. Trash Wheel" in Baltimore, Maryland, which collected approximately 1 million pounds of waste over the course of 2.5 years by funneling trash from harbors and rivers, pre-venting loss to the sea (Campbell 2016; Schmaltz et al. 2020).[29] Adopting these technologies has been very important in cleaning that city's waterways, and similar technologies (such as the Watergoat Trash Traps in Augusta, Georgia) have been used elsewhere (Savannah Riverkeeper 2018; Schmaltz et al. 2020). Increased media attention to marine pollution in the past few years has triggered even more the development of technology projects and increased expectations. But although these efforts to collect plastic pollution are laudable, their current capacity and widespread implementation are still limited in scope, and most of them are in the trial phase.

Local beach cleanups have become a common local response to marine-plastic pollution. Usually, these are largely conducted by volunteers concerned about the crisis who react by gathering together with like-minded people to clean specific hot spots. Through these actions, coastal communities practice environmental stewardship. One example of beach cleanup coordination at the global level is the Ocean Conservancy International Coastal CleanUp campaign, which encourages individuals and groups to organize their own local volunteer cleanups as part of a global campaign.

Voluntary beach cleanups should not crowd out actions taken by industry or government. One common concern about framing volunteer beach cleanups as a solution to the plastic-pollution crisis is that such framing shifts the "narrative of responsibilization" away from industry and government to consumers and civil society (Jorgensen, Krasny, and Baztan 2021). Cleanup methods are helpful to reduce the impacts on marine life and restore degraded ecosystems and should be used in tandem with other preventive solutions, such as the ones mentioned in the previous sections.

To briefly summarize the key message presented in this policy-review section on the marine-plastic pollution of the Middle East and North Africa's seas: The environment, its residents, and their national economy are not independent entities; they are very much interdependent. Each can—and each does—very substantially affect the well-being of the other two.

As the examples in this section have shown, strategies are available to limit marine-plastic pollution contemporaneously and synergistically in the Middle East and North Africa region while protecting and enhancing residents' well-being and furthering local and national economic development. The only way to comprehensively achieve these three interrelated goals of protecting and enhancing the environment, human health, and local and national economic development is through a circular-economy approach, which consists of the "3 R's + 2". The 3 R's are *Reduce*, *Reuse*, and *Recycle*, and the "+ 2" are (1) *proper disposal* (including cleanups) of plastics that cannot be reused or recycled, and (2) *awareness-raising* campaigns to inform the public about plastic pollution of the region's seas.

Only if residents are educated about the problems that such pollution causes can public opinion be mobilized to support government policies that advance the principles and practices of a circular economy.

NOTES

1. Macroplastics are plastic items with a size of at least 5 millimeters, and microplastics are smaller than 5 millimeters.
2. The figures estimated by Jambeck et al. (2015) are unique for their global coverage, but this comprehensiveness comes with the caveat that, for some countries, the estimates of marine-plastic debris may be less precise and deviate from estimates in country-specific studies. Still, the Jambeck et al. (2015) figures are widely recognized and often used as a benchmark to compare the discharge of plastic debris into the oceans across countries.
3. Microplastics are divided into two types: primary and secondary. Examples of primary microplastics are found in personal-care products; plastic pellets (or nurdles, which are the size of a lentil) used in industrial manufacturing; and plastic fibers used in synthetic textiles (for example, nylon). Secondary microplastics are ones that result from the fragmentation of larger pieces through exposure to, for example, wave action, wind abrasion, and ultraviolet radiation from sunlight.
4. These fish species are the European pilchard (*Sardina pilchardus*) and the European anchovy (*Engraulis encrasicolus*).
5. The ROPME refers to the RSA as the sea area at the most northwestern part of the Indian Ocean, surrounded by the eight ROPME member states: Bahrain, Iraq, the Islamic Republic of Iran, Kuwait, Oman, Qatar, Saudi Arabia, and the United Arab Emirates.
6. The five major gyres—large systems of rotating ocean currents—are the North and South Pacific Subtropical Gyres, the North and South Atlantic Subtropical Gyres, and the Indian Ocean Subtropical Gyre.
7. The major oceanic zones are the North Pacific, South Pacific, North Atlantic, South Atlantic, and Indian Oceans.
8. In terms of mass, the Mediterranean contains 5–10 percent of global microplastics, with the small average particle size accounting for the differences between counts and mass (van Sebille et al. 2015).
9. Ocean plastic can persist in a sea's surface waters, eventually accumulating in remote areas of the world's oceans. In the Great Pacific Garbage Patch (GPGP)—a major plastic-accumulation zone formed in subtropical waters between California and Hawaii—at least 46 percent of the GPGP mass consisted of fishing nets (Roman et al. 2020).
10. The reasons for this level of human ingestion are (a) the amount of plastic produced in recent decades, which has been increasing at a steady rate of 4 percent a year; and (b) the primary use of plastic as a disposable material, with large amounts of plastic waste not adequately managed, as noted earlier.
11. The midpoint of the 4–8 percent range is taken as the plastic industry's share of global oil production and growth rates of consumption, in line with projected industry growth of 3.8 percent annually 2015–30 and 3.5 percent annually 2030–50. Increases in efficiency are limited (EMAF 2016).

12. For more information, see the SwitchMed website: https://switchmed.eu/.

13. The estimate is based on analysis of five Asian countries: China, Indonesia, the Philippines, Thailand, and Vietnam (Engel, Stuchtey, and Vanthournout 2016).

14. Countries consulted were Egypt, Jordan, Lebanon, Morocco, and all of the GCC countries. Issues with data availability for SUP items and their alternatives restricted the inclusion of other economies in the analysis. The websites consulted were https://www.luluhypermarket.com/; https://www.desertcart.ae/; https://egypt.souq.com/eg-ar/; https://www.noon.com/; and https://www.ubuy.com.kw/. The data were obtained on December 2, 2020.

15. Prices were standardized to relate to a single unit of each good; however, batch sizes may vary where that variation was unavoidable. Such differences could distort per unit prices. Furthermore, purchases from wholesale distributors by commercial clients, such as restaurants or catering services, most likely involve varying prices (and price discrepancies) and are not reflected in this analysis.

16. Data are from the EUROMAP (2016) comparison of production, consumption, and imports of plastic resins for the Middle East region. Changes in inventories produced in a previous year are not considered, leading to potential deviations. The EUROMAP report considers the Middle East only in the aggregate—that is, it does not include North African countries, Egypt, or Djibouti. However, it is reasonable to assume that much of the plastic consumed in these countries is imported from Middle East producers in intraregional trade.

17. For ethane and natural gas, these prices are significantly lower than ones from comparable global markets, with prices of the latter 50–300 percent higher. Furthermore, fixing the prices of inputs gives companies an advantage by easing their planning of production costs and capacities.

18. Globally, around one-third of produced food ends up as waste (FAO 2011). This problem is likely accentuated in GCC countries: for example, survey respondents in Saudi Arabia stated that more than 75 percent of food purchased is discarded every week to make room for new groceries (Baig et al. 2019).

19. Coated plastic excludes sterilized aseptics, beverage cartons, and coffee cups for which the lamination weight or double-sided application mean they are recyclable only in a specialized recycling facility.

20. For example, the European industrial composting standard EN 13432 could be mandated where industrial-equivalent composting is available and effective.

21. Saudi Standards, Metrology and Quality Organization (SASO), "Technical Regulation for Degradable Plastic Products," No. M.A-156-16-03-03, *Official Gazette*, October 14, 2016.

22. Emirates Authority for Standardization and Metrology (ESMA), "Standard & Specification for Oxo-biodegradation of Plastic Bags and Other Disposable Plastic Objects"; "Specific Requirements for the Registration of Oxo-Biodegradable Plastic Objects according to UAE Standard 5009: 2009" (Revision 1, March 1, 2014).

23. See "Regular Members," ESYF website: http://esyf.gr/o-syndesmos/esyf-members/.

24. For more information, see the Bee'ah website: https://www.beeahgroup
.com.

25. See "Sorting and Recycling within Organizations and Municipalities,"
Sustainable Agriculture and Environment Projects, arenciel: https://www
.arcenciel.org/projects/sorting-and-recycling-in-municipalities/.

26. For more information, see the Ervis Foundation website: https://www
.ervisfoundation.org/; and "Educating the Youth for Responsible Plastic
Consumption and Disposal," UN Sustainable Development Goals website:
https://sustainabledevelopment.un.org/partnership/?p=32210.

27. For more information, see The Ocean Cleanup website: https://www
.theoceancleanup.com.

28. For more information, see the Seabin Project website: https://seabinproject
.com/.

29. For more information, see the Mr. Trash Wheel website: https://www
.mrtrashwheel.com/technology/.

REFERENCES

Abejón, R., J. Laso, M. Margallo, R. Aldaco, G. Blanca-Alcubilla, A. Bala, and
P. Fullana-i-Palmer. 2020. "Environmental Impact Assessment of the
Implementation of a Deposit-Refund System for Packaging Waste in Spain:
A Solution or an Additional Problem?" *Science of the Total Environment* 721:
137744.

Abu Zaid, M. 2020. "Egypt Issues First Green Bonds in MENA." *Arab News*,
September 29. https://arab.news/5junu.

Adyel, T. M. 2020. "Accumulation of Plastic Waste During COVID-19." *Science*
369 (6509): 1314–15.

Ahmad, A. 2020. "UAE Environmentalists Welcome Abu Dhabi's New Single-
Use Plastic Policy." *Gulf News*, March 10. https://gulfnews.com/uae
/uae-environmentalists-welcome-abu-dhabis-new-single-use-plastic
-policy-1.70283337.

Al Ahad, M. A., A. Chalak, S. Fares, P. Mardigian, and R. R. Habib. 2020.
"Decentralization of Solid Waste Management Services in Rural Lebanon:
Barriers and Opportunities." *Waste Management & Research:* 38 (6):
639–48.

Alessi, E., and G. Di Carlo. 2018. "Out of the Plastic Trap: Saving the
Mediterranean from Plastic Pollution." Report, World Wide Fund for Nature
(WWF), Rome.

Aljaradin, M., K. M. Persson, and H. I. Al-Itawi. 2011. "Public Awareness and
Willingness for Recycle in Jordan." *International Journal of Academic Research*
3 (1): 508–10.

Alkalay, R., G. Pasternak, A. Zask. 2007. "Clean-Coast Index—A New Approach
for Beach Cleanliness Assessment." *Ocean & Coastal Management* 50: 352–62.

Al-Sadoun, A. 2018. "Sustainable Plastics Solutions Crucial for Global
Challenges." *Saudi Gazette*, September 4. https://saudigazette.com.sa/
article/542715.

Al-Sadoun, A. 2019. "From Waste to Value: The GCC Chemical Industry's Contribution to Circular Economy." Message from the Secretary General, Gulf Petrochemicals & Chemicals Association (GPCA), Dubai, United Arab Emirates.

Al Sarihi, A. 2019. "The Pros and Cons of Expanding Plastics Production in the GCC." Gulf Monitor, September 27. Castlereagh Associates, Surrey, UK.

Aly, B. 2020. "Egypt: Waste Management Upgrade." Ahram Online, January 24. https://english.ahram.org.eg/NewsContent/50/1201/360035/AlAhram -Weekly/Egypt/Egypt-Waste-management-upgrade.aspx.

Anouti, Y., A. Klat, J, Bertal, and M. Bejjani. 2019. "Putting GCC Cities in the Loop: Sustainable Growth in a Circular Economy." Report published for the World Government Summit by the Ideation Center of Strategy & in the Middle East, PricewaterhouseCoopers, Beirut, Lebanon.

APICORP (Arab Petroleum Investments Corporation). 2016. "Energy Price Reform in the GCC: Long Road Ahead." APICORP Energy Research: 1 (4): 1–4.

Baalkhuyur, F. M., El-J. A. Bin Dohaish, M. E. A. Elhalwagy, N. M. Alikunhi, A. M. AlSuwailem, A. Røstad, D. J. Coker, M. L. Berumen, and C. M. Duarte. 2018. "Microplastic in the Gastrointestinal Tract of Fishes along the Saudi Arabian Red Sea Coast." Marine Pollution Bulletin 131: 407–15.

Baig, M. B., K. H. Al-Zahrani, F. Schneider, G. S. Straquadine, and M. Mourad. 2019. "Food Waste Posing a Serious Threat to Sustainability in the Kingdom of Saudi Arabia–A Systematic Review." Saudi Journal of Biological Sciences 26 (7): 1743–52.

Bakr, A. 2020. "Saudis May Stop Subsidizing Petrochemicals." Energy Intelligence, June 23. https://www.energyintel.com/0000017b-a7da -de4c-a17b-e7dae8d60000.

Bappenas and UNDP (Ministry of National Planning and Development Indonesia and the United Nations Development Programme). 2021. "The Economic, Social, and Environmental Benefits of a Circular Economy in Indonesia." Report, Kementerian PPN/Bappenas and UNDP, Jakarta.

Barbuscia, D., and A. Ramnarayan. 2020. "Egypt Becomes First Arab Country to Issue Green Bonds with $750 Million Deal." Reuters, September 29.

Barrowclough, D., and C. D. Birkbeck. 2020. "Transforming the Global Plastics Economy: the Political Economy and Governance of Plastics Production and Pollution." Working Paper No. 142, Global Economic Governance Programme, University College and Blavatnik School of Government, University of Oxford.

Barrowclough, D., and R. Kozul-Wright. 2018. "Institutional Geometry of Industrial Policy in Sustainable Development." In Industrial Policy and Sustainable Growth, edited by M. A. Yülek. Singapore: Springer.

Benali, L. R., and R. Al-Ashmawy. 2020. "MENA Gas & Petrochemicals Investment Outlook 2020-24." Report, APICORP (Arab Petroleum Investments Corporation), Dammam, Saudi Arabia.

Boskovic, G., N. Jovicic, S. Jovanovic, and V. Simovic. 2016. "Calculating the Costs of Waste Collection: A Methodological Proposal." Waste Management & Research 34 (8): v775–83.

Boucher, J., and G. Billard. 2020. "The Mediterranean: Mare Plasticum." Report, International Union for Conversation of Nature, Gland, Switzerland.

Boucher, J., and D. Friot. 2017. *Primary Microplastics in the Oceans: A Global Evaluation of Sources.* Gland, Switzerland: International Union for Conservation of Nature.

BP. 2015. "Energy Outlook 2015." https://www.bankofcanada.ca/wp-content/uploads/2015/05/bp-energy-outlook-2035.pdf.

Calabrese, A., R. Costa, N. Levialdi Ghiron, T. Menichini, V. Miscoli, and L. Tiburzi. 2021. "Operating Modes and Cost Burdens for the European Deposit-Refund Systems: A Systematic Approach for their Analysis and Design." *Journal of Cleaner Production* 288: 125600.

Campbell, C. 2016. "Rank Record: Mr. Trash Wheel Gathers 1 Millionth Pound of Trash from Jones Falls." *Baltimore Sun*, October 20.

CEDARE (Center for Environment and Development for the Arab Region and Europe). 2020. "Measures to Address Single-Use Plastic Bags (SUPBs) in Egypt." Policy brief, CEDARE, Cairo.

CIEL (Center for International Environmental Law). 2019. "Plastic & Climate: The Hidden Costs of a Plastic Planet." Washington, DC: CIEL.

Claro, F., M. C. Fossi, C. Ioakeimidis, M. Baini, A. L. Lusher, W. McFee, R. Ruth McIntosh, et al. 2019. "Tools and Constraints in Monitoring Interactions Between Marine Litter and Megafauna: Insights from Case Studies around the World." *Marine Pollution Bulletin* 141: 147–60.

CMFA (California Municipal Finance Authority). 2020. "CMFA Completes the Issuance of $150,000,000 in Bonds for Waste Management." Press release, October 1.

Cogut, A. 2016. "Open Burning of Waste: A Global Health Disaster." R20 Regions of Climate Action.

Convery, F., S. McDonnell, and S. Ferreira. 2007. "The Most Popular Tax in Europe? Lessons from the Irish Plastic Bags Levy." *Environmental and Resource Economics* 38 (1): 1–11.

Cózar, A., F. Echevarría, J. I. González-Gordillo, X. Irigoien, B. Úbeda, S. Hernández-León, Á. T. Palma, et al. 2014. "Plastic Debris in the Open Ocean." *Proceedings of the National Academy of Sciences* 111 (28): 10239–44.

CTBH (CTBH Partners LLC). 2020. "CTBH Partners LLC Announces Successful Issuance of $40.0 Million of Solid Waste Industrial Revenue Bonds for Casella Waste Systems Inc." Press release, September 3.

Dalberg Advisors. 2019a. *No Plastic in Nature: Assessing Plastic Ingestion from Nature to People.* Gland, Switzerland: World Wide Fund for Nature (WWF).

Dalberg Advisors. 2019b. "Stop the Flood of Plastic: A Guide for Policy-Makers in Morocco." Report for the World Wide Fund for Nature (WWF), Gland, Switzerland.

Dalberg Advisors. 2019c. "Stop the Plastic Flood: A Guide for Policy-Makers in Tunisia." Report for the World Wide Fund for Nature (WWF), Gland, Switzerland.

Dalberg Advisors. 2019d. "Stop the Plastic Flood: How Mediterranean Countries Can Save their Sea." Report for the World Wide Fund for Nature (WWF), Gland, Switzerland.

Dalberg Advisors. 2019e. "Plastic Pollution in Greece: How to Stop It. A Guide for Policy-Makers." Report for the World Wide Fund for Nature (WWF), Gland, Switzerland.

D'Angelo, S., and R. Meccariello. 2021. "Microplastics: A Threat for Male Fertility." *International Journal of Environmental Research and Public Health* 18 (5): 2392.

Defra and EA (Department for Environment, Food and Rural Affairs and Environment Agency, Government of the United Kingdom). 2018. "Our Waste, Our Resources: A Strategy for England." Policy paper, Defra and UK Government, London.

Dinan, T. M. 1993. "Economic Efficiency Effects of Alternative Policies for Reducing Waste Disposal." *Journal of Environmental Economics and Management* 25 (3): 242–56.

EC (European Commission). 2018. "A European Strategy for Plastics in a Circular Economy." Strategy document approved January 2018, EC, Brussels.

EC (European Commission). 2020. *Circular Economy Action Plan: For a Cleaner and More Competitive Europe.* Luxembourg: Publications Office of the European Union.

EcoConServ. 2010. "Consultancy for Up Stream Poverty and Social Impact Analysis (PSIA) for Egypt's Solid Waste Management Reform – Final Report." EcoConServ Environmental Solutions. Cairo, Egypt.

ECOELASTIKA. 2014. "Annual Report 2014." ECOELASTIKA, Marousi, Greece.

Egypt Today. 2020. "Electricity Min. to Stop Collecting Cleaning Fees Starting July." *Egypt Today*, January 14. https://www.egypttoday.com /Article/3/79625/Electricity-Min-to-stop-collecting-cleaning-fees -starting-July.

El-Katiri, L., and B. Fattouh. 2017. "A Brief Political Economy of Energy Subsidies in the Middle East and North Africa." In *Combining Economic and Political Development: The Experience of MENA*, edited by G. Luciani, 58–87. Leiden, the Netherlands: Brill.

Elliott, T., H. Xirou, V. Stergiou, A. Bapasola, and H. Gillie. 2020. "Policy Measures on Plastics in Greece: A Report for WWF Greece." Eunomia Research & Consulting. Athens, Greece.

EMAF (Ellen MacArthur Foundation). 2016. "The New Plastics Economy: Rethinking the Future of Plastics." Project MainStream report, EMAF, Cowes, UK.

EMAF (Ellen MacArthur Foundation). 2017. "The New Plastics Economy: Catalysing Action." Report for the New Plastics Economy initiative, EMAF, Cowes, UK.

EMAF (Ellen MacArthur Foundation). 2020. "Financing the Circular Economy: Capturing the Opportunity." Report, EMAF, Cowes, UK.

Emirate of Abu Dhabi. 2020. "Abu Dhabi Emirate Single Use Plastic Policy." Abu Dhabi.

Engel, H., M. Stuchtey, and H. Vanthournout. 2016. "Managing Waste in Emerging Markets." Article, McKinsey Sustainability, McKinsey & Co., Frankfurt, Munich, and Geneva.

EUROMAP (European Plastics and Rubber Machinery). 2016. "Plastics Resin Production and Consumption in 63 Countries Worldwide, 2009–2020." Survey report, EUROMAP General Secretariat, Frankfurt, Germany.

FAO (Food and Agriculture Organization of the United Nations). 2011. "Global Food Losses and Food Waste: Extent, Causes and Prevention." Study for the International Congress, "Save Food!" FAO, Rome.

Fattouh, B., and L. El-Katiri. 2013. "Energy Subsidies in the Middle East and North Africa." *Energy Strategy Reviews:* 2 (1): 108–15.

Galgani, F., G. Hanke, and T. Maes. 2015. "Global Distribution, Composition and Abundance of Marine Litter." In *Marine Anthropogenic Litter*, edited by M. Bergmann, L. Gutow, and M. Klages, 29–56. Cham, Switzerland: Springer.

Gallo, F., C. Fossi, R. Weber, D. Santillo, J. Sousa, I. Ingram, A. Nadal, and D. Romano. 2018. "Marine Litter Plastics and Microplastics and Their Toxic Chemicals Components: The Need for Urgent Preventive Measures." *Environmental Sciences Europe* 30 (1): 1–14.

Geissdoerfer, M., P. Savaget, N. Bocken, and E. Hultink. 2017. "The Circular Economy – A new Sustainability Paradigm?" *Journal of Cleaner Production:* 143 (1): 757–68.

GPCA (Gulf Petrochemicals & Chemicals Association). 2014. "GCC Plastic Industry Indicators 2014." Statistical report, GPCA, Dubai, United Arab Emirates.

GPCA (Gulf Petrochemicals & Chemicals Association). 2016. "GCC Plastic Industry Indicators 2016." Statistical report, GPCA, Dubai, United Arab Emirates.

GPCA (Gulf Petrochemicals & Chemicals Association). 2017. "GPCA Study Reveals Lack of Awareness about Recycling among GCC Public." Press release, April 16.

GPCA (Gulf Petrochemicals & Chemicals Association). 2018. "Sustainable Plastic Innovation: Closing the Loop." Post-event report, GPCA PlastiCon 9th edition, Dubai, United Arab Emirates, March 14–15. https://www.gpca.org.ae/pdfuploads/2018/9th-gpcaplasticon-post-event-report.pdf.

GPCA (Gulf Petrochemicals & Chemicals Association). 2019. "Innovative Plastic Designs: Sustainability for Future Generations. 10th Edition." Post-event report, GPCA PlastiCon, Amwaj Islands, Bahrain, March 11–12.

Gulf News. 2017. "Dubai Municipality to Shift 75 Per Cent of Waste from Landfills." *Gulf News*, February 12. https://gulfnews.com/going-out/society/dubai-municipality-to-shift-75-per-cent-of-waste-from-landfills-1.1977096.

Hamdallah, D. 2020. "Producing Medical Masks and Coveralls in Jordan during the Coronavirus Pandemic." European Bank for Reconstruction and Development (EBRD) News, April 22.

HERRCO (Hellenic Recovery Recycling Corporation). 2015. "Packaging Recycling: A Project for All of Us." Annual report, HERRCO, Athens.

Hicks, R. 2020. "Cheap Virgin Plastic Is Being Sold as Recycled Plastic—It's Time for Better Recycling Certification." Eco-Business, May 28. https://www.eco-business.com/news/cheap-virgin-plastic-is-being-sold-as-recycled-plastic-its-time-for-better-recycling-certification/.

Hogg, D., D. Fletcher, T. Elliot, and M. von Eye. 2010. "Have We Got the Bottle? Implementing a Deposit Refund Scheme in the UK." A report for the Campaign to Protect Rural England, Eunomia Research & Consulting, Bristol, UK.

Hopewell, J., R. Dvorak, and E. Kosior. 2009. "Plastics Recycling: Challenges and Opportunities." *Philosophical Transactions of the Royal Society B: Biological Sciences* 364 (1526): 2115–26.

Houssari, N. 2020. "Lebanon Divided Over Face Masks in Virus Battle." *Arab News*, April 5. https://arab.news/w47um.

Ibrahim, A. A. 2020. "6 Ways that Gulf Cities Can Turn Waste into Wealth." World Economic Forum article, June 17. https://www.weforum.org/agenda/2020/06/6-ways-that-gulf-cities-can-turn-waste-into-wealth/.

Ibrahim, Y. S., S. T. Anuar, A. A. Azmi, W. M. A. W. M. Khalik, S. Lehata, S. R. Hamzah, D. Ismail, et al. 2021. "Detection of Microplastics in Human Colectomy Specimens." *JGH Open* 5 (1): 116–121.

IEA (International Energy Agency). 2018. "The Future of Petrochemicals: Towards More Sustainable Plastics and Fertilisers." Report, IEA, Paris.

Jambeck, J. R., R. Geyer, C. Wilcox, T. R. Siegler, M. Perryman, A. Andrady, R. Narayan, and K. Lavender Law. 2015. "Plastic Waste Inputs from Land into the Ocean." *Science* 347 (6223): 768–71.

Jorgensen, B., M. Krasny, and J. Baztan. 2021. "Volunteer Beach Cleanups: Civic Environmental Stewardship Combating Global Plastic Pollution." *Sustainability Science* 16 (1): 153–67.

Kaandorp, M. L. A., H. A. Dijkstra, and E. van Sebille. 2020. "Closing the Mediterranean Marine Floating Plastic Mass Budget: Inverse Modeling of Sources and Sinks." *Environmental Science & Technology* 54 (19): 11980–89.

Kaza, S., L. Yao, P. Bhada-Tata, and F. Von Woerden. 2018. *What a Waste 2.0: A Global Snapshot of Solid Waste Management to 2050.* Washington, DC: World Bank.

Laglbauer, B. J. L., R. M. Franco-Santos, M. Andreu-Cazenave, L. Brunelli, M. Papadatou, A. Palatinus, M. Grego, and T. Deprez. 2014. "Macrodebris and Microplastics from Beaches in Slovenia." *Marine Pollution Bulletin* 89 (1–2): 356–66.

Lam, C. S., S. Ramanathan, and M. Carbery. 2018. "A Comprehensive Analysis of Plastics and Microplastic Legislation Worldwide." *Water, Air, & Soil Pollution* 229 (11): 345.

Lavee, D. 2010. "A Cost-Benefit Analysis of a Deposit–Refund Program for Beverage Containers in Israel." *Waste Management* 30 (2): 338–45.

Law, K. L., N. Starr, T. R. Siegler, J. R. Jambeck, N. J. Mallos, and G. H. Leonard. 2020. "The United States' Contribution of Plastic Waste to Land and Ocean." *Science Advances* 6 (44): eabd0288.

Lebreton, L., and A. Andrady. 2019. "Future Scenarios of Global Plastic Waste Generation and Disposal." *Palgrave Communications* 5 (1): 1–11.

Lebreton, L., B. Slat, F. Ferrari, B. Sainte-Rose, J. Aitken, R. Marthouse, S. Hajbane, et al. 2018. "Evidence that the Great Pacific Garbage Patch is Rapidly Accumulating Plastic." *Scientific Reports* 8 (1): 1–15.

Lebreton, L., J. van der Zwet, J. W. Damsteeg, B. Slat, A. Andrady, and J. Reisser. 2017. "River Plastic Emissions to the World's Oceans." *Nature Communications* 8: 15611.

Lebreton, L., M. Egger, and B. Slat. 2019. "A Global Mass Budget for Positively Buoyant Macroplastic Debris in the Ocean." *Scientific Reports* 9 (1): 1–10.

Li, K., Y. Yan, J. Yin, Y. Tan, and L. Huang. 2016. "Seasonal Occurrence of *Calanus sinicus* in the Northern South China Sea: A Case Study in Daya Bay." *Journal of Marine Systems* 159: 132–41.

Lim, X.-Z. 2021. "Microplastics Are Everywhere—But Are They Harmful?" *Nature News Feature*: 22–25.

Linderhof, V., F. H. Oosterhuis, P. J. H. Van Beukering, and H. Bartelings. 2019. "Effectiveness of Deposit-Refund Systems for Household Waste in the Netherlands: Applying a Partial Equilibrium Model." *Journal of Environmental Management* 232: 842–50.

Lindo, N. 2020. "Dubai Acts to Stem the Tide of Single Use Plastic." Euronews report, January 24. https://www.euronews.com/next/2020/01/24/dubai -acts-to-stem-the-tide-of-single-use-plastic.

López-Martínez, S., C. Morales-Caselles, J. Kadar, and M. L. Rivas. 2021. "Overview of Global Status of Plastic Presence in Marine Vertebrates." *Global Change Biology* 27 (4): 728–37.

Loukil, F., and L. Rouached. 2012. "Modeling Packaging Waste Policy Instruments and Recycling in the MENA Region." *Resources, Conservation, and Recycling* 69: 141–52.

Mahjoub, O., A. Jemai, and I. Haddaoui. 2020. "Waste Management in Tunisia— What Could the Past Bring to the Future?" In *Waste Management in MENA Regions*, edited by A. M. Negm and N. Shareef, 35–69. Cham, Switzerland: Springer.

Makki, F., A. Lamb, and R. Moukaddem. 2020. "Plastics and the Coronavirus Pandemic: A Behavioral Science Perspective." *Mind & Society* 20 (2): 1–5.

Malek, C. 2020. "How Middle East's Coronavirus Crisis Threatens the Environment Too." Arab News, June 15. https://arab.news/2mv52.

Marazzi, L., S. Loiselle, L. G. Anderson, S. Rocliffe, and D. J. Winton. 2020. "Consumer-Based Actions to Reduce Plastic Pollution in Rivers: A Multi-Criteria Decision Analysis Approach." *PLOS ONE* 15 (8): e0236410.

Market Data Forecast. 2021. "Middle East and Africa Biodegradable Plastics Market, by Type, Application & Region: Industry Analysis on Size, Share, Growth, Trends & Forecast Report, 2021–2026." Research report, Market Data Forecast, Hyderabad, India. https://www.marketdataforecast.com /market-reports/mea-biodegradable-plastics-market.

Martin, C., S. Parkes, Q. Zhang, X. Zhang, M. F. McCabe, and C. M. Duarte. 2018. "Use of Unmanned Aerial Vehicles for Efficient Beach Litter Monitoring." *Marine Pollution Bulletin* 131: 662–73.

Martinho, G., N. Balaia, and A. Pires. 2017. "The Portuguese Plastic Carrier Bag Tax: The Effects on Consumers Behavior." *Waste Management* 61: 3–12.

McElearney, R., and J. Warmington. 2015. "Carrier Bag Charge 'One Year On'." Zero Waste Scotland, Stirling.

Menachery, M. 2020. "The $6bn GCC Plastic and Metal Waste Recycling Opportunity." *Refining & Petrochemicals Middle East*, October 7. https://www .refiningandpetrochemicalsme.com/petrochemicals/29169-the-6bn-gcc -plastic-and-metal-waste-recycling-opportunity.

Moerenhout, T., and T. Irschlinger. 2020. "Exploring the Trade Impacts of Fossil Fuel Subsidies." Global Subsidies Initiative Report, International Institute for Sustainable Development, Winnipeg, Canada.

Mostafa, N. 2020. "Logistics of Waste Management with Perspectives from Egypt." In *Waste Management in MENA Regions*, edited by A. M. Negm and N. Shareef, 171–91. Cham, Switzerland: Springer.

Munari, C., C. Corbau, U. Simeoni, and M. Mistri. 2016. "Marine Litter on Mediterranean Shores: Analysis of Composition, Spatial Distribution and Sources in North-Western Adriatic Beaches." *Waste Management* 49: 483–90.

Muscat Daily. 2018. "88% Vote to Regulate Use of Plastic Bags in Malls, Markets, Says Ministry of Environment and Climate Affairs." *Muscat Daily*, July 13.

Nachite, D., F. Maziane, G. Anfuso, and A. T. Williams. 2019. "Spatial and Temporal Variations of Litter at the Mediterranean Beaches of Morocco Mainly Due to Beach Users." *Ocean & Coastal Management* 179: 104846.

Naji, A., Z. Esmaili, and F. R. Khan. 2017. "Plastic Debris and Microplastics along the Beaches of the Strait of Hormuz, Persian Gulf." *Marine Pollution Bulletin:* 114 (2): 1057–62.

Nor, M., N. Hazimah, M. Kooi, N. J. Diepens, and A. A. Koelmans. 2021. "Lifetime Accumulation of Microplastic in Children and Adults." *Environmental Science & Technology* 55 (8): 5084–96.

OECD (Organisation for Economic Co-Operation and Development). 2015. "Creating Incentives for Greener Products: A Policy Manual for Eastern Partnership Countries." OECD Green Growth Studies. Paris: OECD.

OSPAR (OSPAR Commission). 2020. *OSPAR Scoping Study on Best Practices for the Design and Recycling of Fishing Gear as a Means to Reduce Quantities of Fishing Gear Found as Marine Litter in the North-East Atlantic.* London: OSPAR Commission.

Pasternak, G., D. Zviely, C. A. Ribic, A. Ariel, and E. Spanier. 2017. "Sources, Composition and Spatial Distribution of Marine Debris along the Mediterranean Coast of Israel." *Marine Pollution Bulletin* 114 (2): 1036–45.

Pew Charitable Trusts and SYSTEMIQ. 2020. "Breaking the Plastic Wave: A Comprehensive Assessment of Pathways Towards Stopping Ocean Plastic Pollution." Report, The Pew Charitable Trusts, Washington, DC.

Plastics Europe. 2015. "Plastics – The Facts 2015: An Analysis of European Plastics Production, Demand and Waste Data." Report, Plastics Europe, Brussels.

Portman, M. E., and R. E. Brennan. 2017. "Marine Litter from Beach-Based Sources: Case Study of an Eastern Mediterranean Coastal Town." *Waste Management* 69: 535–44.

Portman, M. E., G. Pasternak, Y. Yotam, R. Nusbaum, and D. Behar. 2019 "Beachgoer Participation in Prevention of Marine Litter: Using Design for Behavior Change." *Marine Pollution Bulletin* 144: 1–10.

Powell, D. 2018. "The Price Is Right… Or Is It? – The Case for Taxing Plastic." Paper by the New Economics Foundation for Zero Waste Europe, Ixelle, Belgium.

Ragusa, A., A. Svelato, C. Santacroce, P. Catalano, V. Notarstefano, O. Carnevali, F. Papa, et al. 2021. "Plasticenta: First Evidence of Microplastics in Human Placenta." *Environment International* 146: 106274.

Randone, M., G. Di Carlo, and M. Constantini. 2017. "Reviving the Economy of the Mediterranean Sea: Actions for a Sustainable Future." Report, World Wide Fund for Nature (WWF) Mediterranean Marine Initiative, Rome.

Reloop. 2020. "Global Deposit Book 2020: An Overview of Deposit Systems for One-Way Beverage Containers." Report, Reloop, Brussels.

Ribeiro-Broomhead, J., and N. Tangri. 2021. "Zero Waste and Economic Recovery: The Job Creation Potential of Zero Waste Solutions." Report, Global Alliance for Incinerator Alternatives (GAIA), Berkeley, CA.

Roman, L., F. Kastury, S. Petit, R. Aleman, C. Wilcox, B. D. Hardesty, and M. A. Hindell. 2020. "Plastic, Nutrition and Pollution; Relationships between Ingested Plastic and Metal Concentrations in the Livers of Two *Pachyptila* Seabirds." *Scientific Reports* 10 (1): 1–14.

Rubeis, M., M. Jędrzejewski, A. Rajamani, and U. Jung. 2016. "Why the Middle East's Petrochemical Industry Needs to Reinvent Itself." Report, Boston Consulting Group, Boston, MA.

Ruth, John. 2017. "Doing More with Dollars." *Waste Today*, April 6.

SABIC (Saudi Basic Industries Corporation). 2019. "SABIC First in Industry to Launch Polycarbonate Based on Certified Renewable Feedstock." Press release, October 17.

Savannah Riverkeeper. 2018. "'Watergoat' Trash Traps Helping Curb Litter in Augusta." *The Blue Heron* (blog), November 30. https://www .savannahriverkeeper.org/blue-heron-blog/watergoat-trash-traps-helping -curb-litter-in-augusta#/.

Scheinberg, A., and R. A. Savain. 2015. *Valuing Informal Integration: Inclusive Recycling in North Africa and the Middle East.* Bonn: German Agency for International Cooperation (GIZ).

Schlüter, B. A., and M. B. Rosano. 2016. "A Holistic Approach to Energy Efficiency Assessment in Plastic Processing." *Journal of Cleaner Production* 11: 19–28.

Schmaltz, E., E. C. Melvin, Z. Diana, E. F. Gunady, D. Rittschof, J. A. Somarelli, J. Virdin, and M. M. Dunphy-Daly. 2020. "Plastic Pollution Solutions: Emerging Technologies to Prevent and Collect Marine Plastic Pollution." *Environment International* 144: 106067.

Senathirajah, K., and T. Palanisami. 2019. "How Much Microplastics Are We Ingesting?: Estimation of the Mass of Microplastics Ingested." Unpublished paper, University of Newcastle, Australia.

Shen, L., and E. Worrell. 2014. "Plastic Recycling." In *Handbook of Recycling: State-of-the-Art for Practitioners, Analysts, and Scientists,* edited by E. Worrell and M. A. Reuter, 179–90. Waltham, MA: Elsevier.

Skovgaard, J., and H. van Asselt. 2019. "The Politics of Fossil Fuel Subsidies and their Reform: Implications for Climate Change Mitigation." *Reviews: Climate Change:* 10 (4): 1–12.

Staub, C. 2020. "Low Virgin Plastics Pricing Pinches Recycling Market Further." *Plastics Recycling Update*, May 6. https://resource-recycling.com /plastics/2020/05/06/low-virgin-plastics-pricing-pinches-recycling-market -further/.

SwitchMed. n.d. "Jordan Country Hub." SwitchMed Programme of the European Union, Barcelona. https://switchmed.eu/country-hub/jordan/.

Thiel, M., I. A. Hinojosa, L. Miranda, J. F. Pantoja, M. M. Rivadeneira, and N. Vásquez. 2013. "Anthropogenic Marine Debris in the Coastal Environment: A Multi-Year Comparison between Coastal Waters and Local Shores." *Marine Pollution Bulletin* 71 (1–2): 307–16.

Tobin, D. 2012. Pricing Reforms and Capacity Constraints in China's Petrochemical Sector." *Oxford Energy Forum:* 88: 15–16.

Tsakana, M., and I. Rucevska. 2020. "Baseline Report on Plastic Waste." Report commissioned by the Secretariat of the Basel Convention for the first meeting of the Basel Convention Plastic Waste Partnership, Beau Vallon, Seychelles, March 2–5. United Nations Environment Programme (UNEP), Nairobi.

Uddin, S., S. W. Fowler, and T. Saeed. 2020. "Microplastic Particles in the Persian/Arabian Gulf–A Review on Sampling and Identification." *Marine Pollution Bulletin* 154: 111100.

UNEP (United Nations Environment Programme). 2015. *Marine Litter Assessment in the Mediterranean.* Athens: UNEP.

UNEP (United Nations Environment Programme). 2016. *Marine Plastic Debris & Microplastics: Global Lessons and Research to Inspire Action and Guide Policy Change.* Nairobi: UNEP.

UNEP (United Nations Environment Programme). 2018a. "'One Dead Sea Is Enough': Jordan Takes Action on Marine Litter." News story, March 13. UNEP, Nairobi.

UNEP (United Nations Environment Programme). 2018b. *Single-Use Plastics: A Roadmap for Sustainability.* Nairobi: UNEP.

UNEP/MAP and Plan Bleu (United Nations Environment Programme/ Mediterranean Action Plan and Plan Bleu). 2020. *State of the Environment and Development in the Mediterranean.* Nairobi: UNEP.

van den Oever, M., K. Molenveld, M. van der Zee, and H. Bos. 2017. "Bio-Based and Biodegradable Plastics: Facts and Figures: Focus on Food Packaging in the Netherlands." Wageningen Food & Biobased Research Working Paper 1722, Wageningen University.

van Sebille, E., C. Wilcox, L. Lebreton, N. Maximenko, B. D. Hardesty, J. A. van Franeker, M. Eriksen, D. Siegel, F. Galgani, and K. L. Law. 2015. "A Global Inventory of Small Floating Plastic Debris." *Environmental Research Letters* 10 (12): 124006.

Veiga, J. M., T. Vlachogianni, S. Pahl, R. C. Thompson, K. Kopke, T. K. Doyle, B. L. Hartley, et al. 2016. "Enhancing Public Awareness and Promoting Co-Responsibility for Marine Litter in Europe: The Challenge of MARLISCO." *Marine Pollution Bulletin* 102 (2): 309–15.

Verisk Maplecroft. 2020. "Recycling Index." Database, Verisk Maplecroft Limited, Bath, UK.

Verma, S. 2020. "Ahmedabad Issues Municipal Bonds to Implement Green Projects." Case Study for the Climate and Development Knowledge Network (CDKN), ICLEI South Asia, New Delhi.

Vié, J.-C., C. Hilton-Taylor, and S. N. Stuart, eds. 2009. *Wildlife in a Changing World: An Analysis of the 2008 IUCN Red List of Threatened Species.* Gland, Switzerland: International Union for Conversation of Nature (IUCN).

WAM (Emirates News Agency). 2018. "Dubai Municipality Promotes Smart Sustainability Oasis Project at Dubai Customs." WAM, March 1. http://wam.ae/en/details/1395302657974.

WAM (Emirates News Agency). 2020. "Dubai Sharpens Focus on Circular Economy in Preparation for the Next 50 Years." WAM, December 22. https://www.wam.ae/en/details/1395302897217.

World Bank. 2021a. "Market Study for Malaysia: Plastics Circularity Opportunities and Barriers." Marine Plastics Series, East Asia and Pacific Region.

World Bank. 2021b. "Market Study for Thailand: Plastics Circularity Opportunities and Barriers." Marine Plastics Series, East Asia and Pacific Region.

World Bank. 2021c. "Market Study for the Philippines: Plastics Circularity Opportunities and Barriers." Marine Plastics Series, East Asia and Pacific Region.

World Bank. 2021d. "Stratégie du Maroc 'Littoral Sans Plastique' LISP: Réduction de la Pollution Marine par le Plastique et Promotion des Approches de l'économie Circulaire." World Bank, Washington, DC.

World Bank. 2021e. "Stratégie du Maroc 'Littoral Sans Plastique' LISP: Réduction de la Pollution Plastique Marine et Promotion des Approches de l'économie Circulaire. Rapport de Benchmark International. Version Définitive." World Bank, Washington, DC.

World Bank. 2021f. "Stratégie de la Tunisie 'Littoral Sans Plastique' LISP: Diadnostic de la Situation et èbauche de plan d'action." Unpublished manuscript, World Bank, Washington, DC.

World Bank and IMF (International Monetary Fund). 2021. "From COVID-19 Crisis Response to Resilient Recovery: Saving Lives and Livelihoods while Supporting Green, Resilient and Inclusive Development (GRID)." Document No. DC2021-0004 for the April 9, 2021, Meeting of the Development Committee (Joint Ministerial Committee of the Boards of Governors of the Bank and the Fund on the Transfer of Real Resources to Developing Countries), Washington, DC. https://www.devcommittee.org/sites/dc/files/download/Documents/2021-03/DC2021-0004%20Green%20Resilient%20final.pdf.

Wright, S., and F. Kelly. 2017. "Plastic and Human Health: A Micro Issue?" *Environmental Science and Technology* 51 (12): 6634–47.

YouGov. 2018. "Are UAE Residents Aware of Plastic Legislations in the Country?" Survey report, November 6, YouGov, London.

Zhang, Q., E. G. Xu, J. Li, Q. Chen, L. Ma, E. Y. Zeng, and H. Shi. 2020. "A Review of Microplastics in Table Salt, Drinking Water, and Air: Direct Human Exposure." *Environmental Science & Technology* 54 (7): 3740–51.

Zhou, G., Y. Gu, Y. Wu, Y. Gong, X. Mu, H. Han, and T. Chang. 2020. "A Systematic Review of the Deposit-Refund System for Beverage Packaging: Operating Mode, Key Parameter and Development Trend." *Journal of Cleaner Production* 251: 119660.

CHAPTER 5

Blue Seas: Fighting Coastal Erosion

OVERVIEW

Beaches are retreating and coastal areas are eroding fast in parts of the Middle East and North Africa. Especially in the Maghreb subregion, they are already in an advanced stage of degradation, having eroded at alarming rates in recent decades. With uncontrolled human development and the intensification of climate change impacts, pressures are likely to increase. Managing coasts sustainably is a critical part of the green, resilient, and inclusive development (GRID) paradigm for the region's economies to adopt, because most of their people live in coastal areas—and many, especially in the poorer strata of society, depend on intact coasts for their livelihoods.

Coastal erosion in the region already entails substantial costs, as this chapter shows in an economic quantification exercise for four Maghreb countries.[1] Those estimated direct costs are conservative since they incorporate costs only for lost land and destroyed buildings and do not take into account the forgone revenues from tourism or fishing activities or damages to marine infrastructure and ecosystems that are important for the region's biodiversity. These indirect costs are most likely significant, especially for economies heavily dependent on tourism, because tourists are less willing to return when beaches are gradually disappearing. Nonetheless, the estimated annual costs due to losses in land and destroyed buildings alone already amount to large sums, in both absolute terms and as shares of annual economic output.

The reasons for eroding beaches are manifold, but knowledge is often sparse about specific drivers of coastal erosion in the region's hot spots.

This chapter discusses the general drivers of coastal erosion; however, comprehensive and granular assessments of the Middle East and North Africa's coasts are limited, which undermines the effective control of erosive processes. The next section presents findings from two novel such analyses of the coasts of Morocco and Tunisia, conducted for this report in cooperation with the National Oceanography Centre (NOC) of the United Kingdom. However, even here, detailed studies on the balance between sediment added to and removed from the coastal system and its transport, prevalent wave dynamics, and the possible effects of control measures would be necessary to effectively combat the disappearance of the region's beaches.

While most measures for combating coastal erosion require knowledge about local characteristics, starting a comprehensive integrated coastal zone management (ICZM) scheme is a crucial step to ensure sustainable development of the region's coasts in the future by avoiding coastal erosion where possible and mitigating erosion where it is not. Such an initiation includes identifying and involving all relevant stakeholders; increasing capacity for data, monitoring, and analysis; and planning prospective and reactive measures. The inclusion of all parties interested in sustainable coastal development—not only the local population and environmental nongovernmental organizations (NGOs) but also public authorities and private sector players—is crucial and also underlines the importance of managing coasts for development within a GRID framework.

A key part of ICZM is segmenting the coast into different zones for which different development policy strategies ought to be formulated, making the inclusion of different interests possible. Another priority recommendation is the increased use of nature-based solutions (NBS) like mangroves or marshes to protect existing human structures. NBS solutions, as opposed to other possibilities, have the advantage of conserving or restoring biodiversity and habitats for flora and fauna while also offering job opportunities for the local population, irrespective of gender, with knowledge of the local characteristics, in line with a GRID framework.

The chapter concludes with a summary of various other potential ways to combat coastal erosion under the tenets of an ICZM scheme. International and regional examples illustrate the different possibilities for defending the coast and its assets. In addition to NBS, these include "hard-defense" structures such as groins or seawalls and "soft-defense" measures such as beach nourishment—adding sand to eroding shorelines. Finally, it discusses some regulations regarding dam construction and sand mining.

Targeted policies that include suitable defense strategies and regulations in a comprehensive scheme for managing the Middle East and

North Africa's coasts are necessary to protect these assets so future generations can benefit from intact beaches. Such policies are crucial for the economic prosperity of countries, the integrity of their natural habitats and biodiversity, and the resilience of coastal assets and communities to the challenges ahead posed by climate change.

HOW ERODED IS THE COAST?

The coastlines of the Middle East and North Africa, which stretch for nearly 25,000 kilometers, are central to the region's economy and history. These coastal and marine environments vary widely.

The Red Sea has an average depth of nearly 1,500 meters and a maximum depth of 2,600 meters.[2] It has a rich marine life and many shorelines that are still sparsely populated except in its far northern reaches and near the cities of Jeddah and Yanbu, Saudi Arabia; Port Sudan, Sudan; and Al-Hodeidah, Republic of Yemen. In contrast, the inner RSA—that is, the Regional Organization for the Protection of the Marine Environment (ROPME) Sea Area, which extends over 1,000 kilometers from the Strait of Hormuz to the northern coast of the Islamic Republic of Iran—has an average depth of only 35 meters. The inner RSA is characterized by heavy ship traffic and several large, rapidly growing, and densely populated cities. The Mediterranean Sea has an average depth of 1,500 meters with a relatively narrow continental shelf across its southern and eastern shores. The western and eastern Mediterranean coasts have been populated for millennia and include some of the world's most ancient cities, while the North Atlantic has Morocco's largest city on its shoreline.

Although none of the Middle East and North Africa economies is landlocked, the lengths of their coastlines vary considerably. Jordan has less than 30 kilometers of shoreline on the Gulf of Aqaba, which provides it with access to the Red Sea, whereas Saudi Arabia has more than 7,500 kilometers of shoreline; the Arab Republic of Egypt, 2,900 kilometers; and the Islamic Republic of Iran, 2,800 kilometers.

Significance of the Coasts to the Region's Populations and Economies

Coasts are home to large parts of the Middle East and North Africa's population. Almost 50 percent of the region's population lived within 100 kilometers of the coast in 2020. The total coastal population is expected to increase by about 18 percent by 2035 (Maul and Duedall 2019). In 13 of the region's economies, over 60 percent of the population lives within

100 kilometers of the coast, ranging from around 6 percent in Iraq to over 30 percent in Saudi Arabia, more than 80 percent in Oman and Tunisia, and 100 percent in Kuwait, Lebanon, Malta, and Qatar.

Moreover, 10 of the region's capital cities are located on or near coastlines. Other major cities and economic centers on the coasts include Alexandria in Egypt and Jeddah in Saudi Arabia, but smaller towns and villages cluster along the coast as well. Population growth has increased pressure on the natural environment of the coasts and coastal wetlands as well as on urban water and wastewater systems and marine-water quality.

Given the concentration of settlements, the Middle East and North Africa's coasts are centers for economic activity, as in the following examples:

- *In Morocco*, two-thirds of the population and 90 percent of industries are located on or near the coast, and coastal tourism, largely dependent on intact beaches, is a major contributor to the national economy (Snoussi et al. 2009).

- *In Tunisia*, more than 83 percent of industrial firms occupy specialized industrial zones along the Sahel coast (Tunisia's central coast), and 90 percent of the country's total economic output is achieved in near-shore areas (Albrecht-Heider 2020).

- *In Djibouti*, the city-state's US$2 billion economy is driven by a state-of-the-art port complex that serves as the principal port of entry to Ethiopia. Given its location at the southern entrance to the Red Sea, it hosts military bases and supports global antipiracy efforts.

- *In the Gulf Cooperation Council (GCC) countries*, several city-states are dependent for their existence on their coastal locations (Tolba and Saab 2009).

- *In Egypt*, the Mediterranean city of Alexandria hosts about 40 percent of the country's industrial capacity and is an important summer resort (El-Raey 2010).

Moreover, regional transportation hubs such as ports and other economic centers occupy low-lying coastal areas throughout the Middle East and North Africa (Schäfer 2013)—many of which are vulnerable to sea level rise (SLR) and coastal erosion.

In some of the region's economies, much of the population depends on marine resources that are threatened by further environmental degradation. The fishing industry is economically significant in several, including Morocco, Oman, and the Republic of Yemen. In Morocco, for instance, fisheries and connected industries contributed around 2.3 percent to national gross domestic product (GDP) and created employment for almost 700,000 people (directly and indirectly) in 2014 (FAO 2020). Morocco is also developing its aquaculture industry. Its fish production in capture fisheries dominates the region, representing about 40 percent of more than 4 million tons of fish landed every year in the Middle East and North Africa (OECD and FAO 2018). Furthermore, coastal areas are important to the livelihoods of hundreds of thousands of people in the region—many of them among the vulnerable and poor—who work in small-scale fisheries and aquaculture, both of which have been increasing substantially in recent years (OECD and FAO 2018; Sieghart, Mizener, and Gibson 2019).

Tourism is a major pillar of the economy in many of the region's economies, accounting for a substantial portion of GDP in several—and in some, adding up to well over 10 percent of economic output. For example, in Morocco, tourism and connected value chains contributed almost 19 percent of GDP, with employment in these sectors making up 19.6 percent of total employment in 2017 (Kasmi et al. 2020). Similarly, in Tunisia, the tourism sector offered jobs for around 2.3 million people, and economic activity amounted to 14.2 percent of GDP in 2018 (Saidani 2019). In recent years, many of the region's economies, including Egypt, Jordan, and Tunisia, have experienced rebounds from previous slumps in tourist arrivals following political incidents. However, the COVID-19 crisis led to sharp reductions in tourist numbers throughout the Middle East and North Africa (box 5.1). It is important that the tourism sector—which is also dependent on intact, clean coasts—emerge from this crisis and recover.

For several of these countries, intact beaches and coastlines are vital to their attractiveness to tourists. For example, Djerba Island off the coast of Tunisia accounts for about one-fourth of all of Tunisia's international tourist arrivals (Carboni, Perelli, and Sistu 2014), and tourism in Tunisia as a whole is focused strongly on coastal areas (Widz and Brzezińska-Wójcik 2020). Similarly, in Egypt, coastal tourism has grown significantly in recent decades, bringing economic benefits to host communities but also increasing environmental pressures (Abdel-Latif, Ramadan, and Galal 2012). Losing the natural assets represented by beaches can severely harm the tourism sector by diminishing the number of tourists or threatening tourism-related infrastructure.

Coastal Erosion and Beach Loss: Global Comparisons

Coastal erosion and beach loss are global phenomena with regional hot spots, including the southern Mediterranean. Historically, significant shares of sandy beach areas around the world have eroded (lost area), while other have accreted (gained area) or stayed stable (Luijendijk et al. 2018).

BOX 5.1

Tourism in the Middle East and North Africa and the Impact of COVID-19

Tourism is one of the Middle East and North Africa region's most important economic and employment sectors. Before the COVID-19 pandemic hit, the tourism sector was growing. Tourists are attracted to the region's spectacular landscapes and beaches, its cultural heritage, its entertainment and shopping opportunities, and its mild winter climate. Ecotourism is also a growing area, while religious tourism is important especially in the Islamic Republic of Iran and Saudi Arabia. Oil exporting countries such as Oman and Saudi Arabia are recognizing the potential of increasing tourism to diversify their economies and are investing heavily to increase their attractiveness for international tourists.

According to the United Nations World Tourism Organization, almost 90 million international tourist arrivals were recorded in the Middle East and North Africa in 2018—about 6 percent of the world's total arrivals and about 10 percent more than in 2017 (UNWTO 2019). In the North African countries of Morocco and Tunisia, which rely to a large extent on beach tourism, solid growth was recorded: In 2018, Tunisia's tourism arrivals experienced double-digit growth. Morocco increased its arrivals by around 8 percent, exceeding 12 million visitors (UNWTO 2019). In 2019, the tourism sector also showed strength elsewhere in the region, as follows:[a]

- Tourism accounted for over 45 percent of export revenues in Lebanon and over 40 percent in Jordan.

- More than 20 million tourists visited the United Arab Emirates and Saudi Arabia.

- Between 10 million and 15 million tourists visited Bahrain, the Arab Republic of Egypt, and Morocco.

- Between 5 million and 10 million tourists visited the Islamic Republic of Iran, Kuwait, Lebanon, and Tunisia.

Domestic tourism was also growing before the pandemic. The sector and its value chains accounted for about 20 percent of employment and gross domestic product (GDP) in Jordan and Lebanon in 2019; over 15 percent in Morocco; nearly 15 percent in Tunisia; and nearly 10 percent in Bahrain, Egypt, Qatar, Saudi Arabia, and the United Arab Emirates. Tourist arrivals are also highly affected by security and geopolitical concerns, but by 2019, tourist arrivals had largely recovered following several terrorist attacks on tourists in Egypt and Tunisia between 2014 and 2016.

The COVID-19 pandemic has had severe negative impacts on the Middle East and North Africa's tourism sector. In 2020, international tourist arrivals globally decreased by more than 74 percent, according to UNWTO data. Regional impacts were as follows:[b]

- In April 2020, international tourist arrivals in the Middle East were down 90 percent compared with April 2019 and, for the year as a whole, they were 76 percent lower than in 2019.

(continued)

BOX 5.1

Tourism in the Middle East and North Africa and the Impact of COVID-19 (*Continued*)

- In Tunisia, tourism revenues dropped by around 60 percent in the first three quarters of 2020 compared with 2019, with almost 80 percent fewer international visitors arriving in 2020. In April to June 2020, the reduction in visitors reached almost 100 percent.

- In Egypt, tourism receipts in the first two quarters of 2020 were only one-third of those in 2019, with reductions in tourist arrivals reaching almost 100 percent in April to August 2020.

- In Morocco, international tourism receipts were down 93 percent in the third quarter of 2020 compared with 2019.

- In Algeria, the tourism receipts for the second quarter were down 82 percent compared with 2019.

- Similarly, receipts and arrivals of international tourists in Bahrain, Kuwait, Oman, Qatar, Saudi Arabia, and the United Arab Emirates have been hit hard by the COVID-19 crisis, almost obliterating international tourism in the months after the pandemic started in the GCC countries.

With the onset of second and third waves of COVID-19 in the region and around the world, the tourism sector is expected to begin making a real recovery only after the pandemic is contained and the countries are able to welcome tourists again.

a. Tourism data are from the World Development Indicators database.
b. Data on COVID-19 impact from the World Tourism Organization of the United Nations (UNWTO), "Global Tourism Dashboard," https://www.unwto.org/international-tourism-and-covid-19.

Map 5.1 illustrates these processes in some of the most affected areas between 1984 and 2016, displaying severe coastal erosion as red dots and accretion as green dots. The areas shown on the southern Mediterranean shore are mostly eroding.

The Maghreb subregion (Algeria, Libya, Morocco, and Tunisia) is the second fastest coastally eroding area in the world (by 15 centimeters per year), exceeded by only South Asia (86 centimeters per year), as shown in figure 5.1. The Mashreq subregion's shorelines also retreated, albeit more slowly (around 7 centimeters per year). In contrast, the GCC's coasts accreted substantially, by almost 70 centimeters per year, owing in part to large-scale coastal reclamation and development projects.

MAP 5.1

Average Annual Erosion and Accretion of Selected Beaches Worldwide, 1984–2016

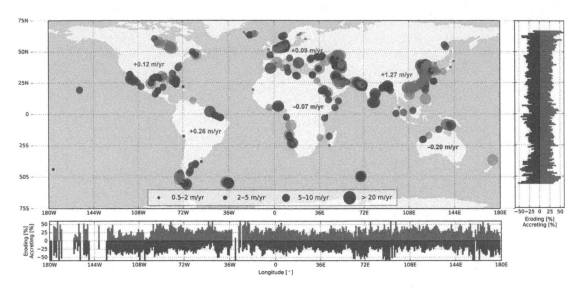

Source: Luijendijk et al. 2018. Map available under Creative Commons Attribution license (CC BY 4.0).
Note: The selected beaches are those considered "fast changing"—that is, showing more than 0.5 meters (m) accretion or erosion per year.
Red dots designate beach areas that display coastal erosion (lost area), and green dots those that have accreted (gained area). Ancillary graphs
show the annual percentages of eroding and accreting beach areas by longitude (bottom graph) and latitude (right graph). m/yr = meters
per year.

Coastal Erosion and Beach Loss: Intraregional Variations

Coastal erosion varies substantially across the Middle East and North Africa and even among the Maghreb countries, with Tunisia facing the highest rates of coastal erosion. For example, Libya, Morocco, and Tunisia face net erosion of their beaches, while in Algeria accretion dominates (figure 5.2). GCC countries such as Bahrain, Oman, Qatar, and the United Arab Emirates have experienced accretion, owing largely to land reclamation and coastal development projects (Luijendijk et al. 2018), while Saudi Arabia's vast coastline has been rather stable. The coasts of two of the poorest countries in the region, Djibouti and the Republic of Yemen, have been retreating on average by 30–50 centimeters per year. Overall, Tunisia faced the highest net rates of coastal erosion.

At the site (hot spot) level, coastal erosion varies significantly. The dynamics of erosive processes and beach retreat can vary significantly for individually affected beaches. For instance, photo 5.1 shows shoreline changes between Chekka and El Heri in Lebanon. Most of the shoreline has experienced severe erosion, with a maximum retreat of 81 meters

FIGURE 5.1

Average Annual Net Coastal Accretion or Erosion, Global Regions and Middle East and North Africa Subregions, 1984–2016

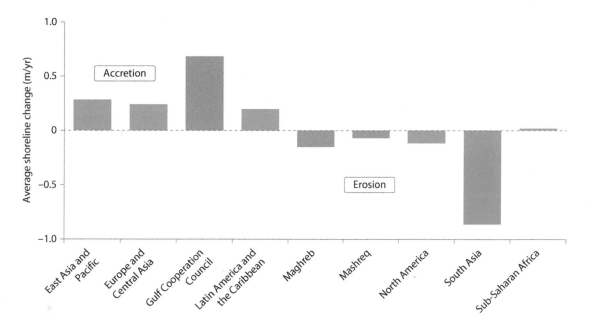

Source: Based on Luijendijk et al. 2018.
Note: Positive values indicate accretion, and negative values, erosion. Orange bars designate the Middle East and North Africa region or its subregions. Middle East and North Africa subregions are as follows: (a) Maghreb, including Algeria, Libya, Malta, Morocco, and Tunisia; (b) Mashreq, including Djibouti, the Arab Republic of Egypt, the Islamic Republic of Iran, Iraq, Jordan, Lebanon, the Syrian Arab Republic, West Bank and Gaza, and the Republic of Yemen; and (c) Gulf Cooperation Council, including Bahrain, Kuwait, Oman, Qatar, Saudi Arabia, and the United Arab Emirates. "North America" includes Canada and the United States. m/yr = meters per year.

between 1962 and 2007, totaling 79 hectares over the years (Abou-Dagher, Nader, and El Indary 2012). Reasons for the erosion of this part of the shoreline are large-scale extraction of sediments (sand and pebble) and the influence of the jetty that was added to service one of the major cement factories near the coast.

Other parts of the beach have been accreting during this period, mostly because of blocked sediment transport caused by the jetty built perpendicular to the sandy beach. Around the connection of the jetty to the mainland, accretion reached a maximum of 94 meters, and the accreted area amounted to around 25 hectares during that time, illustrating the effect that human intervention can have on beach erosion.

Among the most eroded areas in Lebanon is the Akkar shoreline north of Tripoli, which is 13 kilometers long and has been subject to high erosion. With beach retreats of up to 150 meters and an area of 635 hectares lost to coastal erosion, it accounts for 68 percent of the

FIGURE 5.2

Average Annual Net Coastal Accretion or Erosion in the Middle East and North Africa, by Economy, 1984–2016

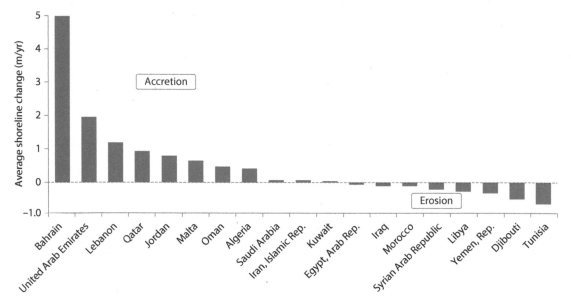

Source: Based on Luijendijk et al. 2018.
Note: Positive values indicate accretion, and negative values, erosion. Data on West Bank and Gaza are unavailable. m/yr = meters per year.

total eroded coastal area in northern Lebanon (Abou-Dagher, Nader, and El Indary 2012).

Close Analyses of Tunisian and Moroccan Shorelines

Small-scale differences are important; hence higher-resolution analyses of coastlines are preferable. Using the data of Luijendijk et al. (2018), the average shoreline retreat of sandy beaches was calculated to be 12 centimeters per year on the Atlantic coast of Morocco, 14 centimeters on the Mediterranean coast of Morocco, and 70 centimeters on the coast of Tunisia (Heger and Vashold 2021). However, these aggregate national indicators hide heterogeneity within countries, where some local areas are eroding as others are accreting.

To better understand the erosion rates in Morocco and Tunisia—whose coasts are threatened by erosion and where beach tourism and coast-related activities play a large role in the economic mix—the authors of this report partnered with NOC of the United Kingdom and the European Space Agency (ESA) to better understand the distribution of erosion along the coast. To carry out a more detailed analysis in

PHOTO 5.1

Total Shoreline Accretion or Erosion of Chekka and El Heri, Lebanon, 1962–2007

Source: Nader 2015. © Manal Nader. Used with the permission of Manal Nader. Further permission required for reuse.
Note: "Sea filling" (yellow area) designates a jetty built to service a cement factory near the coast. km = kilometers ; m² = square meters.

these two countries, only photos made by satellites launched from 2000 onward were used to divide the coasts into finer segments, allowing for more precise interpretations.

Coastal Changes in Tunisia

Shoreline changes vary greatly, even in Tunisia, where 85 percent of the coastline is identified as sandy. The fastest-accreting areas are along the coasts of the Sfax, Gabés, and Médenine Governorates (map 5.2). Accretion rates of 2 meters per year or more are occurring in 13 governorates. Intensive erosion is concentrated in seven main areas, exceeding 2 meters per year in Utiquere in Bizerte and Korba in Nabeul. Overall, there is more erosion in the north and more accretion in the south.

More than one-third of Tunisia's sandy beaches (about 35 percent) are eroding at a rate of more than 0.5 meters per year, with some eroding by several meters per year. One declared erosion hot spot is Hammamet Bay along the Mediterranean coast, south of the capital in northeast Tunisia. Coastal erosion of Hammamet's beach resulted in the loss of 24,000 square meters of beach area in only 13 years (2006–19), at a rate of 3–8 meters per year, as shown in photo 5.2 (Heger and Vashold 2021). This is mostly caused by rapid urbanization on the coast of Hammamet, hindering the natural sediment flow to the shoreline (Amrouni, Hzami, and Heggy 2019). Urbanization, coastal erosion, and associated vegetation loss have also exposed aquifers to seawater intrusion and salinization.

Coastal Changes in Morocco

Coastal erosion and accretion processes in Morocco also vary widely. On Morocco's Mediterranean coast, intensive accretion is occurring in Fahs Anjra and Tétouan Provinces (map 5.3). Developments such as North Africa's largest port, Tanger-Med—a strip of development 1.6 square kilometers long on the Strait of Gibraltar, at Morocco's northern tip—are likely to significantly affect shoreline-change rates. Other sources of change relate to the hydrological cycle and the rates of river-flow deposition at estuarine locations.

Farther east along Morocco's Mediterranean coast, erosion processes become more dominant, particularly on either side of Driouch Province in the Port of Al Hoceima Bay and surrounding the Nador West Med Port project. There are severe accretion rates within the lagoon at Nador, whereas the outer coast of Nador has intensive erosion. Numerous studies on the threat of SLR to the Moroccan coastline have been carried out (Kasmi et al. 2020; Snoussi, Ouchani, and Niazi 2008; Snoussi et al. 2009).

MAP 5.2

Annual Average Coastal Erosion and Accretion in Tunisia, 2000–20

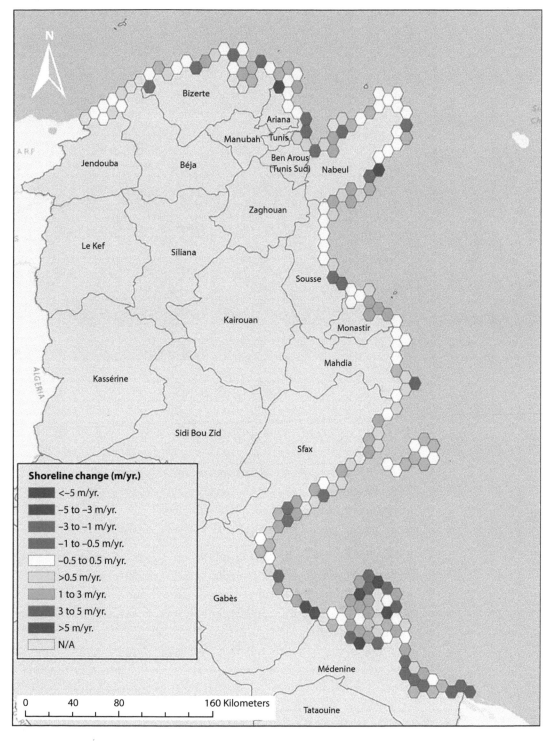

Source: NOC 2020. © World Bank. Further permission required for reuse.
Note: Tunisia shoreline-change rates are aggregated within areas of about 50 square kilometers, designated by the hexagons. m/yr. = meters per year; N/A = not applicable.

PHOTO 5.2

Coastal Erosion at Hammamet Beach, Tunisia, 2006 vs. 2019

Source: © 2020 CNES/Airbus, courtesy of Google Earth.
Note: Shades superimposed on the satellite image show the beach area in 2019 (orange) and the beach area in 2006 (light green).

THE ECONOMIC IMPACTS OF ERODED COASTS

If no measures to combat erosion are undertaken, sandy beaches will inevitably be lost, with cascading effects on the economy and the well-being of the local populations, particularly those dependent on tourism.
—Snoussi et al. (2017, 30) on coastal erosion in Morocco

Coastal erosion processes are already severe and will be exacerbated with climate change and without remedial action—threatening the well-being of coastal communities. This section assesses the human exposure to coastal erosion and estimates the economic costs in the Middle East and North Africa region.

These estimates are conservative, not taking into account the indirect impacts on the tourism industry. As noted earlier, much of this sector is significantly threatened by coastal erosion and contributes, with its value chains, more than 18 percent of value added to GDP in Morocco and about 14 percent of value added to GDP in Tunisia (Heger and Vashold 2021). International experiences have shown that, in some

MAP 5.3

Annual Average Coastal Erosion and Accretion, Mediterranean Coast of Morocco, 2000–20

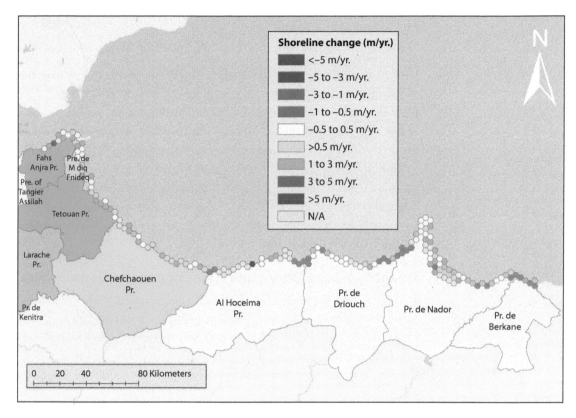

Source: NOC 2020. © World Bank. Further permission required for reuse.
Note: Morocco Mediterranean shoreline-change rates are aggregated within areas of about 15 square kilometers, designated by the hexagons. m/yr = meters per year; N/A = not applicable. Pr. = Province; Pre. = Prefecture.

cases, the majority of certain tourist groups reported they would not return if the beaches disappeared or were reduced (Raybould et al. 2013 for Australia; Tarui, Peng, and Eversole 2018 for Hawaii; Uyarra et al. 2005 for Barbados). Thus, unchecked coastal erosion is more than just worrisome, because tourism (of which beach recreation is an important part) is a main sector of many Middle East and North Africa economies.

Furthermore, SLR linked to climate change will accelerate erosion processes. Even a modest SLR of only 0.35 meters—consistent with a global temperature rise of 1.5 degrees Celsius (the most optimistic scenario)—would have substantial impacts. Alexandria, the delta coastal cities, and Port Said, Benghazi, and Algiers on the Mediterranean are particularly vulnerable to flooding (Elsharkay, Rashed, and Rached 2009). An SLR of only 0.30 meters would flood 30 percent of metropolitan

Alexandria, forcing about 545,000 people to abandon their homes and land and leading to the loss of 70,500 jobs. Other vulnerable cities include Muscat in Oman; Dubai in the United Arab Emirates; Aden, Republic of Yemen; and Basra, Iraq (El-Raey 2009).

Costs from Destruction of Land Values and Built Assets

Coastal erosion poses a particular threat to Middle East and North Africa economies because of their high economic exposure to sectors that derive their income from coast-related services, like beach tourism. Many of the region's largest urban centers as well as many smaller towns and villages are on the shore, which concentrates economic activity in these areas.

In this section, the results of a coastal-erosion monetization exercise show that coastal erosion presents a challenge, particularly to the four selected countries: Algeria, Libya, Morocco, and Tunisia. These contemporary developments emphasize the need for timely collective action on this issue to mitigate future impacts.

Quantifying the Extent and Rates of Coastal Degradation

In this exercise, the average erosion rates along a country's eroding coasts are extracted from the global dataset (mentioned earlier) on historical shoreline changes from 1984 to 2016 (Luijendijk et al. 2018). Following previous monetization studies, this analysis focuses on gross coastal erosion, not net coastal erosion (gross erosion plus accretion) (see, for example, Croitoru, Miranda, and Sarraf 2019).

This focus on gross coastal erosion also leads to erosion values that differ from those reported in table 5.1 and depicted in figure 5.2. In other words, this analysis does not include the value of the accretion of areas. This is largely because coastal erosion's economic effects are always

TABLE 5.1

Extent and Rates of Coastal Erosion in Selected North African Countries

Country	Share of coastline subject to erosion (%)	Share of coastline urbanized (%)	Long-term erosion Rate(m/yr)	Area(ha/yr)
Algeria	29	14.5	−2.1	−90.5
Libya	55	7.0	−0.9	−100.1
Morocco	54	6.6	−0.9	−139.9
Tunisia	59	15.0	−2.4	−247.3

Source: Heger and Vashold 2021, based on Luijendijk et al. 2018.
Note: The shares of urbanized coastline are from 2017 data. All other values are annual averages from a dataset on historical shoreline changes from 1984 to 2016, based on Luijendijk et al. (2018).
ha/yr = hectares per year; m/yr = meters per year.

negative (because territory is lost), whereas coastal accretion effects can be negative or positive. For example, accretion near a harbor's entrance may hinder ships from entering the port and lead to siltation of pathways, necessitating dredging to maintain them. Similarly, when rivers are used for shipping from the ocean to inland destinations, accreted areas can block river entrances. Furthermore, even when accreted land is not detrimental by itself, it remains unclear whether it can be used for development or recreational purposes.

In Libya, Morocco, and Tunisia, more than half the coastline is subject to coastal erosion, while lower shares are reported for Algeria (table 5.1). The rightmost column of table 5.1 shows yearly land loss from coastal erosion in the four countries—ranging from about 90 hectares in Algeria to almost 250 hectares in Tunisia. The subsequent calculations of yearly direct costs arising from shore retreat are based on these values.

Quantifying the Costs of Coastal Degradation to Land Values and Built Assets

To quantify the value of coastal land lost, the unit price of land per square meter in these countries was assessed, based on market data and, where available, official statistics.[3] Prices of land near the coast differ substantially based on location. Urban land prices considerably exceed rural land prices.[4]

These assessments show that urban land is most expensive in Morocco and least expensive in Algeria (table 5.2). Rural land prices vary less between countries, ranging from US$20 per square meter in Algeria and Morocco to US$30 in Libya. The present value of annual rents for the next 30 years is then used as an estimate of the value of land. Some assumptions had to be made to calculate the present value of

TABLE 5.2

Average Coastal Land Prices in Selected North African Countries, 2020

US$ per square meter

Country	Urban	Rural[a]
Algeria	350	20
Libya	480	30
Morocco	650	20
Tunisia	450	25

Source: Heger and Vashold 2021.
Note: Estimates are based on rapid price assessment from a combination of online property portals and official sources (where available).
a. Because of data constraints, the average rural land prices do not distinguish between land used explicitly for agriculture and land that may be a building plot. Hence, they are composite prices for rural areas.

land. The rent-to-price ratio of land was assumed to be 8 percent, with rents increasing by 8 percent and 5 percent for urban and rural land, respectively. To account for agglomeration effects in coastal regions, average urbanization rates as estimated by the United Nations (UN) for the 30 years were used. Finally, a standard rate of 3 percent was used for discounting future rents forgone by erosion of coastal areas.

The distinction between land prices due to location necessitates the classification of eroded land by land use. The ESA's Global Land Cover database was used to determine the share of urban areas on the total coastline (shown in table 5.1).[5] More than 15 percent of Tunisia's coasts and 14.5 percent of Algeria's coasts are urbanized. This share is lower in the other countries: 7 percent in Libya and 6.6 percent in Morocco.[6]

To estimate the value of built housing assets destroyed per year, the estimated average replacement costs of buildings in coastal districts were adapted from data for 12 Middle East and North Africa economies (Dabbeek and Silva 2020). The numbers of dwellings, buildings, and population are downscaled to a fine grid (1 square kilometer) to estimate the economic value based on geographical location and physical characteristics. Aggregating these data for coastal districts in the four countries being considered here enables an estimate of the value of assets lost because of coastal erosion.[7]

The land and built-asset destruction costs of coastal erosion in these four Maghreb countries are high, especially in Tunisia. They range from US$273 million per year in Libya to more than US$1.1 billion per year in Tunisia (table 5.3). Annual losses are equivalent to about 0.2 percent of GDP in Algeria, 0.4 percent in Morocco, 0.7 percent in Libya, and 2.8 percent in Tunisia.

The estimates are conservative; they do not take into account losses in adjacent properties. Near-shore properties derive part of their value from their proximity to the sea and hence may be affected indirectly; their

TABLE 5.3

Direct Economic Costs of Coastal Erosion in Selected North African Countries

Cost metric	Algeria	Libya	Morocco	Tunisia
Buildings lost (US$, millions)	3	1	8	29
Land lost (US$, millions)	310	272	425	1,078
Total losses (US$, millions)	313	273	434	1,107
Total losses (% of GDP)	0.2	0.7	0.4	2.8

Source: Heger and Vashold 2021.
Note: The building-loss estimates are based on the replacement costs of dwellings and other buildings and do not take damages to infrastructure or reductions in value of undestroyed buildings and land explicitly into account. Values are averages computed using the average erosion rates shown in table 5.1.

value will be reduced even if erosive forces do not directly destroy them (Fraser and Spencer 1998; Pompe and Rinehart 1995; Scott, Simpson, and Sim 2012). The negative effect of beach retreat on property values diminishes with distance, implying that properties near but not necessarily bordering the shore can be affected through negative spillover effects arising from erosion (Rinehart and Pompe 1994). These effects are not included in the estimates of direct costs; hence, the estimates should be viewed as rather conservative assessments of the overall costs due to coastal erosion in the countries discussed in this section.

The effects on developments such as ports or industrial sites as well as on ecosystems are substantial but hard to quantify. The analysis above does not capture these specific effects, which would require detailed modeling of effects and costs of coastal erosion linked to these developments—a task that is hardly possible at a national or regional scale.

SLR and greater frequency of extreme weather events driven by climate change will increase coastal erosion and its costs. Coastal flooding is exacerbated by shoreline retreat and causes significant losses for major cities in the Middle East and North Africa, increased by socioeconomic changes. For example, in Alexandria, a projected SLR of 20–40 centimeters, subsidence, and measures to keep the flooding probability constant could lead to annual losses of US$504–US$581 million in 2050 because of coastal floods (Hallegatte et al. 2013).

Lost Tourism Revenue from Coastal Areas

Coastal erosion is an existential threat to tourism, a sector that contributes significantly to GDP in many Middle East and North Africa economies. Probably the largest share of the costs of coastal erosion, especially in the long term, will be indirect by reducing revenues resulting from tourism in affected areas.[8] Forgone revenues from tourism are a severe threat, especially for countries where "blue" tourism represents a large part of their revenues.

As noted earlier, the tourism sector accounts for more than 10 percent of GDP In several of the region's economies (Heger and Vashold 2021). For example, Morocco and especially Tunisia depend heavily on tourism, which in turn largely depends on their beaches. In Morocco, over 12 million international visitors were recorded in 2018, with receipts totaling more than US$9.5 billion (around 8 percent of GDP), according to data from the UN World Tourism Organization (UNWTO 2019), and tourism activities account for more than half of the country's export services. Considering indirect economic impacts as well, tourism accounted for 18.6 percent of Morocco's GDP in 2017 and 16.4 percent of employment (Kasmi et al. 2020).

In Tunisia, tourism-related activity accounted for 14.2 percent of GDP in 2018 and employed more than 2 million (Saidani 2019). International tourists alone contributed over US$2.3 billion to the economy, representing around 6 percent of GDP in 2018 (based on UNWTO data). Given that more than 90 percent of the country's recorded tourist bed nights were spent in coastal areas (Jeffrey and Bleasdale 2017), the economic threat posed by the disappearance of beaches due to coastal erosion should be recognized.

Beachgoer Surveys: Willingness to Return, Willingness to Pay

Coastal erosion would discourage tourists from visiting the region's coasts. International evidence, whereby tourists are asked whether they would return to an area if the coast were eroded, shows that coastal erosion significantly affects tourism. However, the propensity of tourists to visit a certain location does not decrease in a linear fashion with advancing beach retreat. For example, in a survey carried out at beaches in the US state of Delaware, around two-thirds of visitors stated that a reduction of a beach's width to a quarter of its current size would worsen their experience, and one-third indicated that they would reduce their number of visits (Parsons et al. 2013). For California beaches narrower than 20 meters, a reduction in width is associated with much larger decreases in the propensity of recreational visitors to come back than when initial beach width is larger than 20 meters (Pendleton et al. 2012). In Barbados, tourists' aversion to returning is especially strong if beaches are less than 8–10 meters wide (Schuhmann et al. 2016). A similar nonlinear relation can be found regarding the speed of erosion, where surveys reveal that faster beach retreat led to a disproportionate reduction in consumer surplus, and hence propensity to revisit, compared with slower retreat (Huang et al. 2011).

In the extreme case, disappearance of beaches could lead to total losses if tourists decide not to visit the affected areas anymore, as several surveys showed:

- Most respondents to a survey conducted in a Hawaiian town stated that they would not consider staying in a hotel should the nearby beach completely erode (Tarui, Peng, and Eversole 2018).

- More than three-quarters of surveyed tourists in Barbados were unwilling to return for the same price should beaches largely disappear; this was associated with a 46 percent decrease in tourism revenues (Schuhmann et al. 2016).

- Along two stretches of the Australian coast, large shares (exceeding 50 percent) of surveyed tourists stated that major erosion events would lead them to switch to other destinations. The associated losses would equate to more than US$75 million per year (Raybould et al. 2013).

Hence, the retreat of beaches could lead to severe economic losses, especially in regions that are primarily visited for their beaches, as is the case in many coastal-tourism destinations in the Middle East and North Africa. These findings imply that the relationship between tourists' unwillingness to return and beach width can be thought of as having a shape like the one depicted in figure 5.3.

In coastal districts, a higher share of open or flat coastlines (that is, intact beaches) is a significant determinant of higher accommodation prices (Hamilton 2007). In Spain's northeast territory of Catalonia, hotels along the Costa Brava that are near a beach with "Blue Flag certification" have room prices that are more than 10 percent higher on average (Rigall-I-Torrent et al. 2011) than hotels without such a nearby beach.

FIGURE 5.3

Stylized Relationship between Beach Width and Visitors' Unwillingness to Return

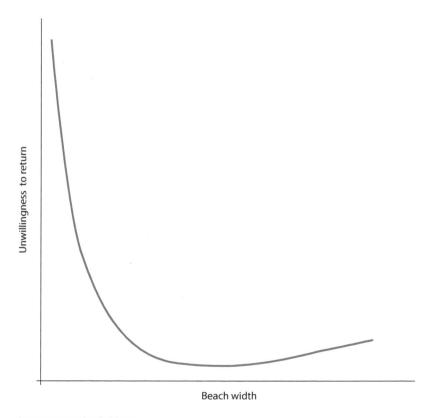

Source: Heger and Vashold 2021.

Blue Flag certification indicates that the beach and seawater quality meet certain standards and that beach management also meets specific environmental standards. Both Morocco and Tunisia are part of this program, and efforts to extend its scope are ongoing to accommodate tourists' changing attitudes. The revenues of hotels near retreating beaches are likely to decrease because the amenity value for tourists decreases if the beach quality deteriorates.

One way to estimate the indirect cost of coastal erosion is to ask individuals their preferences about avoiding eroded beaches and asking them to monetize those preferences. This "willingness-to-pay" measure reflects the amount of money that residents and tourists would be willing to pay for beach-saving initiatives such as beach nourishment or building offshore safeguards. The nature of these surveys, which build to a certain extent on respondents' attachment to the region under consideration, limits their practical use to small geographical units such as individual beaches or beaches on a certain island. However, their importance for informing policy makers about possible sources for financing such precautionary measures is not negligible, and some lessons can be drawn from such analyses.

Interventions that increase beach width (such as beach nourishment) are often viewed more positively than other interventions that mitigate coastal erosion. Beach nourishment refers to the filling of a certain beach with sand or similar sediment to restore the width (or area) lost to coastal erosion. Studies have found that large portions of survey participants respond positively to such interventions. A large-scale survey in North Carolina found that almost half the respondents were in favor of beach nourishment to increase beach width, which (using an econometric model) is estimated to lead to more beach trips (Whitehead et al. 2008, 2010). These results have been confirmed by other studies (Landry and Liu 2009, 2011).

On the other hand, some interventions may decrease the propensity to return. Beachgoers tend to take fewer trips to beaches that have sand dunes and jetties. A study on beaches in New Hampshire and Maine found that erosion leads to an average of 1.36 fewer trips per resident. Where an erosion-control program prevents erosion, the impact was attenuated; however, respondents would still take 1.01 fewer trips. Hence, erosion control can be desirable, but the potential negative impacts on the beach environment can offset the benefits of decreased erosion (Huang et al. 2011).

Studies have shown the potentially high losses of tourist revenue stemming from coastal erosion in the Middle East and North Africa. For example, tourists visiting North Africa's largest island—Djerba Island in

Tunisia—would be willing to contribute over €5 million a year for a project to reduce coastal erosion (Dribek and Voltaire 2017). (For the effects of coastal erosion on a Djerba Island hotel, see photo 5.3.) This figure implies that beach preservation measures to reduce coastal erosion may be financed largely by contributions from tourists[9] and also measures

PHOTO 5.3

Effects of Coastal Erosion on Hotel les Sirenes, Djerba Island, Tunisia, 1992–2019

Sources: Oueslati, Labidi, and Elamri 2015; Heger and Vashold 2021. 1992 and 2013 photos: © Ameur Oueslati / Agence de protection et d'aménagement du littoral (APAL). 2016 and 2019 photos: © 2020 CNES/Airbus, courtesy of Google Earth. Permission required for reuse.

the value that tourists attribute to intact beaches. Moreover, it implies a modest contribution of under US$3 per tourist per year. Djerba hosted about 25 percent of the 9 million tourists who visited Tunisia in 2019, or about 2 million tourists.

For a similar region in Morocco—the Tetouan coast with its main tourist beaches—total eroded surface in the period 1958–2018 amounted to approximately 490 hectares (Benkhattab et al. 2020). Drawing on these numbers and projections about future beach retreat, lost revenues to the economy due to the retreat of these beaches alone could total US$190 billion in the next few decades (Flayou et al. 2017).[10]

Other Analyses

Other approaches that quantify the economic effects of coastal erosion also highlight large potential losses to the local economy. They may use information on beach attributes such as their width, land value, and characteristics of nearby hotels (for example, room price) to directly estimate reductions of land values and revenues due to coastal erosion in the framework of hedonic price regressions.

Using such an approach, the decrease in beach width due to coastal erosion in Rethymnon on the Greek island of Crete could lead to revenue losses amounting to around €18.5 million in the next 10 years because of the progressing retreat of a single beach (Alexandrakis, Manasakis, and Kampanis 2015). The impact of SLR on beach tourism in Sahl Hasheesh and Makadi Bay on the Red Sea in Egypt is expected to lead to losses in revenues that could exceed US$350,000 per day in 2050 (Sharaan, Somphog, and Udo 2020). Similar studies show beach-surface reduction to have a decisive negative impact on the overall image of tourist destinations, decreasing the number of arrivals and hence reducing receipts from them (Bitan and Zviely 2019; Raybould et al. 2013; Scott, Simpson, and Sim 2012).

Therefore, direct losses due to coastal erosion are only a fraction of total losses to the economy, not considering the impact on tourism and other development activities (such as ports and so on) as well as ecosystems. Considering the evidence on indirect losses attributable to coastal erosion, the analysis of direct costs of coastal erosion presented in the previous section ("Costs from Destruction of Land Values and Built Assets") provides only a conservative estimate of the total costs of this phenomenon to Middle East and North Africa economies.

Accounting for lost future revenues is crucial for assessing the real costs of coastal erosion. Furthermore, coastal erosion processes can lead to lost jobs and reduced tax revenues from ports and near-shore industries dependent on intact coasts. Another important point to consider is that coastal erosion may permanently damage ecosystems that can be important income sources, such as for fisheries. However, quantifying the effects on these sectors is extremely challenging and out of the scope of this report.

POLICY REVIEW: HOW TO COMBAT COASTAL EROSION

This section first discusses the principles for sustainably managing the coastline and effectively avoiding coastal erosion in the Middle East and North Africa. It then highlights priority recommendations for policies and actions that the region's economies can take to combat coastal erosion. Managing and effectively avoiding coastal erosion necessitates detailed information on coastal dynamics. Therefore, the section next reviews the general drivers of coastal erosion and highlights the paucity of this information in the Middle East and North Africa, arguing that to plan interventions effectively, such gaps in the evidence basis must be filled.

Finally, the section presents a comprehensive review of actions and recommendations for reducing the effects of coastal erosion, including ICZM planning; forward (prospective) management strategies; reactive mitigation (such as defensive investments) including NBS; and preventive policies (such as banning sand mining). The objective is not to present a detailed pathway with concrete recommended options for each city and country—which would be a futile task given the paucity of evidence on the sources—but rather to present a menu of options with priority recommendations that can be beneficial to many countries.

ICZM: Principles and Benefits of Sustainable Coastline Management

To be successful, ICZM requires continuous assessments of the coast as well as engagement of all stakeholders at all stages (figure 5.4). Given the proper identification of the sources of coastal erosion (with the aid of data, monitoring, and analysis) and consideration of different stakeholders' claims, suitable strategies can be formulated to tackle the challenge posed by coastal erosion. These include prospective management, such as risk assessment for different parts of the shoreline, and appropriate zoning. Participation of local communities and other stakeholders is a key element in designing successful strategies.

Once the coastal erosion problems are identified, the question becomes one of adapting to the challenge, presented in figure 5.4 under "reactive management and control measures." The adaptation options include either (a) defending developments where possible (using hard defenses, soft defenses, or both, including natural solutions and NBS); or

FIGURE 5.4

Key Elements of Sustainable Coast Management to Mitigate Coastal-Erosion Effects

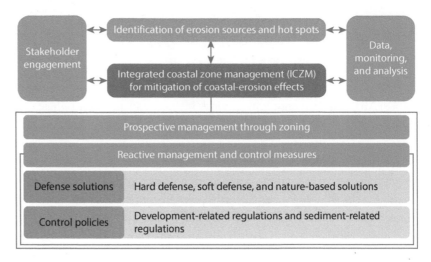

Source: Original figure prepared for this publication.

(b) avoiding impacts where such defenses are not possible (by managing the retreat).

Complementary work on river-basin management may also be necessary because interference in natural processes—through reduction in river flow from abstraction for upstream water use, urbanization, loss of natural flooding regimes, and dam regulation—may all affect the flow of sediment to the coast. Regulatory controls may vary from banning sand mining to modifying dam operations in order to limit losses from trapped sediment. The "Comprehensive Policies and Actions for Reducing Coastal Erosion" section later reviews the experiences of Middle East and North Africa economies as well as international best-practice examples in coastal management and combating coastal erosion.

Managing coastal development will bring important cross-benefits for coastal issues beyond mitigation of coastal erosion. Coastal zones are major contributors to the continued flow of plastics into the Middle East and North Africa's seas, together with plastic waste deposited by rivers. Hence, setting up a comprehensive ICZM scheme that includes improved waste management processes (for both solid waste and effluent discharge) will reduce the amount of plastics ending up in marine spaces. Furthermore, clear development plans in such schemes could involve coordinated measures to reduce plastic waste in coastal zones, whether it has been dumped there or deposited by waves from the open sea. Similarly, setting up a comprehensive development plan is crucial to ensure that coasts are used in the most productive and most sustainable fashion. This, together with the increased protection from coastal erosion, would lead to more efficient use and valorization of these assets while also protecting them for future generations.

Setting in motion sustainable management schemes for coastal development will contribute to greener, more resilient, and more inclusive development (the GRID path) in the Middle East and North Africa. Coastal erosion comes with large adverse effects on human development (that is, built infrastructure) as well as on local ecosystems in the region (as discussed earlier in the chapter). Long-term management of coastal development under the tenets of an ICZM scheme will hence benefit coastal communities and local biodiversity in various ways:

- Limiting uncontrolled development will reduce conflicts among stakeholders and ensure that the gains from using coastal assets will be distributed more justly and sustainably, providing income sources for those depending on intact coasts.

- Restoring ecosystems using NBS will not only reduce coastal erosion but also provide habitats for local flora and fauna, greening the region's coasts.

- Actively involving local communities in such revitalization efforts implies additional income-generating activities for these communities, especially for the poor (given the high labor intensity of such work).

- Promoting more environmentally friendly coasts and beaches can maintain tourism while also encouraging development of new ecotourism ventures.

Thus, successfully managing the coasts and mitigating their erosion will contribute to putting Middle East and North Africa's economies on a GRID path and ensure that these important natural assets remain the boon to future generations that they have been to others for so long.

Priority Recommendations for Combating Coastal Erosion

Better management of coastal-erosion risks will benefit the Middle East and North Africa region. Climate change exacerbates these risks. This section discusses three priorities, based on their effectiveness in combating coastal erosion, their economic and social benefits, and their level of interdependence with other policies and interventions.

No. 1: Understand the Drivers of Coastal Erosion and Improve Access to This Data

Understanding the drivers of coastal erosion is a prerequisite for many interventions. First, identifying and monitoring the erosion hot spots are critical preconditions (as further discussed later in this section). In the selected hot spots, identifying the drivers of coastal erosion is critical to inform the planning for interventions to reduce erosion. Because the local drivers of coastal erosion are site-specific, the most effective and efficient solutions also vary across these sites.

Source-identification analysis includes monitoring and computational modeling of coastal morphology, sediment flows, and fluid mechanics as well as the impact of coastal development. This entails analysis of river flows and the impacts of (a) upstream river-basin development schemes on coastal wetlands, (b) sediment flows on shoreline replenishment processes, and (c) particular activities such as extraction of gravel and sands from rivers and beaches.

Finally, the data and related information must be made publicly available to build public engagement, awareness, and consensus for future policy reforms. Data-based impact analysis is critical for designing solutions for coastal-defense investments and for environmental impact assessments of ICZM plans, including new developments. The transboundary nature of coastal erosion also makes regional cooperation on data and information sharing important to identify drivers of the phenomenon across country borders.

No. 2: Engage Stakeholders in ICZM Planning at All Levels

Multistakeholder participation in ICZM planning at the municipal, national, and regional (subnational and multinational) levels is key for the design of effective solutions for combating coastal erosion. Coastal-zone management requires a holistic approach that includes all stakeholders—for example, port authorities, fishers, hotels, utility service providers, nature conservation specialists, cultural heritage authorities, restaurants, technical specialists, and representatives of local communities (more on this in the "Comprehensive Policies and Actions for Reducing Coastal Erosion" section). Thus, joining forces is an important step to take in parallel with launching the data collection, analysis, and monitoring effort. Doing so across country borders would be an important step to address joint efforts and maximize the effectiveness of ICZM schemes. ICZM also needs to use multistakeholder processes to identify hazardous and ecologically vulnerable areas, areas designated for development, and measures for reducing erosion.

Comprehensive ICZM planning, both spatial and temporal, will help ensure coastal sustainability and combat erosion—all while identifying economic opportunities and supporting efficient coastal development plans, which differ across coastal areas and shape the site-specific responses to coastal erosion. A comprehensive ICZM plan will help inform policies through not only *prospective* management and development but also *reactive* management and control interventions, as discussed in detail in the "Comprehensive Policies and Actions for Reducing Coastal Erosion" section. As mentioned earlier, implementing such an ICZM scheme will also bring cross-benefits by reducing marine-plastic pollution of the seas and coasts. Through the consultation with all relevant stakeholders and consideration of their needs, the introduction of ICZM schemes also directly contributes to more inclusive development in general.

No. 3: Use Nature-Based Solutions on Land and Sea

Nature-based solutions (NBS) can combat coastal erosion and restore coastal and marine ecosystems. Using NBS in the seas, coastal wetlands, and along shorelines can increase coastline resilience by weakening incoming waves and inland winds and by retaining and stabilizing sediments. (For more on these approaches, see the later subsection on soft defenses under "Reactive Management and Control Measures.")

Increasing vegetation cover through the planting or restoration of (a) seagrass fields within marine ecosystems, mangroves, and other natural vegetation in coastal wetlands and shorelines; and (b) marshes and other flora on coastal dunes will help improve habitats and stabilize sand dunes, beaches, wetlands, and natural coastal protection processes. Restoring coral reefs in the Red Sea and RSA can help combat erosion by diminishing wave energy. Importantly, these efforts should take into account the preexisting natural landscape to avoid disturbing the local flora and fauna. And all these natural interventions require supporting environmental policies.

NBS offer multiple benefits in addition to coastal resilience, including carbon sequestration, biodiversity, and coastal ecosystem restoration. They also can be used for educational purposes and provide lasting benefits to the tourism sector. And, given the labor-intensive nature of these interventions, the environmental benefits are accompanied by economic opportunities. Moreover, relying on the knowledge of the local population regardless of gender or educational background is key for a successful implementation of NBS. Hence, their broader use also contributes directly to longer-term, inclusive economic development through sustainable management of natural resources—very much in line with a GRID framework.

One Must Measure What One Would Manage: The Sources of Coastal Erosion

Before one can effectively manage coastal erosion and ensure the sustainability of the coastline, one must first understand the spatial patterns (*where*) of coastal erosion (hot spots) and the reasons why (*how*) the erosion is occurring. Knowing where and how coastal erosion occurs requires a holistic approach using data, monitoring, and analysis as well as stakeholder engagement to sustainably plan ahead for mitigating the risks of coastal erosion. These elements, as well as the evaluation of potential solutions, are essential before putting the necessary policies into place.

Although the particular sources of coastal erosion along an individual part of the coast are very site-specific, some general drivers are known to cause changes in the coastal landscape, not only in the Middle East and North Africa but globally. This section identifies these general drivers and how they might be affecting coastlines. However, for a specific part of the coast, multiple sources may be acting together to shape the coastal landscape, with specific dynamics and their sources varying across the coastline. It is hence important to account for local characteristics to effectively combat coastal erosion at a specific site, necessitating detailed antecedent analyses.

Furthermore, as this section lays out, human interventions at one site may influence erosive processes at others. Hence, it is also important to consider coastal dynamics (such as sediment transport across local hot spots) in an integrated manner. For the Middle East and North Africa, such detailed studies of flow dynamics along the coast and site-specific sources of coastal erosion are limited, increasing the difficulties of tackling coastal erosion effectively.

Overview of Coastal Erosion Drivers in the Middle East and North Africa

Various factors drive changes in the coastal landscape and coastal erosion—some occurring naturally and others induced by human activities (figure 5.5). Direct anthropogenic drivers include coastal subsidence (due to heavy infrastructure near the coast or aquifer-water extraction), coastal infrastructure (such as ports and marinas), defense developments, and land reclamation. Natural physical forcing elements such as storms, SLR, and currents are exacerbated by human-induced climate change (Sytnik et al. 2018).

Coastal areas with different tidal dynamics and wave-energy incidences demonstrate unique coastal morphologies (Hayes and FitzGerald 2013). Fluvial and alongshore sediment transport are major morphodynamic processes that determine the shape of the coastline (Sytnik et al. 2018) and often determine how the coast is divided into management cells. Human interventions, such as ports or groins, intervene in the hydrodynamic processes along the shoreline. Often, these interventions

FIGURE 5.5

Major Factors Affecting Coastal Morphology, Including Coastal Erosion

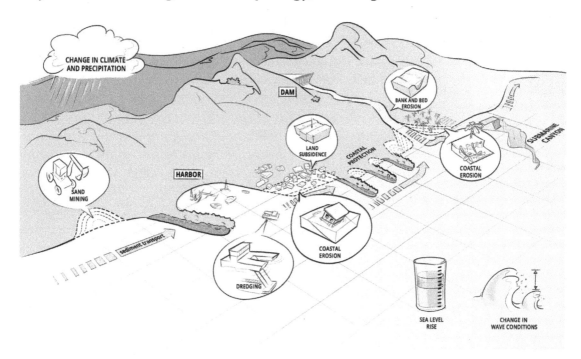

Source: Giardino et al. 2018. Available under Creative Commons Attribution license (CC BY-NC-ND).

stabilize areas through sediment buildup in one location, but they starve the sediment and intensify the erosion rate at other locations; in extreme instances, poorly informed adaptation measures may cause more damage than doing nothing (Hoggart et al. 2014).

Coastal and inland development can exacerbate coastal erosion locally and at other sites, as shown by examples in the Middle East and North Africa. Because of the blocking of sediment transport, structures further upstream can cause or accelerate erosion of downstream coastlines. Although the 2010 expansion of the commercial port Tanger-Med in northern Morocco included an environmental impact assessment (EIA) with a desk-based assessment of known archeological sites, the port's presence has affected the coastline at the nearby archeological site of Ksar es-Seghir (Trakadas 2020). In addition, there is a lack of environmental and social impact assessment (ESIA) regarding coastal erosion in the region. These circumstances highlight potential insufficiencies in the understanding of erosion processes and their proper incorporation into EIAs and ESIAs.

As illustrated in figure 5.5, the construction of a dam inland can have negative effects on sediment transport by the river on which it is built. While this alone has adverse effects on coastal regions at the mouth, coastal protection measures along the coast can also trap sediments that would otherwise feed the coastline. River deltas can be starved of replenishment by upstream construction; combined with coastal construction, the impacts on coastlines can be severe. An example is at the Rosetta promontory in the Nile delta, where the construction of 15 groins has exacerbated the erosive forces caused by the building of the Aswan High Dam. These groins have led to a reversal from accretion to fast erosion along the leeside of the promontory, with some segments exhibiting erosion rates as high as 30.8 meters per year (Ghoneim et al. 2015).

The following brief discussion of key drivers of coastal erosion in the Middle East and North Africa provides a concise view of the issues the region faces. It includes drivers stemming from both anthropogenic sources and natural forces that have been, and will be, exacerbated by climate change.

Dam Construction

Many rivers in the region have been dammed to provide water supplies, control flooding, and produce hydroelectric power, with the unintended consequences of reducing flows of sediment to the coast (Syvitski et al. 2005). Diminishing sediment flows can be important drivers of coastal erosion, reducing beach areas and eroding shorelines, as seen with Egypt's Aswan High Dam (Masria et al. 2015b) and with the Moulouya River in Morocco, where dams capture over 90 percent of the sediment supply (Snoussi, Haïda, and Imassi 2002). Such sediment-flow issues also threaten the dam's functions because as sediment is trapped, the dam's capacity decreases. Thus, it is crucial to use various techniques proven internationally to manage sediment flows for dams (Kondolf et al. 2014).

Subsidence

Ground deformation is a severe geological hazard in the Middle East and North Africa. It results mainly from anthropogenic activities such as fluid extraction or injection, underground excavations, and construction expansion. The Islamic Republic of Iran is one of the most-affected countries in the Middle East because of groundwater overexploitation and 30 years of drought affecting large cities such as Mashhad, Neyshabour, Rafsanjan, and Tehran (Fattahi 2019; Khorrami et al. 2020). Coastal subsidence over time negatively affects farmland, urban areas, and wastewater infrastructure, and it can cause cracks in roads and water and natural-gas pipes.

The United Arab Emirates is also at great risk because its groundwater use is 20 times higher than the natural recharge rate, where 60 percent of its consumption comes from aquifers, compared with 29 percent from desalination plants and 6 percent from water recycling (*Construction Week* 2015). And Alexandria, Egypt, is a well-known example of sediment compaction in the Nile delta that provokes land subsidence, thereby enhancing the effects of climate-induced SLR, as further described in the SLR subsection below (Syvitski et al. 2009).

Dredging

The Gulf region has been active in dredging for both navigation (that is, channels for ships) and land reclamation in the past few years (Kloosterman 2010). Efforts to facilitate commercial, recreational, and navigational activities have important effects on coastal waterways, inlets, and bays—altering currents and wave patterns, creating hydraulic mining, and moving sediment. Shallow deltas and inlets that once constrained the outward flow of freshwater and held the salt sea at bay are now hydraulic superhighways because of dredging. This process also fosters the loss of coastal freshwater aquifers to saltwater infiltration, affecting water treatment and distribution plants and potable water sources for human consumption.

Sand Mining

Sand mining—extraction of sand from sandy areas on the coasts or along riverbanks, mostly for construction and industrial uses—has been a problem in the Middle East and North Africa, particularly in Morocco. In Morocco, coastal sand mining, often illegal, has been rising with the increase in construction and real estate development (Aldar.ma 2019) and has increased erosion along vulnerable coastlines. Around half of the sand used annually in Morocco (about 10 million cubic meters) is from illegal coastal sand mining.

Illegal sand mining operations can, at the extreme, leave behind bare beaches such as a large beach between Safi and Essaouria (along the Atlantic coast of Morocco) that became rocky terrain (UNEP 2019). Coastlines weakened by mining are also more vulnerable to the impact of storms, further accelerating beach erosion. The coastline of Asilah, Morocco, has been severely damaged because of the increased demand for sand to mix with cement for urban construction. Similar operations have been documented along the Atlantic shoreline in Larache or Kenitra and farther south in Morocco, as well as in Algeria (Coastal Care 2009, 2020; Greene 2016). Sand mining can also affect dune replenishment as well as the integrity of coastal marshes.

Storm Surge

Storm-induced erosion and coastal flooding are the two most important natural hazards to coastal systems worldwide and are interdependent (Kron 2013). In some cases, storms and storm surges, exacerbated by SLR, are the principal cause of erosion (Katz and Mushkin 2013; Nicholls et al. 2007).

Storm surges severely affect coastal regions in the southern Mediterranean, and low-lying areas are especially vulnerable (Satta et al. 2017). Middle East and North Africa countries exposed directly to the Indian Ocean (for example, Djibouti, Oman, and the Republic of Yemen) are also regularly exposed to tropical storms, whereas the west coast of Morocco is exposed to Atlantic storms (Becker et al. 2013; Knapp et al. 2010).

By 2100, SLR and storm surges in Tangier, Morocco, will affect a projected 34.8 percent of its urban area, 99.9 percent of its port infrastructure, and 36 percent of its roads (Snoussi et al. 2009; World Bank 2014, 127). Storm-surge zones are also projected to increase this century by 84 percent in Egypt, 57 percent in Algeria, 54 percent in Libya, 30 percent in Morocco, and 27 percent in Tunisia (Dasgupta et al. 2009).

Sea Level Rise

Rising seas contribute to faster coastline retreat, particularly in low-lying areas (Stive, Ranasinghe, and Cowell 2010). Since 1990, the rate of SLR in the Mediterranean has been above the global average (Tsimplis and Baker 2000). Atmospheric influence is thought to be the primary driver: pressure and wind variations associated with the North Atlantic Oscillation control water flow through the Straits of Gibraltar (Gomis et al. 2006; Landerer and Volkov 2013; Tsimplis et al. 2013). Compared with the well-studied Mediterranean, tide-gauge records in the Red Sea and the RSA are much sparser because they are noncontinuous and limited to a few years.[11]

A stronger rise in the Arabian Sea than in the Mediterranean is projected under both a Representative Concentration Pathway (RCP) 2.6 scenario (a 1.5 degrees Celsius world) and an RCP 8.5 scenario (a 4 degrees Celsius world).[12] Under the latter scenario, Muscat is expected to experience a median SLR of 0.64 meters, and Tunis an SLR of 0.56 meters, by the last decades of the century (2081–2100). The most-affected cities in the region include Muscat (12.0 millimeters per year), Alexandria (10.9 millimeters per year), Tangier (10.2 millimeters per year), and Tunis (10.1 millimeters per year) under an RCP 8.5 scenario (World Bank 2014). Among these cities, Alexandria is projected to lose the most local value added as a result of damages from SLR by 2050 (Hallegate et al. 2013). These changes in the sea level are a major potential threat to coastal areas because of the cumulative effect of SLR and long-term shoreline retreat.

How to Understand These Sources of Coastal Erosion

As discussed above, data, monitoring, and analysis form the cornerstone of effective coastal erosion mitigation and management, specifically for ICZM plans. Data on factors such as sediment flows, erosion rates, coastal and marine physical processes, and infrastructure and development are critical for monitoring, analysis, and determination of informed engineered and other solutions and interventions. These data will help all stakeholders to understand areas of accretion, areas of erosion, and sediment feed, which in turn help to identify hot spots, erosion sources, and zones for intervention.

Monitoring the coasts for risk management, identifying threat levels, and implementing immediate interventions are also important. Monitoring also aids the process of keeping records that could also be used for analysis. Both data and monitoring aid the analysis process through various tools and computational modeling to identify sources, risks, hot spots, and potential solutions.

First, it is necessary to understand the degree of coastal erosion in a location-specific way—by identifying and analyzing the hot spots. Coastal erosion is highly dependent on localized physical processes such as fluid mechanics and sediment balance and flows, meaning that although there may be severe erosion in one area, the adjacent area can behave very differently. Analyzing hot spots at a comprehensive yet granular scale (as done in the close analyses of Moroccan and Tunisian coastlines, elaborated earlier in the chapter) is an important first step. Extending the hot-spot analysis to cover as much of the Middle East and North Africa's coasts as possible is desirable. With such analyses in hand, sites where actions are needed in a timely manner can be identified, and policy makers and researchers could prioritize these sites.

Second, why does coastal erosion occur in a selected hot spot? Once the hot spots of erosion along the coast have been identified, more-detailed analyses of prevalent dynamics (for example, the morphological cycle or sediment transport) of these sites is crucial. In addition, anthropogenic perturbations, such as those highlighted earlier (dam construction, harbor construction, sand mining, coastal subsidence, SLR, and so forth) must be modeled covering larger spatial scales because, for example, changes up-current can have down-current effects. Implementing measures without precise knowledge of coastal dynamics (spatially and temporally) can worsen the effects of coastal erosion. This was the case for the Rosetta promontory in Egypt's Nile delta region, where the construction of defense structures has exacerbated the effects of coastal erosion at other sites (Ghoneim et al. 2015). It is important to study the effects of different measures at different sites to avoid unintended side effects.

Measures to combat coastal erosion also require knowledge of geomorphological characteristics. In addition to knowing how sediment is transported along the coast, it is important to know what types of sediments are predominant in certain areas such as beaches. For some defense solutions, it is important to consider existing structure and granularity. Using sand that is too different in its characteristics (for example, desert sand versus coastal sand) to refill beaches will undermine these efforts and will get washed away quickly.

Erosion Data Gathering, Monitoring, and Modeling within the Region

Detailed sediment budgets and numerical models are important tools to understand coastal changes in the Middle East and North Africa. Such analyses have the potential to uncover important sediment dynamics along coasts influenced not only by coastal developments such as ports but also by inland structures such as dams. Although such models have been employed at some specific sites along the region's coasts, a comprehensive analysis of these dynamics would allow for the incorporation of transboundary effects of sediment transport and its effects on coastal erosion. In West Africa, such modeling exercises were recently undertaken (box 5.2). Such analysis provides policy makers with a comprehensive and easy-to-use tool to simulate the impact of different coastal developments and climate change on coastal sediment transport and in turn on coastal erosion.

In the Middle East and North Africa, monitoring and analyses of sediments, coastal changes, and human intervention relevant to coastal erosion have been carried out only for some areas, but comprehensive modeling for the entire coast is needed. For example, a 25-year study (1990–2014) investigated coastal erosion rates and sources along the Nile delta coast (Ali and El-Magd 2016).

Tools such as geographic information systems (GIS) and sediment-budget computational modeling are normally used for such analysis, including statistical and physical data. Studies that investigate the behavior of sediment transport on a large scale have the advantage of incorporating and allowing for evaluation of effects that various changes in certain parts of the coastal landscape have on other parts of the shoreline. Identifying sources and sediment flows can inform the analysis to identify potential solutions and analyze their potential to prevent coastal erosion, mitigate its impacts, or both. The sources and hot spots of coastal erosion become part of the input that aids such analysis to come up with sustainable measures, policies, and designs of engineered solutions. The information from sediment studies provides useful inputs in

BOX 5.2

Sediment Budgets and Numerical Modeling in West Africa

West African coasts have undergone high rates of coastal retreat because of changes in the wave-driven alongshore transport of sand. These changes are largely caused by human developments such as river dams or harbor jetties and have led to severe erosion on the order of 10 meters per year at some locations. In a large-scale project, the World Bank's West Africa Coastal Areas Management Program (WACA) partnered with the Dutch research company Deltares[a] to set up a coastal sediment budget for Benin, Côte d'Ivoire, Ghana, and Togo. This involved estimating annual alongshore sediment transport capacity using numerical modeling to quantitatively assess the effects of different human interventions.

Illustrating the implications of interventions on the evolution of the coast and transboundary effects, together with the effects of climate change, the model provided an important awareness-raising tool for decision-makers. The project developed a digital coastal viewer to facilitate communications to the different stakeholders who were both affected by changes in the coastal landscape and responsible for these changes.

The numerical approach in the modeling exercise developed three submodels in an integrated framework:

- *A large-scale wave model* was employed to simulate wave propagation from offshore to nearshore, taking into account wind generation, dissipation, and nonlinear wave-wave interactions. It covered the complete West African coast and included 15 spatially more explicit models to ensure sufficient spatial resolution.

- *A hydrological model* that takes into account land use and soil maps as well as meteorological data was used to compute the annual runoff that influences the sediment yield from West Africa's major rivers.

- *A shoreline-evolution model* was employed to simulate coastal evolution. It calculates the magnitude of alongshore sediment transport for specific locations along the coast, considering wave-induced drift, tidal flow, or both. It used information on coastal orientation and local wave angles to carry out long-term simulations at large spatial scales.

The analysis allows simulation of the impacts of different changes in the coastal landscape under different climatic or meteorological conditions. Scenarios that can be assessed include major anthropogenic developments such as ports or dams, rising sea levels, and changes in wave height and direction due to climate change. Hence, it provides researchers and policy makers with an efficient tool to model and investigate the effects of changes in coastal development and measures deemed to reduce sediment blockage or coastal erosion. Including the effects of climate change allows for the evaluation of shoreline changes in the long term. Given its large scale, the simulations lend themselves to the analysis of possible transboundary effects of coastal interventions at specific hot spots.

Source: Giardino et al. 2018.
a. For more information, see the Deltares website: https://www.deltares.nl/en/.

planning new development. Extensive simulation studies have been used, for example, in the design stage of the Al-Faw Grand Port in southeast Iraq (box 5.3).

Numerical analyses can be used for assessing the potential effectiveness of different protective measures. For example, the Rosetta promontory on Egypt's Nile delta coast has been the subject of several scientific studies investigating the hydrodynamics and sediment-flow dynamics

---

BOX 5.3

Iraq's Al-Faw Grand Port: Computational Modeling to Eliminate Coastal Erosion

Computational modeling of hydrodynamics and sediment processes were important tools in the design and assessment stages of development and construction of the Al-Faw Grand Port. The port will be located in southeast Iraq, along the Kawr Abdallah Channel, near the mouth of the Shatt Al-Arab waterway that runs between Iraq and the Islamic Republic of Iran. Its design takes into account not only technical navigational aspects but also other surrounding ongoing project interferences and morphological and hydrodynamic effects, thus optimizing the port's location and layout (Technital, n.d.).

The master plan considered many technical aspects, including the results of geotechnical and hydrodynamic studies. It included the construction of two large breakwaters, at the east and the west of the port, the former being around 8 kilometers long and the latter almost 14 kilometers long. The plans account for possible quays and yards extensions in the future and included large dredging works and rock revetments of the reclaimed port embankments (approximately 13 kilometers

long) as well as additional ones placed along the north and south boundaries of the navy base embankment (Technital, n.d.).

Morphological studies were also conducted for the final design of Al-Faw Grand Port to assess the effects of the port's construction on the morphology of the surrounding coastline, on the Shatt Al-Arab delta, and on erosion and accretion rates surrounding the port. The computational modeling assessing these various effects predicted that there will be no impacts overall (for example, no local coastline changes surrounding the port and no impacts on the Shatt Al-Arab delta). Nonetheless, the model did predict the accretion of muddy sediment into the port, thereby aiding the design process to account for this and optimize the design, considering hydrodynamic processes to reduce port siltation and move the sediment out of the port.[a]

The Al-Faw Grand Port illustrates the value of computational modeling in predicting, designing, and assessing in a circular approach to arrive at an optimal design and avoid coastal erosion from coastal projects of this kind.

a. For a more detailed description of the computational modeling, see the Deltares project page, "Morphological Studies for Al Faw Port": https://www.deltares.nl/en/projects/morphological-studies -al-faw-port/.

of different protective measures (box 5.4). Such simulations allow for closer inspection of different measures and their effects and can hence aid decision-making in an evidence-based manner before implementation. These studies are sparse, however, often because comprehensive analyses of sediment budgets and coastal-flow dynamics are lacking. Hence, prospective investigations lack the foundational information they

---

BOX 5.4

Rosetta Promontory: Computational Modeling of Solutions to Fight Coastal Erosion

The use of data, monitoring, and analysis through computational modeling of hydrodynamics and sediment processes is essential in assessing solutions for combating coastal erosion. An example of this praxis is the evaluation of different solutions to stabilize the Rosetta promontory at the northern end of Egypt's Nile delta coast. Morphological and hydrodynamic studies and computational modeling assessed the effects of different solutions (such as hard and soft measures as well as nearshore and beach nourishment) on the stability of the Rosetta promontory. This is important in understanding which solutions can work most effectively while accordingly increasing the efficiency of the planning and management process.

Different nourishment interventions (and at different volumes) were assessed regarding their impacts on the stabilization of the Rosetta promontory. One set of scenarios is for nearshore nourishment and another set for beach nourishment. The latter is the focus here as an example of the use of Coastal Modeling System software to analyze morphological changes of potential beach nourishment interventions.

One solution set compares the results of different nourishment-placement scenarios around the river mouth (involving a nourishment volume of 300,000 cubic meters) with those of a no-action scenario. The analysis predicted that introducing beach nourishment on the western headland is the optimal solution. It would help reduce coastal erosion on the western side of the promontory while decreasing the accretion inside the outlet of the river mouth; however, the eastern side of the promontory would still be subject to coastal erosion (Masria, Abdelaziz, and Negm 2015).

The modeling also assessed the impacts of various soft and hard measures to arrive at an optimal solution for the stability of the Rosetta promontory. The impact of beach nourishment as a soft-defense solution in combination with different hard-defense solutions were simulated using computational modeling. The resulting optimal solution is to provide 300,000 cubic meters of beach nourishment on the western side and nearshore nourishment on the east side and a 360-meter jetty on the east side of the river mouth. The analysis predicted this scenario to be the optimal solution because it could combat coastal erosion threatening the stability of the seawalls, reduce accretion inside the outlet of the river mouth, and decrease wave height at the outlet as well (Masria, Abdelaziz, and Negm 2015).

require—a situation that emphasizes the need for such comprehensive analyses for the various shorelines of Middle East and North Africa economies as well as for specific hot spots of coastal erosion.

Making data and information openly available and easily accessible for academic research and for development purposes is of paramount importance to spur research on coastal erosion and its drivers. The complexity of this task requires a deep understanding of the various dynamics and characteristics of the Middle East and North Africa's shores—including, as already mentioned, local geomorphological specificities, wave dynamics, and sediment transport, among others. Increasing research—both academic and applied (for example, for the establishment of ICZM schemes)—and strengthening the knowledge base about these issues is crucial for choosing appropriate policy responses and implementing a holistic ICZM scheme.

These efforts should identify not only the most effective strategies to combat coastal erosion at specific hot spots but also the root causes of the erosion, requiring a broad set of different skills. This necessitates capacity building in a broad range of specialized fields such as coastal engineering, fluid mechanics, and urban planning. Furthermore, wide availability of data and easy access to these data—such as through websites showing the degree and evolution of coastal degradation and its causes—can increase public awareness about the issue of coastal erosion. This in turn will help build a broad base of support for some measures (such as strict zoning laws and potentially necessary plans for managed retreat) that may otherwise be met with some degree of discontent.

Comprehensive Policies and Actions for Reducing Coastal Erosion

ICZM, with its subcomponents, represents the overarching approach to mitigating coastal erosion effects and conducting broader coastal zone planning. It is an integrative approach that involves stakeholder participation, along with assessments of both existing development and future proposals, to identify hazards that potentially lead to coastal erosion. This section discusses ICZM and experience with it in the Middle East and North Africa and elsewhere. The issues discussed below are (a) *prospective* (forward-looking) management by identifying parts of the coast for strategic intervention (that is, zoning); and (b) *reactive* management and control measures, including hard and soft defenses as well as control policies such as managed retreat and dam regulations.

ICZM continuously assesses changes in the coastal landscape, identifies their drivers, and uses this information to plan and manage coastal development. Identifying the sources of coastal erosion is a necessary prerequisite for planning and managing changes in the coastal land-scape, whether they are caused by human development or natural forces. ICZM begins with (a) identifying stakeholders whose actions may influence these processes (for example, port authorities, fishers, hotels, utility service providers, nature conservation specialists, cultural herit-age authorities, restaurants, technical specialists, and representatives of local communities); and (b) assessing the initial situation with the help of data collection, monitoring, and analysis. However, an ICZM scheme is premised on *continuous* integration of affected stakeholders and *ongoing* assessment of changes that certain management practices might have had on the coastal landscape. In the absence of a holistic planning process that incorporates the various involved parties, which ICZM represents, coastal erosion is likely to increase as development continues along the Middle East and North Africa's coasts.

A multisector approach is crucial for long-term optimal socio-economic outcomes and mutually beneficial trade-offs. ICZM is hence important in creating an integrated policy and technology intervention plan from the local to the regional level, and ideally the national or even international level, that optimizes the outcomes of combating coastal erosion and enabling coastal zones' sustainability and economic oppor-tunities for all stakeholders.

In the Middle East and North Africa, ICZM efforts regarding stake-holder inclusion started decades ago, with many lessons learned and areas for growth. Of the 22 countries bordering the Mediterranean Sea (9 of which are Middle East and North Africa economies), 10 have not yet begun to enforce the 2008 ICZM Protocol 10 from the Barcelona Convention, with Malta last to enter it into force in 2019 (UNEP 2020). Of these 10 countries, four are in the Middle East and North Africa: Algeria, Egypt, Libya, and Tunisia.

Most of the published work on ICZM in countries sharing the Red Sea and Gulf of Aden coasts (Djibouti, Egypt, Jordan, Saudi Arabia, Somalia, and the Republic of Yemen) has focused only on capacity building and ecosystem-based management, including coral reef management—which helps combat coastal erosion but is not sufficient.

ICZM stakeholder engagement in Egypt began in 1996, resulting in the establishment of a national ICZM committee (including stakehold-ers) as well as a department of coastal and marine-zone management including divisions for the Mediterranean and the Red Sea coasts (Abul-Azm, Abdel-Gelil, and Trumbic 2003). Additional recent initiatives in the region, such as Morocco's ICZM scheme, are in the implementation stage (box 5.5).

To tackle issues related to climate change and its impact on the coasts, Egypt has taken steps to enhance the resilience of coastal settlements and

BOX 5.5

Integrated Coastal Zone Management in Morocco

After ratifying the Barcelona Convention's ICZM Protocol 10 in 2012, the Moroccan Parliament approved a new Coastal Law (Law No. 81-12) in June 2015. The aim was to balance the protection and promotion of natural assets along the coast with economic, social, and cultural development. Some of the explicit goals of Law No. 81–12 are "(i) Preserve the coast's biological and ecological balance, natural and cultural heritage, while combating coastal erosion; (ii) prevent and reduce pollution and the coast's degradation, while rehabilitating polluted and damaged areas; (iii) improve planning, by means of a national plan for the coast and compatible regional spatial planning documents; […] (vi) advance research and innovation promoting the coast and its resources" (Trakadas 2020, 3; based on World Bank reports). With it, Morocco became one of the first countries in the region to have a legal instrument dedicated for the integrated management of its coastal resources.

Currently, the government of Morocco is implementing the mandate for regional coastal management plans (Schémas Régionaux d'Aménagement du Littoral, or SRLs). With support from the World Bank, the National Integrated Coastal Management Plan (NLP) was developed incorporating some of the main principles of ICZM.

In the Rabat-Salé-Kénitra region of northern Morocco, an SRL was implemented and launched in early 2021 (World Bank 2021). Set up with technical and analytical support from the World Bank and the Italian government, the SRL represents the first regional strategy arising from the NLP that aims to reconcile environmental protection and economic activity under one umbrella. Human activities will be coordinated under the guiding principles of a comprehensive ICZM scheme, part of the 2040 road map for sustainable development of the coastline in the Rabat-Salé-Kenitra region.

The SRL includes setting up a regional zoning plan and incorporates participation of the various stakeholders affected. Some planned investments aim to "green" the activities scattered along the coast, strengthening the resilience of the coastline. An important part of these investments aims to address coastal erosion by embedding safeguard measures in the development plans of urban centers—including, for example, the biological stabilization of dunes. Other investments within the long-term coastline development scheme include construction of wastewater treatment programs, rehabilitation of coastal wetlands, increasing capabilities for recycling and recovery of plastic waste, and organizing and training fishers for sustainable fishing along the coast.

infrastructure through stakeholder engagement and capacity building. Egypt's Nile delta is considered one of the areas most vulnerable to the adverse effects of climate change, including the threat of coastal erosion. Coastal lakes are key ecosystems acting as protective zones for inland economic activities, but they are separated from the Mediterranean Sea only by dune systems that are already eroding and retreating. Rising sea levels and accompanying intensification in erosive forces threaten these ecosystems.

Egyptian authorities together with the UN Development Programme (UNDP) established the project, "Enhancing Climate Change Adaptation in the North Coast and Nile Delta Regions in Egypt," to efficiently manage the risks posed by SLR and accompanying coastal erosion (box 5.6). It aims to integrate the management of these risks into the development of low-elevation coastal zones in the Nile delta that play an important role in the regional economy. The lessons learned in this project can be important for future projects trying to implement ICZM principles in the Middle East and North Africa.

Prospective Management through Integrated Planning

Under the ICZM umbrella, analysis can inform the identification of zones for different interventions, the formulation of plans with targeted interventions, and the development of other policies and institutional measures. Zoning is an effective approach that—using data, monitoring, and analysis for risk assessment and solution—identifies areas with a set of action plans for coastal development to address challenges (including coastal erosion) and opportunities in a comprehensive and sustainable way. As such, coastal zone planning is one of the most important cornerstones of an effective ICZM scheme. These plans usually designate specific interventions for different parts of the coast and divide the coastline into zones based on current usage and, perhaps more importantly, possible future usage. Strategies will differ depending on natural characteristics as well as priorities expressed by different stakeholders.

Where adaptive actions to address coastal erosion are part of an ICZM process, they generally involve three options: protection, accommodation, and retreat (Few, Brown, and Tompkins 2007). Accordingly, in an ideal ICZM scheme for combating coastal erosion, the spatial, temporal, and stakeholder engagement aspects are all integrated and balanced against one another to achieve a solution acceptable to all involved parties.

Zoning regulation based on considerations of hazard areas and existing human structures are crucial in combating coastal erosion and mitigating its effects. Coasts have different hazard areas such as tidal inlets, swashes, and permanent overwash passes; these may have different effects on human structures based on characteristics like critical sediment-balance zones, elevated sites, and low-density development zones (Rangel-Buitrago, Anfuso, and Williams 2015). Identifying such hazard areas and considering the claims of stakeholders is a crucial step in mitigating coastal erosion risks in planning and management. Suitable policies and

BOX 5.6

Enhancing Climate Change Adaptation in the North Coast and Nile Delta Regions, the Arab Republic of Egypt

The UN Intergovernmental Panel on Climate Change (IPCC) has identified the Nile delta as one of the world's three areas most vulnerable to climate change. It faces the threat of flooding of low-lying coastlines from sea-level rise (SLR) and increased storm intensity and frequency. Salinization of land and water resources is expected to have significant impacts on agriculture, fishing, and the availability of freshwater resources.

To improve coastal zone management in the context of SLR, the Arab Republic of Egypt is working with the UN Development Programme (UNDP) on the "Enhancing Climate Change Adaptation in the North Coast and Nile Delta Regions in Egypt" project. Supported by the Green Climate Fund and approved in 2017, the project aims to reduce coastal flooding risks to Egypt's northern coast due to SLR and extreme storms. The project will address barriers including a lack of high-quality data to inform planning decisions; the absence of a suitable framework for integrated approaches to coastal adaptation; weak institutional coordination to build coastline resilience; issues with the disposal of dredge material that would otherwise be disposed of in the marine environment; and low institutional capacity to anticipate and manage the expected impacts of SLR.

To facilitate transformational change by reducing coastal flooding threats and laying the framework for more sustainable coastal development, the project has two key components:

- *Development of 69 kilometers of sand-dune dikes along five vulnerable hot spots within the Nile delta.* The dikes have been designed to mirror natural coastal features and/or sand dunes and will transform the areas from high-risk to low-risk zones for flooding. They will be stabilized with a combination of rocks and local vegetation species to encourage dune growth by trapping and stabilizing blown sand. Importantly, the coastal protection measures will provide beneficial reuse of dredge material that would otherwise be disposed of in the marine environment.

- *A climate change risk-informed integrated coastal zone management (ICZM) plan.* The plan will enable high-resolution diagnosis of coastal threats, updated regulatory and institutional frameworks to manage SLR, and a coastal observation system for ongoing data collection and analysis.

Altogether, the program will benefit nearly 800,000 people directly and up to 14 million indirectly in the coastal governorates of Beheira, Dakhalia, Kafr El Sheikh, and Port Said.

The project's costs will total an estimated US$105 million. Key central government implementing agencies in Egypt include the Ministry of Environment, the Ministry of Agriculture and Land Reclamation, and the Egyptian Meteorological Authority. Research institutes and universities are also involved.

Source: "Enhancing Climate Change Adaptation in the North Coast and Nile Delta Regions in Egypt," Project FP053, Green Climate Fund, Incheon, Republic of Korea: https://www.greenclimate.fund/project/fp053.

regulations, including zoning regulations and implementation measures, must also be identified.

Principles of zoning for coastal erosion management have been adopted in large-scale projects. A recent example is the "Shoreline Management Sub-Plan for Odisha Coast," published in July 2018 as part of India's ICZM scheme (supported by the World Bank), which included a comprehensive and integrated analysis and implementation plan. Here, for the sake of combating coastal erosion, the coastlines were divided into nine different zones with accompanying policies. These zone-specific policies include combinations of abstaining from intervention (doing nothing), limited interventions, holding the line, managed realignment, and stabilizing sea walls (map 5.4).

MAP 5.4

Shoreline Management Sub-Plan for Odisha Coast, India, 2018

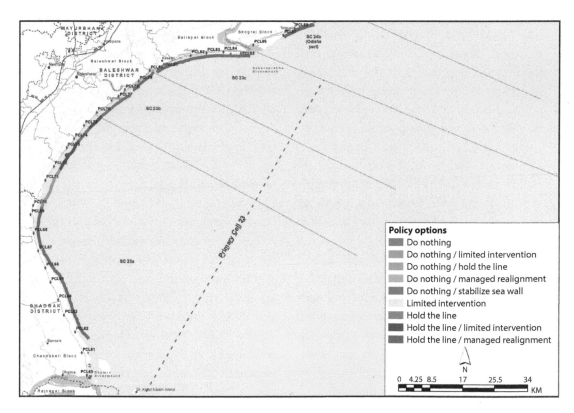

Source: IPE Global 2018. © PD, ICZMP, Odisha. Used with the permission of PD, ICZMP, Odisha. Permission required for reuse.
Note: Colors designate nine coastline zones corresponding to zone-specific policy categories, as follows: "Do nothing" applies to zones where natural shore evolution is preferred and no investment in coastal defense is recommended. Under "managed realignment," natural shoreline changes are allowed, and investments may involve moving human settlements beyond the predicted shoreline change if they are threatened by erosion. "Limited interventions" involve working with natural coastal changes while reducing risk through measures to slow rather than stop coastal erosion, ranging from vegetative measures to appropriate location of dredging disposal at river mouths for beach nourishment. "Hold the line" options include soft-defense and hard-defense solutions (with a preference for the former in combination with vegetation) that aim to mitigate or stop erosion. It is assumed that managed realignment or retreat is not an option in the "hold the line" areas.
KM = kilometers; PCL = Permanent Coastal Location; SC = subcell.

The ICZM scheme in India involves multiple states and is one of the most recent attempts to implement a holistic plan for large shorelines in regions severely stressed by the consequences of coastal erosion compounded by climate change. Its initial implementation in the states of Gujarat, Orissa, and West Bengal has been successful. Box 5.7 describes the overall ICZM scheme in more detail.

BOX 5.7

India's ICZM Project: A Comprehensive Approach for Combating Coastal Erosion

The World Bank supported the government of India in building capacity for the implementation of a comprehensive coastal management approach beginning in 2010. The first component of the project included mapping, delineation, and demarcation of hazard lines together with the delineation of coastal-sediment cells along India's main coast. The other three components were (a) piloting of integrated coastal zone management (ICZM) approaches in the states of Gujarat, Orissa, and West Bengal; (b) building the capacity of agencies and institutions at the state level; and (c) training technical and administrative staff in ICZM planning and implementation. The project included pilot investments to protect coastal assets while also protecting and enhancing biodiversity at various sites (for example, through mangrove-shelterbelt plantations) as well as investments in better waste management and improvement in the livelihoods of coastal communities (for example, by promoting small-scale ecotourism).

In the course of the project, more than 7,800 kilometers of India's mainland coast were mapped, delineated, and classified in the coastal hazard line—that is, the line of likely impact from natural hazards—to implement policies to protect coastal assets and communities from the adverse effects of disaster risks, including coastal erosion. The example shown in map 5.4 involves identifying different policy options for different parts of the shoreline together with detailed recommendations on the most promising interventions. For example, measures in intervention zones—that is, Permanent Coastal Location (PCL) 72 to PCL 77—included a combination of mangrove afforestation, installing low-slope revetment walls behind the shore, and beach nourishment to reduce the pressure of erosive forces and limit the risk of inundation during storm events (IPE Global 2018).

The preparation of plans and pilot activities in the states of Gujarat, Orissa, and West Bengal was highly participatory. Nearly 72,000 local inhabitants and 2,500 representatives of stakeholder groups were engaged, including government agencies, the private sector, fishing communities, and other members of civil society as well as tourists (World Bank 2020a). Emphasis was given to activities with the dual advantage of reducing erosive processes

(continued)

BOX 5.7

India's ICZM Project: A Comprehensive Approach for Combating Coastal Erosion (*Continued*)

while benefiting biodiversity. About 19,500 hectares of mangroves were restored or planted, acting as coastal carbon sinks while protecting coastal assets and enriching local biodiversity.

A project in Pentha, Odisha, demonstrated the merits of Geotube technology (further described in the subsection on hard defenses below). It involves 505 meters of geotextile tubes along the Pentha village coast that act as a bund against tidal wave actions and mitigate erosive forces. It protects the lives and livelihoods of more than 40,000 residents as well as around 250,000 tourists who visit annually (World Bank 2020a). In addition to helping the village withstand the severe cyclones Phailin, Hud Hud, and Fani that hit Odisha in 2019, it

has contributed to the restoration of the beachfront (Technical Textiles 2020).

Following the successes in developing the ICZM plan for India, the World Bank in April 2020 approved a multiyear US$400 million financing envelope to further support coastal states in India in enhancing the resilience of their coastal resources and populations. In the first phase, around US$180 million was provided to the Enhancing Coastal and Ocean Resource Efficiency (ENCORE) project, which will cover eight coastal states (Andhra Pradesh, Goa, Gujarat, Karnataka, Kerala, Odisha, Tamil Nadu, and West Bengal) as well as three coastal union territories (Daman and Diu, Lakshadweep, and Puducherry) where coastal resources are under significant pressure (World Bank 2020b).

Zoning plans as part of ICZM schemes have also become an important part of regional initiatives to efficiently address more-local coastal erosion issues. The ICZM plan for the beaches near Mundesley Beach in Norfolk, UK, is an example of such an intervention scheme. The region and its beaches are severely affected by coastal erosion, and local authorities have been planning and implementing erosion protections in various forms since the 1940s. In-depth studies have assessed the risks of coastal erosion on local municipalities (Dawson et al. 2009; Dickson, Walkden, and Hall 2007).

The ICZM plan divides the coast into three types of zones, each with a dominant policy intervention: doing nothing, holding the line, or maintaining existing defenses (but once the defenses are projected to fail, then strategically retreating). It also assigns different priorities to different localities—for example, the protection of the Bacton gas terminal, compared with the longer-term protection of Mundesley itself. Furthermore, because of the national importance of the Bacton gas terminal, both the short-term (0–20 years) and medium-term (20–50 years) policies are to hold the line (because the terminal's lifetime can reach up to 50 years) along with other interventions such as maintaining existing defense infrastructure and introducing a large volume of sediments starting in 2018 (Norfolk Vanguard Limited 2018).

Even when there are no plans to protect, other strategies should be considered. For example, the ICZM zoning plan for Mundesley Beach considers no protective measures for the coast along the village of Happisburgh—that is, the plan is to do nothing. However, this does not necessarily imply that no strategies are under consideration to address the coast's erosion and the risk that near-shore properties will be destroyed. Providing inhabitants of these properties with clearly defined adaptation strategies in advance reduces the pain of eventual relocation. A clear strategy for the abandonment of land that cannot be saved provides the various stakeholders with a perspective that enables them to plan accordingly.

Support measures for displaced persons include buyout schemes, subsidies for the relocation process, and purchasing or building new developments elsewhere that are safe. This was done in Happisburgh. Properties considered to be at immediate risk of erosion were purchased from the residents, and they were simultaneously granted an automatic planning right to replace their homes on the landward side of the village. Similarly, the cliff-top caravan park, which is an important source of income, was relocated farther inland (Kerby 2019).

Reactive Management and Control Measures

Stakeholders or policy makers can adapt existing structures to coastal erosion and manage coastal erosion in two ways: (a) defend the coast with "hard solutions" like seawalls or "soft solutions" such as beach nourishment, including natural and NBS such as mangrove restorations; or (b) implement control policies for development purposes or directly for sediment-balance purposes. These two options of defense and control policies are discussed in more detail below.

Defense Solutions

The range of possible measures to address coastal erosion and defend shorelines and settlements can be categorized as hard-defense solutions or soft-defense solutions (including NBS). All these measures aim to reduce or absorb the erosive forces of currents and waves but differ in their integration with the natural environment. The specific implementation of one measure or another is highly case-specific and depends on local characteristics but should be coordinated, ideally in a comprehensive action plan designed under the tenets of ICZM.

Hard defenses. Hard-defense solutions often are interventions including gray infrastructure that affect wave intensity and sediment transport. These include artificial headlands, groins, offshore structures such as breakwaters, and seawalls and revetment rock armors. They are often made of natural stones, concrete, or a combination, and their main objectives are to absorb and disperse wave intensity, influence sediment transport, or affect both wave intensity and sediment transport. Box 5.8 provides an overview of some of the most often used hard-defense structures.

BOX 5.8

General Overview of Hard-Defense Options

Hard-defense solutions include a variety of options, some of which are placed foreshore (figure B5.8.1), and others are located on the shore (figure B5.8.2). The figures show both the intended effects of such protection solutions and their unintended, possibly disadvantageous, effects. For some examples of such solutions that were used in the Arab Republic of Egypt, see photo 5.4.

Foreshore Structures

Groins. Foreshore structures such as *groins* (row 1 of figure B5.8.1) are structures built approximately perpendicular to the shoreline outward to locally trap transported sediments by shoreline drift. The objective is to stabilize the shoreline by decreasing sediment transport away from certain sites and encouraging buildup of beaches there; there may again be adverse effects on nearby areas. They can be built out of different materials, including wood poles, timber, or rocks.

Jetties. Built in a similar fashion, jetties (row 2 of figure B5.8.1) are typically installed for the protection of navigation channels used by ships. Their interference with natural sediment transport can also cause erosion in and along other coastal areas downdrift.

Breakwaters. Breakwaters (row 3, figure B5.8.1) can be built offshore and reduce the impact of waves on the coast to minimize coastal erosion and inundation along the coast locally to protect strategic points of the shoreline. When connected to the shore, they are also often called artificial headlands.

Temporary breakwaters or headlands can be formed of gabions or sandbags, but those will dissipate faster and are used for short-term protection.[a] These structures hinder the process of coastal erosion primarily by reducing the energy with which waves arrive at the shore. Artificial headlands additionally encourage the buildup of beaches by trapping sediment brought in by waves. However, this trapping interferes with the natural flow of sediments and could cause increased erosion at other parts of the coast.

In addition, geotextile tubes are increasingly used in civil and environmental

(continued)

BOX 5.8

General Overview of Hard-Defense Options (*Continued*)

FIGURE B5.8.1

Foreshore Hard-Defense Structures to Combat Coastal Erosion

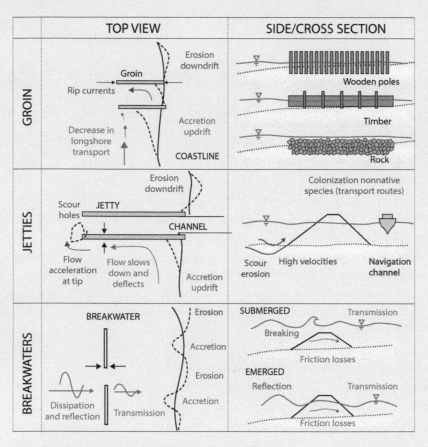

Source: Adapted from Schoonees et al. 2019.
Note: Text in blue designates intended (advantageous) effects of the structure, and text in red designates unintended (disadvantageous) effects.

applications to combat the effects of coastal erosion. They consist of geotextiles filled with soil, sand, or other material and present an alternative to traditional forms of coastal structures, which may be expensive to build and maintain because of shortages of suitable building materials (Shin and Oh 2007).

Onshore Structures

Other hard-defense solutions are more focused on protection behind the shore without directly interfering in natural sediment transport alongshore. They are typically placed onshore to dissipate wave energy and reflect it back.

(continued)

BOX 5.8

General Overview of Hard-Defense Options (*Continued*)

Seawalls. Seawalls (row 1 of figure B5.8.2) are vertical structures built along the coastline to protect the coast from the impacts of coastal erosion such as exposure to waves and tides. Their goal is to protect backshore infrastructure by absorbing shock waves and hence hindering erosion of the natural shoreline, helping to stabilize it.

Although they do not usually block natural sediment transport along the shoreline, they block natural sediment replenishment from the shore (and of course interfere with the natural appearance of the shoreline itself).

Construction of seawalls is typically quite costly and is often considered only

FIGURE B5.8.2

Onshore Hard-Defense Structures to Combat Coastal Erosion

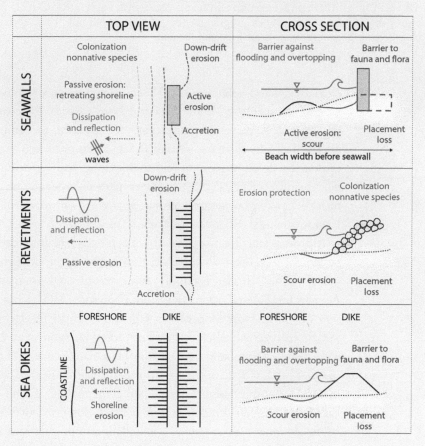

Source: Adapted from Schoonees et al. 2019.
Note: Text in blue designates intended (advantageous) effects of the structure, and text in red designates unintended (disadvantageous) effects.

(continued)

BOX 5.8

General Overview of Hard-Defense Options (*Continued*)

for parts of the shoreline where the assets to be protected have substantial value. Furthermore, seawalls themselves are subject to erosion and hence must be maintained regularly and strengthened periodically.

Revetments. Revetment rock armors, sometimes also called riprap seawalls (row 2 of figure B5.8.2), are often used in combination with seawalls to reduce the intensity of waves before they hit seawalls;

however, they are also employed on their own to protect the coastline.

Sea dikes. Built in a similar fashion, sea dikes (row 3 of figure B5.8.2) are often considered a last line of flood defense and are also known as levees or embankments (Schoonees et al. 2019). They are often composed of an earth-filled core with smooth slopes on both seaward and landward sides and protect the low-lying hinterland thanks to their raised ground level.

a. "Artificial Headlands," Resilience-Increasing Strategies for Coasts Toolkit (RISC-KIT) project of the European Union's Seventh Framework Programme (FP7) for Research and Technological Development, https://coastal-management.eu/artificial-headlands.

Hard-defense solutions have been commonly used throughout the world, including the Middle East and North Africa, for coastal protection against erosion and inundation. For instance, to combat negative effects of climate change such as SLR, Egypt has protected its shores from erosion by erecting barriers and maintaining and repairing its coastal defense infrastructure (SIS 2020). These hard-defense solutions (photo 5.4) include seawalls, breakwaters, groins, and revetment rock armor and have been widely used (EEAA 2016; Koraim, Heikal, and Abozaid 2011).

Morocco and Tunisia have also been active in implementing hard structures to protect their coastal assets. Given the importance of the coasts to the Tunisian economy, especially through tourism, Tunisia's Agency for Coastal Protection and Planning (APAL) has initiated a series of projects in cooperation with the German Development Bank (KfW) (Albrecht-Heider 2020). For example, submarine breakwaters were installed to rebuild the beach in Hamman Sousse, and groins were installed in Raf Raf, north of Tunis. The latter project was complemented by adding more than 500,000 cubic meters of sand—a practice called beach nourishment, a soft-defense solution further discussed below. Another measure was the installation of protective seawalls using natural rocks at the Kerkenna Islands near Sfax.

In Morocco, breakwaters have been used heavily to protect several ports, such as North Africa's largest port, Tanger-Med; a 3-kilometer breakwater development at the shore near Safi; and in the harbor of Rabat.

PHOTO 5.4

Hard-Defense Solutions in the Nile Delta Zone in the Arab Republic of Egypt

| a. Mohamed Ali seawall, Abu Qir Bay | b. Revetment, Rosetta Promontory |

| c. Detached breakwaters, El Agami Beach | d. Breakwater, El Arish Harbor |

| e. Groins, El Mandara Beach | f. Jetty, Ras El Bar |

Source: Iskander 2010. © Moheb M. Iskander. Used with the permission of Moheb M. Iskander. Further permission required for reuse.

Groins have been built in several areas, including the port of Tan Tan and the beaches of Tangier, and seawalls have been used across the country such as in Essaouira, Rabat, and Tangier.

Hard-defense solutions such as groins and breakwaters interfere with sediment transport alongshore, sometimes by design. Hence, deep understanding of sediment transport and prevalent coastal-flow dynamics becomes especially important and necessitates detailed analyses, ideally carried out on a large scale to consider transboundary changes. Different hard-defense solutions may have distinct influences on sediment transport, and hence their ability to prevent coastal erosion (box 5.9).

BOX 5.9

Effects of Different Defense Structures in Soliman Beach, Tunisia

The coastal zone of Soliman is located at the Gulf of Tunis, southeast of the city of Tunis. This coastline, in particular, has seen strong rates of coastal erosion through time and has been well studied in the literature (Marzougui and Oueslati 2017; Saïdi, Souissi, and Zargouni 2012). Infrastructure projects in 1989 and 1990 included the erection of breakwaters, which were replaced by a coastal-groin system in 2018. This example reveals interesting patterns in the coastal-sediment transport and erosion patterns at Soliman Beach. Photo B5.9.1 provides views of the breakwaters used before 2018 (panel a) and after their replacement with groins in 2018 (panel b).

Considering the area where the two breakwaters were replaced (at lower-left of Soliman Beach as depicted in photo B5.9.1), figure B5.9.1 reveals the shoreline changes of this particular area in more detail. Figure B5.9.1, panel a, shows the coastline as of May 2017—that is, before the replacement of the breakwaters—overlaid with the position of the shoreline in the past two decades, indicated by the differently colored lines. Figure B5.9.1, panel b, then shows the

erosion and accretion patterns for each of the transects located at different parts of the coast (as marked in figure B5.9.1, panel a).

As can be seen from the bar chart in figure B5.9.1 (panel b), transects 5–9 especially experienced severe erosion over the years up to 2018. Following the replacement of the breakwaters in 2018, transect 2 shows severe erosion that just reflects the removal of the structures. What is interesting are the changes in transects 5–9, where there is an immediate and large reversal of the erosion that had occurred in the years before the introduction of the groins. This demonstrates the development and regeneration of the beach following the introduction of the groin system, which altered sediment transportation and distribution from the upstream beach parts getting trapped because of the perpendicular design of the groins.

The structural groin also had a positive effect farther downstream (that is, to the left of the breakwaters in photo B5.9.1, panel a), leading to less coastal erosion, with sediments building up as well. These processes show that different structures

(continued)

BOX 5.9

Effects of Different Defense Structures in Soliman Beach, Tunisia (*Continued*)

PHOTO B5.9.1

Soliman Beach, Tunisia, before and after Replacement of Breakwaters with Groins

a. Beach with breakwaters (1990–2018)

b. Beach with groin system (2018–present)

Source: © 2020 CNES/Airbus, courtesy of Google Earth.

(continued)

BOX 5.9

Effects of Different Defense Structures in Soliman Beach, Tunisia (*Continued*)

FIGURE B5.9.1

Changes in Erosion at Soliman Beach, Tunisia, after Replacing Breakwaters with Groins

a. Beach in May 2017, showing transects and shorelines, 2000–20[a]

b. Extent of shoreline change, by transect, 2000–20[b]

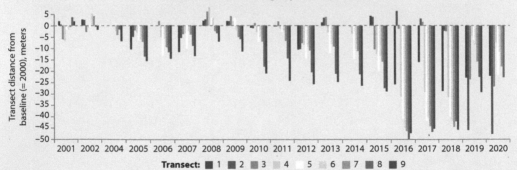

Source: NOC 2020. Panel a: © World Bank. Further permission required for reuse.
a. The photo in panel a shows Soliman Beach as of May 2017, before construction of the groins.
b. The bar chart shows cumulative erosion or accretion at different transects in different years relative to the baseline year (2000).

may have quite dissimilar impacts on the processes affecting the coast and beaches. Careful planning is necessary to incorporate possible effects of different structures and to select the ones that have the desired effects once implemented.

Source: Cooperative analysis for this volume with the National Oceanography Centre (NOC) in the UK using high-resolution satellite photos of Soliman Beach in Tunisia.

Soft defenses. Soft-defense solutions have been used in recent decades. Soft solutions include beach nourishment, wind fences, sand dunes, artificial reefs, and other NBS:

- *Beach nourishment* is the addition of sediments, sand, or both along the shoreline to maintain beach land width and act as a soft-defense mechanism by dissipating wave energy before it reaches the coastline. This aids shoreline stabilization, improves beach quality, and improves storm protection. Beach nourishment can be costly, affect marine ecosystems, and require a constant supply of new sand.

- *Wind fences* are structures that act as barriers against winds to accumulate sand and sediments carried by the winds on the downwind side of the fence as an intervention for shoreline stabilization (Khalil 2008).

- *Sand dunes* are mounds of sand built or maintained to protect the coast from coastal erosion impacts such as exposure to waves and tides. Dunes act as backshore protection, protect the coast from waves, and can be vegetated so they are not easily eroded.

- *Artificial reefs* are objects made of environmentally friendly materials placed on the seabed offshore to help retain sediments and dissipate waves while having other environmental impacts such as creating a thriving marine ecosystem (shoreline stabilization).

- *Nature-based solutions (NBS)* are interventions aimed at safeguarding, sustainably managing, and rehabilitating the ecosystems in addressing societal challenges while benefiting human well-being and biodiversity.

Beach-nourishment projects have been implemented in the Middle East and North Africa and received more attention recently. For example, between 1986 and 1995, five beaches near Alexandria, Egypt, were restored this way. The beaches of El Asafra, El Mandara, El Shatby, Miyami, and Stanley were (re)nourished using sand from the desert near Cairo, with all of these projects meeting or exceeding the expectations for beach restoration (Fanos, Khafagy, and Dean 1995). In an assessment for further protections, beach nourishment under the tenets of an ICZM scheme was considered the most cost-effective measure to combat beach retreat (El-Raey 2009).

In 2014, the United Arab Emirates approved a large-scale project to restore the three Umm Suqeim beaches after they faced severe erosion due to offshore developments. For Umm Suqeim I Beach, installation of five groins and beach nourishment were undertaken, while for the other two, only beach nourishment was used. The works included a total of around 500,000 cubic meters of sand, and around US$9.5 million was approved to implement the measure in multiple phases (*Construction Week* 2014; *Khaleej Times* 2017), which also included a nine-month closing of the beaches (*Khaleej Times* 2015). Innovative approaches for the effective deployment of beach nourishment practices have been used extensively—for example, in the Netherlands (box 5.10).

BOX 5.10

Building with Nature: Approaches for Beach Replenishment from the Netherlands

In the Netherlands, an innovative nourishment process was adopted, pumping sand and gravel onto the beach during high tide and allowing the coastal process to spread the sand, resulting in the restoration of the dunes and beach (de Schipper et al. 2016). The project, called "Sand Motor," is located along the narrow Dutch coast between The Hague and Hook of Holland and was constructed in 2011.[a] The project contains approximately 20 million cubic meters of pumped sand that, in its original form, was placed as a hook-shaped peninsula with a length of 2 kilometers alongshore and extending 1 kilometer into the sea. A series of assessments have found it to be successful in feeding the adjacent coastlines with sediments and aiding beach buildup (de Schipper et al. 2016; Luijendijk et al. 2017; van der Meulen and van der Valk 2019).

The Sand Motor acts as a measure of defense to avoid the erosion of the coast that protects the nearby lands lying below sea level by spreading the sand in a natural way with wind, waves, and currents along the coast. As a result, this peninsula gradually changes its shape, decreasing in size, and over the long term it will fully assimilate with the protected coast. The Sand Motor project is an example not only of beach nourishment but also of an integrated coastal management approach that uses a soft-defense solution, data, monitoring, and analysis along with nature processes and recreation for a comprehensive solution (van der Meulen and van der Valk 2019).

The Spanjaards Dune is part of larger nourishment efforts along the coast (including the Sand Motor project and others) to reduce coastal erosion and ensure coastal safety (van der Meulen et al. 2015). It was constructed in 2008–09 with an original size of around 35 hectares in front of the coast to be protected—the Delfland coast near The Hague—as a compensation for losses of the original dune habitat (van der Meulen and van der Valk 2019). Nourishment through the Spanjaards Dune project required the dredging of 6.5 million cubic meters of sand that was piped to shore from 19 kilometers offshore. Similar to the Sand Motor, natural forces are left free to shape the area further and reinforce dune habitats that need compensation (van der Meulen and van der Valk 2019).

a. For more information, see the Sand Motor website: https://dezandmotor.nl/.

Other soft defenses include the installation of wind fences and artificial reefs. APAL has used several soft measures to combat coastal erosion, including wind fences to trap sand and rebuild dunes. APAL initiated a number of actions, implementing them in a participatory manner involving the local population, while taking into account the ecological, economic, and archeological aspects. These initiatives aim to protect the beaches, with the first three phases of the project costing about €38 million (of which 75 percent was financed with grants and loans from the German government). These initiatives included a series of different structures (for example, submarine breakwaters) as well as installation of over 4 kilometers of pinewood fences to stabilize dunes, which are natural protective barriers (Albrecht-Heider 2020). These protection measures help to withstand erosive forces like waves while having rather few side effects.

A pioneering project in Morocco, although not specifically aiming for protection from coastal erosion, installed artificial reefs in Martil, in the north of the country, in 2012. The project was successful in restoring fish stocks—the main objective—but stabilization of the Martil artificial reef community is a slow and long-term process (El Mdari et al. 2018).

NBS through vegetation and natural coral reef restoration have been increasingly used to retain sediments and restore biodiversity and blue carbon. According to a planning framework by Belize's Coastal Zone Management Authority and Institute, "Terrestrial protected areas provide erosion and flood control, sediment retention, and carbon storage. Marine protected areas which include coral reef, sea grass and mangrove offer a variety of coastal and marine services such as protection against erosion, reduction of damages from storm surge, and protection from sea-level rise" (CZMAI 2016). The use of vegetation—such as through the creation, conservation (or both) of wetlands and mangrove zones— includes rehabilitation or plantation of seawater vegetation along the shores to retain sediments and to act as a natural defense line by dissipating wave energy. This option traps sediments and prevents coastal erosion while also enhancing biodiversity by restoring habitats for land-based creatures and safe havens for fish in the case of mangroves.

Restoring biodiversity and supporting revegetation with native species is an important aspect of NBS. In the Middle East and North Africa, such efforts have included the Corso Commune coastal dune ecosystem rehabilitation project, east of Algiers in Algeria. In the course of it, the Ecological Association of Boumerdes, with cooperation of local public authorities, took steps to stabilize dunes with the aid of vegetation. The project also included action to support rehabilitation of coastal sites against human action by implementing actions for cleaning and development of these spaces (Canals Ventín and Lázaro Marín 2019).

In the Red Sea and the RSA, mangroves and coral reefs provide protection and contribute to combating coastal erosion in a cost-effective
way. When comparing the costs of hard-defense interventions such as
tropical breakwater projects with the costs of NBS such as coral reef
restoration projects, the cost-effectiveness of such NBS becomes clear.
However, mangroves and coral reefs have been under stress in the
Middle East and North Africa from climatic changes as well as human
interventions. For example, along the coasts of Saudi Arabia, mangrove
areas decreased by 75 percent between 1985 and 2013, and reclamation, dredging, and poor fishing practices destroyed coral reefs (MEWA
2017). It is hence important to afforest mangrove forests and restore
coral reefs where possible.

Recently, Egypt announced an ambitious project to plant mangroves
in the Red Sea Governorate. The project, announced in 2020, represents
one of the country's largest environmental projects and is an effort to
combat the effects of coastal erosion and overfishing. More than 200
hectares of land has been set aside for four plant nurseries in the Safaga,
Hamata, and Shalateen areas in the Red Sea Governorate and the Nabaq
Nature Reserve in South Sinai. Besides mitigating coastal erosion and
reviving dwindling fish stocks, the project aims to restore bee populations that feed from mangroves and protect coral reefs (Nile FM 2020).
According to officials, the restoration of coral reefs will also help to turn
the Red Sea coast into one of the most important destinations for environmentally conscious tourism while protecting beaches from erosion
resulting from waves and rising sea levels.

Hybrid defenses. Hybrid defense solutions have been used by
combining hard-defense solutions with soft-defense solutions including
NBS. These mixtures of solutions allow for a combination of the positive aspects of both solutions—for instance, using gray infrastructure for
wave-energy dissipation while also creating natural habitats for species
at or near the coasts, such as in mangrove woods and dune vegetation.
These combinations also allow for the smoother integration of solutions
in landscape planning and enhance policy coherence (Cohen-Shacham
et al. 2019). Hybrid solutions can involve several elements such as
submerged breakwaters, natural defenses, and dune reinforcement and
stabilization through vegetation. (See, for example, figure 3 in Antunes
do Carmo [forthcoming].) Typically, not all of the depicted solutions
are used together but rather as a combination of two or three solutions.

Hybrid schemes have been used internationally and are suitable for
Middle East and North Africa economies. NBS interventions were integrated into the coastal-defense strategy of Medmerry's coastal-defense
management in southeast England (Pearce, Khan, and Lewis 2012).
This strategy incorporated ecological engineering at that site, and by

connecting with similar measures aimed at erosion reduction, those NBS interventions had a large impact in reducing coastal erosion along the coast.

In the Middle East and North Africa, Egyptian authorities considered such a hybrid solution for the development of 69 kilometers of sand dunes that are stabilized by a combination of rocks and local vegetation to enhance resilience to climate change in the Nile delta (as discussed in box 5.6). This combination is set to trap blown sand and encourage dune growth. Given that the shores of the Middle East and North Africa region are habitats for some of the most suitable vegetation for coastal protection (for example, mangroves in the Red Sea), a combination of gray and green measures is a suitable option to consider for coastal protection in the region.

Control Policies

Control policies to complement the discussed defensive interventions and to regulate drivers of coastal erosion are important for sustainably managing the coasts. Such control policies aim to mitigate some of the coastal-erosion impacts of new and existing development, such as dams and coastal infrastructure (ports, jetties, and so forth). Control policies may also target activities that reduce sediment budgets—activities such as sand mining, which is illegal in most countries but unfortunately still practiced in some, such as Morocco.

The construction and development of dams in relevant zones must be regulated to take into account the influence of dams on the natural transport of sediment to the coast. Rivers frequently get dammed for water supplies, flood control, and hydroelectric power throughout the world, with notable projects in the Middle East and North Africa. North African countries like Algeria, Morocco, and Tunisia, and some Middle East countries like the Islamic Republic of Iran operate a large number of dams. For example, Morocco has 140 large dams with an overall capacity of about 17,600 million cubic meters and more than 100 small dam and hill reservoirs (Loudyi et al. 2018). In the Islamic Republic of Iran, over 600 dam projects have been built since 1979 (Shahi 2019). The Nile Dam between Egypt and Ethiopia, one of the largest projects in recent decades, has led to tensions between countries in the region (Mutahi 2020).

Rivers are important sources of sediments for coastal zones, and modifying them can have distinct consequences on their flows to the coasts (Syvitski et al. 2005). This interception of sediment can then significantly affect erosion processes, as was documented for the Aswan Dam (Masria, Nadaoka, Negm, and Iskander 2015) and the Sebou and Ouerrha Rivers in Morocco (Haida, Snoussi, and Probst 2004). For the Mediterranean Maghreb basin, dams are estimated to reduce sediment transport by more than 60 percent, with the highest retention rates in Tunisia (72 percent), Algeria (63 percent), and Morocco (55 percent) (Sadaoui et al. 2018). Hence, it is crucial that future projects involving dams be reviewed in this respect. Suitable regulations that address this issue have to be formulated, and ESIAs and EIAs of new projects should include this dimension.

Existing dams could benefit from retrofitting to allow for sediment flow and less accumulation of sediments upstream of the dam. Various methodologies exist to route sediment around dams, through dams, or to relocate sediment trapped in the reservoir to allow for sediment feed downstream and to sustain reservoir capacity (Kondolf et al. 2014). Some techniques to divert sediment around or through a dam are to (a) use off-channel reservoir storage and bypass the dam through a tunnel or a channel (depending on the geometry of the river and steepness for optimizing the design and cost of such intervention); (b) apply sediment sluicing, which is rapidly discharging sediments during periods of high inflows to the reservoir, allowing fine sediments to flow from top of the dam; (c) apply drawdown flushing (the opposite of sediment sluicing, since the gates are at a low level), allowing the resuspension and transportation of sediments downstream. For a detailed review of these options, see Kondolf et al. (2014).

Regulating coastal sand mining and effectively enforcing laws and regulations that ban sand mining in critical zones for sediment balance is crucial in the Middle East and North Africa. As noted earlier, illegal sand mining at the region's coasts is widespread and has become a serious problem. "Sand mafias" have been established that smuggle sand outside the country, making it a transboundary issue as well as a national and local one (UNEP 2019). Hence, putting an effective ban on illegal sand mining is essential. However, such a ban must go hand-in-hand with a credible enforcement mechanism to ensure compliance. Furthermore, in addition to banning sand mining per se, regulating the use of sand for building purposes and putting in force a strict supervisory mechanism is crucial. For example, in Morocco, half the sand used to build hotels, roads, and other tourism-related infrastructure comes from illegal sources (UNEP 2019). Hence, making a declaration of sources obligatory and introducing hefty fines for noncompliance may be useful in cutting the demand for illegally mined sand.

It is important to start combating coastal erosion by unifying stakeholders' efforts through an ICZM plan that will not only help combat coastal erosion but also help resolve other challenges such as pollution and biodiversity. All the policies mentioned here are important to control and support effective interventions under the umbrella of ICZM. However, to combat coastal erosion effectively and practically, it is critical to start by bringing stakeholders together and putting in place data and monitoring plans. Middle East and North Africa economies will benefit greatly from ICZM plans on both the national and subregional levels, as well as on the basin level with the other countries that share the basin.

NOTES

1. Parts of the next two sections —"How Eroded Are the Coasts?" and "The Economic Impacts of Eroded Coasts"—are adapted from Heger and Vashold (2021) and use similar language.
2. Red Sea, Mediterranean, and other sea area data are from WorldAtlas, https://www.worldatlas.com/.
3. The various sources are either online property portals (such as https://www .avito.ma/, https://www.mitula.ma/, http://www.homeintunisia.com/, and https://www.opensooq.com/ar with various country domains) or official sources where available (mainly in Morocco).
4. It would be preferable to further distinguish between agricultural land and building plots in the case of rural land. However, in addition to the challenges posed by price-data constraints, determining the relative shares of these types of land in total rural coastal areas proved extremely difficult. Although the ESA's land-use dataset distinguishes between already built-up areas and ones that are explicitly used for agricultural purposes, it is not possible to determine whether a certain plot of bare land is dedicated for use as a building plot. Given these difficulties and the often unclear rules for land classification in these countries, the report refrains from drawing such distinctions and uses a composite price for rural areas.
5. For more information on the Global Land Cover database, see the ESA's Climate Change Initiative website: http://www.esa-landcover-cci.org/.
6. The share of urban areas is calculated for the whole coastline and not just for the parts subject to erosion. The limited information on land prices did not allow for such a differentiation.
7. These estimates do not take infrastructure costs explicitly into account.
8. Ghermandi and Nunes (2013) provide estimates for the value of recreational services for near-shore locations on a global scale and show that these vary with accessibility, development, and tourist amenities such as the beaches or coral reefs.
9. The authors also calculate the willingness to pay of Djerba residents and found similar values per capita. Because the number of residents (around 30,000) is tiny relative to the number tourists visiting Djerba (more than 1 million per year), tourists' contributions make up the lion's share of overall contributions.
10. The authors infer these highly detrimental effects on the local tourism sector by means of benefit transfers of willingness-to-pay values from another source and assume complete disappearance of the beaches, a process that varies for the different beaches under scrutiny and may take several decades. Nonetheless, these findings highlight the potentially huge losses in tourist regions that could result from coastal erosion.

11. For the latest Permanent Service for Mean Sea Level (PSMSL) data for the Middle East and North Africa region, see the PSMSL Data Explorer: https://www.psmsl.org/data/obtaining/map.html#metadataTab.

12. The RCP scenario refers to a greenhouse gas concentration (not emissions) trajectory estimated by the UN Intergovernmental Panel on Climate Change (IPCC). Temperatures refer to projected increases in average global surface temperature above preindustrial levels, as also estimated by the IPCC.

REFERENCES

Abdel-Latif, T., S. T. Ramadan, and A. M. Galal. 2012. "Egyptian Coastal Regions Development through Economic Diversity for its Coastal Cities." *HBRC Journal* 8 (3): 252–62.

Abou-Dagher, M., M. Nader, and S. El Indary. 2012. "Evolution of the Coast of North Lebanon from 1962–2007: Mapping Changes for the Identification of Hotspots and for Future Management Interventions." In IV International Symposium "Monitoring of Mediterranean Coastal Areas: Problems and Measurement Techniques." Instituto di Biometeorologia (IBIMET), Consiglio Nazionale Delle Ricerche (CNR), Italy.

Abul-Azm, A. G., I. Abdel-Gelil, and I. Trumbic. 2003. "Integrated Coastal Zone Management in Egypt: The Fuka-Matrouh Project." *Journal of Coastal Conservation* 9 (1): 5–12.

Albrecht-Heider, C. 2020. "Tunisia: More Sand on the Beach." *KfW Stories*, March 9. https://www.kfw.de/stories/environment/climate-change/coastal-protection-tunisia/.

Aldar.ma. 2019. "The 'Sand Mafia' Threatens the Moroccan Coast." Aldar.ma, June 17. https://aldar.ma/32945.html.

Alexandrakis, G., C. Manasakis, and N. A. Kampanis. 2015. "Valuating the Effects of Beach Erosion to Tourism Revenue. A Management Perspective." *Ocean & Coastal Management* 111: 1–11.

Ali, E. M., and I. A. El-Magd. 2016. "Impact of Human Interventions and Coastal Processes along the Nile Delta Coast, Egypt During the Past Twenty-Five Years." *Egyptian Journal of Aquatic Research* 42 (1): 1–10.

Amrouni, O., A. Hzami, and E. Heggy. 2019. "Photogrammetric Assessment of Shoreline Retreat in North Africa: Anthropogenic and Natural Drivers." *ISPRS Journal of Photogrammetry and Remote Sensing* 157: 73–92.

Antunes do Carmo, J. S. Forthcoming. "Coastal Defenses and Engineering Works." In *Life Below Water, Encyclopedia of the UN Sustainable Development Goals*, edited by W. L. Filho, A. M. Azul, L. Brandii, A. L. Salvia, and T. Wall. Geneva: Springer.

Becker, A. H., M. Acciaro, R. Asariotis, E. Cabrera, L. Cretegny, P. Crist, M. Esteban, et al. 2013. "A Note on Climate Change Adaptation for Seaports: A Challenge for Global Ports, a Challenge for Global Society." *Climatic Change* 120 (4): 683–95.

Benkhattab, F. Z., M. Hakkou, I. Bagdanavičiūtė, A. El Mrini, H. Zagaoui, H. Rhinane, and M. Maanan. 2020. "Spatial–Temporal Analysis of the Shoreline Change Rate Using Automatic Computation and Geospatial Tools along the Tetouan Coast in Morocco." *Natural Hazards* 104 (1): 519–36.

Bitan, M., and D. Zviely. 2019. "Lost Value Assessment of Bathing Beaches due to Sea Level Rise: A Case Study of the Mediterranean Coast of Israel." *Journal of Coastal Conservation* 23 (4): 773–83.

Canals Ventín, P., and L. Lázaro Marín. 2019. "Towards Nature-Based Solutions in the Mediterranean: What Does Nature Give Us? 14 Examples Put Forward by IUCN Members and Partners in the Mediterranean Region." Booklet, International Union for Conservation of Nature (IUCN) Centre for Mediterranean Cooperation, Málaga, Spain.

Carboni, M., C. Perelli, and G. Sistu. 2014. "Is Islamic Tourism a Viable Option for Tunisian Tourism? Insights from Djerba." *Tourism Management Perspectives* 11: 1–9.

Coastal Care. 2009. "Mining of Coastal Sand: A Critical Environmental and Economic Problem for Morocco." CoastalCare.org, March 25.

Coastal Care. 2020. "470,000 US Dollars Worth of Illegally Mined Sand Seized in 2019, Algeria." CoastalCare.org, May 17.

Cohen-Shacham, E., A. Andrade, J. Dalton, N. Dudley, M. Jones, C. Kumar, S. Maginnis, et al. 2019. "Core Principles for Successfully Implementing and Upscaling Nature-Based Solutions." *Environmental Science & Policy* 98: 20–29.

Construction Week. 2014. "Umm Suqeim Beach to Receive $9.5mn Renovation." *Construction Week*, June 16.

Construction Week. 2015. "Groundwater Depletion Risks Land Subsidence in UAE." *Construction Week*, April 16.

Croitoru, L., J. J. Miranda, and M. Sarraf. 2019. "The Cost of Coastal Zone Degradation in West Africa: Benin, Cote d'ivoire, Senegal and Togo." Study of the West Africa Coastal Areas Management Program (WACA), World Bank, Washington, DC.

CZMAI (Coastal Zone Management Authority and Institute). 2016. "Belize Integrated Coastal Zone Management Plan: Promoting the Wise, Planned Use of Belize's Coastal Resources." Planning framework, CZMAI, Belize City.

Dabbeek, J., and V. Silva. 2020. "Modeling the Residential Building Stock in the Middle East for Multi-Hazard Risk Assessment." *Natural Hazards* 100 (2): 781–810.

Dasgupta, S., B. Laplante, S. Murray, and D. Wheeler. 2009. "Sea-Level Rise and Storm Surges: A Comparative Analysis of Impacts in Developing Countries." Policy Research Working Paper 4901, World Bank, Washington, DC.

Dawson, R. J., M. E. Dickson, R. J. Nicholls, J. W. Hall, M. J. A. Walkden, P. K. Stansby, M. Mokrech, et al. 2009. "Integrated Analysis of Risks of Coastal Flooding and Cliff Erosion under Scenarios of Long Term Change." *Climatic Change* 95 (1): 249–88.

de Schipper, M. A., S. de Vries, G. Ruessink, R. C. de Zeeuw, J. Rutten, C. van Gelder-Maas, and M. J. F. Stive. 2016. "Initial Spreading of a Mega Feeder

Nourishment: Observations of the Sand Engine Pilot Project." *Coastal Engineering* 111: 23–38.

Dickson, M. E., M. J. A. Walkden, and J. W. Hall. 2007. "Systemic Impacts of Climate Change on an Eroding Coastal Region over the Twenty-first Century." *Climatic Change* 84 (2): 141–66.

Dribek, A., and L. Voltaire. 2017. "Contingent Valuation Analysis of Willingness to Pay for Beach Erosion Control through the Stabiplage Technique: A Study in Djerba (Tunisia)." *Marine Policy* 86: 17–23.

EEAA (Egyptian Environmental Affairs Agency). 2016. "Egypt Third National Communication, under the United Nations Framework Convention on Climate Change." Report submitted to the United Nations Framework Convention on Climate Change (UNFCCC), EEAA, Cairo.

El Mdari, M., M. Idhalla, N. Tamsouri, A. Kaddioui, H. Nhhala, K. Hilmi, and E. D. Rhimou. 2018. "Analysis of Fish Community at the First Artificial Reef in Morocco (Martil, Mediterranean)." *International Journal of Advanced Research* 6 (9): 726–35.

El-Raey, M. 2009. "Impact of Climate Change on Egypt." In "Egypt: Coastal Zone Development and Climate Change," case study, GAIA Environmental Information System, http://80.120.147.2/GAIA/CASES/EGY/impact.html.

El-Raey, M. 2010. "Impacts and Implications of Climate Change for the Coastal Zones of Egypt." In *Coastal Zones and Climate Change*, edited by D. Michel and A. Pandya, 31–50. Washington, DC: Henry L. Stimson Center.

Elsharkay, H., H. Rashed, and I. Rached. 2009. "Climate Change: The Impacts of Sea Level Rise on Egypt." Paper presented at the 45th International Society of City and Regional Planners (ISOCARP) World Planning Congress, Porto, Portugal, October 18–22.

Fanos, A. M., A. A. Khafagy, and R. G. Dean. 1995. "Protective Works on the Nile Delta Coast." *Journal of Coastal Research* 11 (2): 516–28.

FAO (Food and Agriculture Organization of the United Nations). 2020. "Morocco: A Maritime Fishing Nation Works to Develop Its Aquaculture Sector." *Blue Growth* (blog), May 2020 (accessed May 10, 2021), https://web .archive.org/web/20210323053137/https://www.fao.org/blogs/blue-growth -blog/morocco-a-maritime-fishing-nation-works-to-develop-its-aquaculture -sector/en/.

Fattahi, M. 2019. "Giant Sinkholes Form around Tehran due to Drought and Excessive Water Pumping in Iranian Capital." *Independent*, January 24.

Few, R., K. Brown, and E. L. Tompkins. 2007. "Climate Change and Coastal Management Decisions: Insights from Christchurch Bay, UK." *Coastal Management* 35 (2–3): 255–70.

Flayou, L., M. Snoussi, R. Otmane, and O. Khalfaoui. 2017. "Valuing the Economic Costs of Beach Erosion Related to the Loss in the Tourism Industry: The Case of Tetouan Coast (Morocco)." Conference paper, *Euro-Mediterranean Conference for Environmental Integration*: 1633–36.

Fraser, R., and G. Spencer. 1998. "The Value of an Ocean View: An Example of Hedonic Property Amenity Valuation." *Australian Geographical Studies* 36 (1): 94–98.

Ghermandi, A., and P. A. L. D. Nunes. 2013. "A Global Map of Coastal Recreation Values: Results from a Spatially Explicit Meta-Analysis." *Ecological Economics* 86: 1–15.

Ghoneim, E., J. Mashaly, D. Gamble, J. Halls, and M. Abubakr. 2015. "Nile Delta Exhibited a Spatial Reversal in the Rates of Shoreline Retreat on the Rosetta Promontory Comparing Pre- and Post-Beach Protection." *Geomorphology* 228: 1–14.

Giardino, A., R. Schrijvershof, C. M. Nederhoff, H. de Vroeg, C. Brière, P. K. Tonnon, S. Caires, et al. 2018. "A Quantitative Assessment of Human Interventions and Climate Change on the West African Sediment Budget." *Ocean & Coastal Management* 156: 249–65.

Gomis, D., M. N. Tsimplis, B. Martín-Míguez, A. W. Ratsimandresy, J. García-Lafuente, and S. A. Josey. 2006. "Mediterranean Sea Level and Barotropic Flow through the Strait of Gibraltar for the Period 1958–2001 and Reconstructed since 1659." *Journal of Geophysical Research: Oceans* 111 (C11): 1–12.

Greene, M. 2016. "Why Are Beaches Disappearing in Morocco?" Middle East Eye, September 1. https://www.middleeasteye.net/features/why-are -beaches-disappearing-morocco.

Haida, S., M. Snoussi, and J.-L. Probst. 2004. "Sediment Fluxes of the Sebou River (Morocco)." *Le Journal de l'Eau et de l'Environnement* 34: 21–24.

Hallegatte, S., C. Green, R. J. Nicholls, and J. Corfee-Morlot. 2013. "Future Flood Losses in Major Coastal Cities." *Nature Climate Change* 3 (9): 802–06.

Hamilton, J. M. 2007. "Coastal Landscape and the Hedonic Price of Accommodation." *Ecological Economics* 62 (3–4): 594–602.

Hayes, M. O., and D. M. FitzGerald. 2013. "Origin, Evolution, and Classification of Tidal Inlets." *Journal of Coastal Research* 69: 14–33.

Heger, M. P., and L. Vashold. 2021. "Disappearing Coasts in the Maghreb: Coastal Erosion and Its Costs." Maghreb Technical Note No. 4, World Bank, Washington, DC.

Hoggart, S. P. G., M. E. Hanley, D. J. Parker, D. J. Simmonds, D. T. Bilton, M. Filipova-Marinova, E. L. Franklin, et al. 2014. "The Consequences of Doing Nothing: The Effects of Seawater Flooding on Coastal Zones." *Coastal Engineering* 87: 169–82.

Huang, J., G. R. Parsons, P. J. Poor, and M. Q. Zhao. 2011. "Combined Conjoint-Travel Cost Demand Model for Measuring the Impact of Erosion and Erosion Control Programs on Beach Recreation." In *Preference Data for Environmental Valuation: Combining Revealed and Stated Approaches*, edited by T. Haab, J. Huang, and J. Whitehead, 115–38. New York: Routledge.

IPE Global. 2018. "Shoreline Management Sub-Plan (Final) for Odisha Coast." Consultancy support for the World Bank/Government of Odisha II Preparation of Integrated Coastal Zone Management Plan (ICZMP) and Shoreline Management Plan for Odisha, India (2015–17), IPE Global Limited, New Delhi.

Iskander, M. M. 2010. "Environmental Friendly Methods for the Egyptian Coastal Protection." Paper presented at the International Conference on

Coastal Zone Management of River Deltas and Low Land Coastlines (CZMRDLLC), Alexandria, Egypt, March 6–10.

Jeffrey, H., and S. Bleasdale. 2017. "Tunisia: Mass Tourism in Crisis?" In *Mass Tourism in a Small World*, edited by D. Harrison and R. Sharpley, 191–99. Wallingford, UK; and Boston: CAB International.

Kasmi, S., M. Snoussi, O. Khalfaoui, R. Aitali, and L. Flayou. 2020. "Increasing Pressures, Eroding Beaches and Climate Change in Morocco." *Journal of African Earth Sciences* 164: 103796.

Katz, O., and A. Mushkin. 2013. "Characteristics of Sea-cliff Erosion Induced by a Strong Winter Storm in the Eastern Mediterranean." *Quaternary Research* 80 (1): 20–32.

Kerby, M. 2019. "March 2019 Comments." Coastal Concern Action Group (CCAG) Comments, Happisburgh Village Website, February 28. http://happisburgh.org.uk/category/ccag-comment/.

Khaleej Times. 2015. "Umm Suqeim Beach to Be Partially Closed for 9 Months." *Khaleej Times*, April 5.

Khaleej Times. 2017. "Dh35 Million to Improve Dubai Beaches." *Khaleej Times*, February 6.

Khalil, S. 2008. "The Use of Sand Fences in Barrier Island Restoration: Experience on the Louisiana Coast." System-Wide Water Resources Program (SWWRP) Technical Note, Engineer Research and Development Center (ERDC) Environmental Lab, US Army Corps of Engineers, Vicksburg, MS.

Khorrami, M., S. Abrishami, Y. Maghsoudi, B. Alizadeh, and D. Perissin. 2020. "Extreme Subsidence in a Populated City (Mashhad) Detected by PSInSAR Considering Groundwater Withdrawal and Geotechnical Properties." *Scientific Reports* 10 (1): 1–16.

Kloosterman, K. 2010. "Coastal Erosion Threatens Evolutionary Hotspots in Gulf Region." Green Prophet, March 5. https://www.greenprophet.com/2010/03/coastal-erosion-gulf/.

Knapp, K. R., M. C. Kruk, D. H. Levinson, H. J. Diamond, and C. J. Neumann. 2010. "The International Best Track Archive for Climate Stewardship (IBTrACS) Unifying Tropical Cyclone Data." *Bulletin of the American Meteorological Society* 91 (3): 363–76.

Kondolf, G. M., Y. Gao, G. Annandale, G. Morris, E. Jiang, J. Zhang, Y. Cao, et al. 2014. "Sustainable Sediment Management in Reservoirs and Regulated Rivers: Experiences from Five Continents." *Earth's Future* 2 (5): 256–80.

Koraim, A., E. Heikal, and A. Abozaid. 2011. "Different Methods Used for Protecting Coasts from Sea Level Rise Caused by Climate Change." *Current Development in Oceanography* 3: 33–66.

Kron, W. 2013. "Coasts: The High-Risk Areas of the World." *Natural Hazards* 66 (3): 1363–82.

Landerer, F. W., and D. L. Volkov. 2013. "The Anatomy of Recent Large Sea Level Fluctuations in the Mediterranean Sea." *Geophysical Research Letters* 40 (3): 553–57.

Landry, C. E., and H. Liu. 2009. "A Semi-Parametric Estimator for Revealed and Stated Preference Data—An Application to Recreational Beach Visitation." *Journal of Environmental Economics and Management* 57 (2): 205–18.

Landry, C. E., and H. Liu. 2011. "Econometric Models for Joint Estimation of Revealed and Stated Preference Site-frequency Recreation Demand Models." In *Preference Data for Environmental Valuation: Combining Revealed and Stated Approaches*, edited by T. Haab, J. Huang, and J. Whitehead, 115–38. New York: Routledge

Loudyi, D., M. Chagdali, S. Belmatrik, and K. El Kadi Abderrezzak. 2018. "Reservoirs Silting in Morocco." *Hydrolink* 3/2018: 92–94.

Luijendijk, A., G. Hagenaars, R. Ranasinghe, F. Baart, G. Donchyts, and S. Aarninkhof. 2018. "The State of the World's Beaches." *Scientific Reports* 8 (1): 1–11.

Luijendijk, A. P., R. Ranasinghe, M. A. de Schipper, B. A. Huisman, C. M. Swinkels, D. J. R. Walstra, and M. J. F. Stive. 2017. "The Initial Morphological Response of the Sand Engine: A Process-Based Modelling Study." *Coastal Engineering* 119: 1–14.

Marzougui, W., and A. Oueslati. 2017. "Les plages de la côte d'Ejjehmi-Soliman (golfe de Tunis, Tunisie): exemple d'accélération de l'érosion marine dans une cellule sédimentaire artificiellement tronçonnée." *Physio-Géo. Géographie Physique et Environnement* 11: 21–41.

Masria, A., K. Abdelaziz, and A. Negm. 2015. "Testing a Combination of Hard and Soft Measures to Enhance the Stability of Rosetta Outlet." *Journal of Oceanography and Marine Research* 4 (1): 1–9.

Masria, A., K. Nadaoka, Y. Kuriyama, A. Negm, M. Iskander, and O. C. Saavedra. 2015. "Near Shore and Beach Nourishment Effects on the Stability of Rosetta Promontory, Egypt." Paper presented at 18th International Water Technology Conference IWTC18, Sharm Elsheikh, Egypt, March 12–14.

Masria, A., K. Nadaoka, A. Negm, and M. Iskander. 2015. "Detection of Shoreline and Land Cover Changes around Rosetta Promontory, Egypt, Based on Remote Sensing Analysis." *Land* 4 (1): 216–30.

Maul, G. A., and I. W. Duedall. 2019. "Demography of Coastal Populations" In *Encyclopedia of Coastal Science*, 2nd ed., edited by C. W. Finkl and C. Makowski, 692–99. Cham, Switzerland: Springer.

MEWA (Ministry of Environment, Water and Agriculture, Kingdom of Saudi Arabia). 2017. "National Environment Strategy: Executive Summary for the Council of Economic and Development Affairs, December 15, 2017." Vision 2030 strategy document, MEWA, Riyadh, Saudi Arabia.

Mutahi, B. 2020. "Egypt-Ethiopia Row: The Trouble over a Giant Nile Dam." BBC News, January 13. https://www.bbc.com/news/world-africa-50328647.

Nader, M. 2015. "Coastal Zone Management in Lebanon." PowerPoint presentation to Programme Solidarité Eau, Beirut, Lebanon, September 22. https://www.pseau.org/outils/ouvrages/uob_coastal_zone_management_in _lebanon_2015.pdf.

Nicholls, R. J., P. P. Wong, V. R. Burkett, J. Codignotto, J. Hay, R. McLean, S. Ragoonaden, et al. 2007. "Coastal Systems and Low-Lying Areas." In *Climate Change 2007: Impacts, Adaptation and Vulnerability. Contribution of Working Group II to the Fourth Assessment Report of the Intergovernmental Panel on Climate Change*, edited by M. L. Parry, O. F. Canziani, J. P. Palutikof, P. J. van der Linden, and C. E. Hanson, 315–56. Cambridge: Cambridge University Press.

Nile FM. 2020. "Egypt Plants Thousands of the Mangrove Supertree to Revive Its Unique Ecosystem." Nile FM staff report, March 1. https://nilefm.com /digest/article/5062/egypt-plants-thousands-of-mangroves-to-combat -climate-change-over-fishing-and-coastal-erosion.

NOC (National Oceanography Centre). 2020. "Using Optical Satellite Shoreline Detection to Measure Historic and Forecast Future Sandy Shoreline Changes in North Africa." Report for the World Bank's Cleaner Marine and Coastal Ecosystems in North Africa (P170596) and Integrated Marine and Coastal Management in Tunisia (P166339) technical assistance, NOC, Southampton, UK.

Norfolk Vanguard Limited. 2018. "Norfolk Vanguard Coastal Erosion Study." Environmental Impact Assessment Environmental Statement, Document Reference PB4476-005-0043, Norfolk Vanguard Limited, London.

OECD and FAO (Organisation for Economic Co-operation and Development and the Food and Agriculture Organization of the United Nations). 2018. "The Middle East and North Africa: Prospects and Challenges." In *OECD-FAO Agricultural Outlook 2018–2027. Special Focus: Middle East and North Africa*. Paris: OECD; and Rome: FAO.

Oueslati, A., O. Labidi, and T. Elamri. 2015. "Atlas de la vulnérabilité du littoral tunisien à l'élévation du niveau marin." Agency for Coastal Protection and Planning (APAL), Tunis, Tunisia.

Parsons, G. R., Z. Chen, M. K. Hidrue, N. Standing, and J. Lilley. 2013. "Valuing Beach Width for Recreational Use: Combining Revealed and Stated Preference Data." *Marine Resource Economics* 28 (3): 221–41.

Pearce, J., S. Khan, and P. Lewis. 2012. "Medmerry Managed Ralignment–Sustainable Coastal Management to Gain Multiple Benefits." In *Innovative Coastal Zone Management: Sustainable Engineering for a Dynamic Coast*, edited by A. Schofield, 243–52. London: ICE Publishing.

Pendleton, L., C. Mohn, R. K. Vaughn, P. King, and J. G. Zoulas. 2012. "Size Matters: The Economic Value of Beach Erosion and Nourishment in Southern California." *Contemporary Economic Policy* 30 (2): 223–37.

Pompe, J. J., and J. R. Rinehart. 1995. "Beach Quality and the Enhancement of Recreational Property Values." *Journal of Leisure Research* 27 (2): 143–54.

Rangel-Buitrago, N. G., G. Anfuso, and A. T. Williams. 2015. "Coastal Erosion Along the Caribbean Coast of Colombia: Magnitudes, Causes and Management." *Ocean & Coastal Management* 114: 129–44.

Raybould, M., D. Anning, D. Ware, and N. Lazarow. 2013. *Beach and Surf Tourism and Recreation in Australia: Vulnerability and Adaptation*. Gold Coast, Australia: Bond University.

Rigall-I-Torrent, R., M. Fluvià, R. Ballester, A. Saló, E. Ariza, and J.-M. Espinet. 2011. "The Effects of Beach Characteristics and Location with Respect to Hotel Prices." *Tourism Management* 32 (5): 1150–58.

Rinehart, J. R., and J. J. Pompe. 1994. "Adjusting the Market Value of Coastal Property for Beach Quality." *Appraisal Journal* 62 (4): 604–08.

Sadaoui, M., W. Ludwig, F. Bourrin, Y. Le Bissonnais, and E. Romero. 2018. "Anthropogenic Reservoirs of Various Sizes Trap Most of the Sediment in the Mediterranean Maghreb Basin." *Water* 10 (7): 927.

Saidani, M. 2019. "Tunisia: $1.3b of Tourism Revenues Expected in 2019." *Asharq Al-Awsat*, August 2. https://english.aawsat.com//home/article /1840766/tunisia-13b-tourism-revenues-expected-2019.

Saïdi, H., R. Souissi, and F. Zargouni. 2012. "Environmental Impact of Detached Breakwaters on the Mediterranean Coastline of Soliman (North-East of Tunisia)." *Rendiconti Lincei* 23 (4): 339–47.

Satta, A., M. Puddu, S. Venturini, and C. Giupponi. 2017. "Assessment of Coastal Risks to Climate Change Related Impacts at the Regional Scale: The Case of the Mediterranean Region." *International Journal of Disaster Risk Reduction* 24: 284–96.

Schäfer, K. 2013. "Urbanization and Urban Risks in the Arab Region." Presentation at First Arab Region Conference for Disaster Risk Reduction, Aqaba, Jordan, March 19–21.

Schoonees, T., A. Gijón Mancheño, B. Scheres, T. J. Bouma, R. Silva, T. Schlurmann, and H. Schüttrumpf. 2019. "Hard Structures for Coastal Protection, Towards Greener Designs." *Estuaries and Coasts* 42 (7): 1709–29.

Schuhmann, P. W., B. E. Bass, J. F. Casey, and D. A. Gill. 2016. "Visitor Preferences and Willingness to Pay for Coastal Attributes in Barbados." *Ocean & Coastal Management* 134: 240–50.

Scott, D., M. C. Simpson, and R. Sim. 2012. "The Vulnerability of Caribbean Coastal Tourism to Scenarios of Climate Change Related Sea Level Rise." *Journal of Sustainable Tourism* 20 (6): 883–98.

Shahi, A. 2019. "Drought: the Achilles Heel of the Islamic Republic of Iran." *Asian Affairs* 50 (1): 18–39.

Sharaan, M., C. Somphog, and K. Udo. 2020. "Impact of SLR on Beach-Tourism Resort Revenue at Sahl Hasheesh and Makadi Bay, Red Sea, Egypt: A Hedonic Pricing Approach." *Journal of Marine Science and Engineering* 8 (6): 1–13.

Shin, E. C., and Y. I. Oh. 2007. "Coastal Erosion Prevention by Geotextile Tube Technology." *Geotextiles and Geomembranes* 25 (4–5): 264–77.

Sieghart, L. C., J. A. Mizener, and J. Gibson. 2019. "Capturing Opportunities for Integrated Coastal Zone Management and the Blue Economy in MENA." MENA Knowledge and Learning: Quick Notes Series No. 172, World Bank, Washington, DC.

SIS (State Information Service, Arab Republic of Egypt). 2020. "Environment Sector Achievements." June 13 report, SIS, Cairo.

Snoussi, M., S. Haïda, and S. Imassi. 2002. "Effects of the Construction of Dams on the Water and Sediment Fluxes of the Moulouya and the Sebou Rivers, Morocco." *Regional Environmental Change* 3 (1): 5–12.

Snoussi, M., O. Khalfaoui, L. Flayou, S. Kasmi, and O. Raji. 2017. "Can ICZM Help the Resilience of Disappearing Beaches in the Face of Climate Change?" In *Recent Advances in Environmental Science from the Euro-Mediterranean and Surrounding Regions: Proceedings of the Euro-Mediterranean Conference for Environmental Integration (EMCEI-1), Tunisia 2017*, edited by A. Kallel, M. Ksibi, H. Ben Dhia, and N. Khélifi, 29–30. Cham, Switzerland: Springer.

Snoussi, M., T. Ouchani, A. Khouakhi, and I. Niang-Diop. 2009. "Impacts of Sea-Level Rise on the Moroccan Coastal Zone: Quantifying Coastal Erosion and Flooding in the Tangier Bay." *Geomorphology* 107 (1–2): 32–40.

Snoussi, M., T. Ouchani, and S. Niazi. 2008. "Vulnerability Assessment of the Impact of Sea-level Rise and Flooding on the Moroccan Coast: The Case of the Mediterranean Eastern Zone." *Estuarine, Coastal and Shelf Science* 77 (2): 206–13.

Stive, M. J. F., R. Ranasinghe, and P. J. Cowell. 2010. "Sea Level Rise and Coastal Erosion." In *Handbook of Coastal and Ocean Engineering*, edited by Y. C. Kim, 1023–37. New Jersey and Singapore: World Scientific.

Sytnik, O., L. Del Río, N. Greggio, and J. Bonetti. 2018. "Historical Shoreline Trend Analysis and Drivers of Coastal Change along the Ravenna Coast, NE Adriatic." *Environmental Earth Sciences* 77 (23): 1–20.

Syvitski, J. P. M., A. J. Kettner, I. Overeem, E. W. H. Hutton, M. T. Hannon, G. R. Brakenridge, J. Day, et al. 2009. "Sinking Deltas Due to Human Activities." *Nature Geoscience* 2 (10): 681–86.

Syvitski, J. P. M., C. J. Vörösmarty, A. J. Kettner, and P. Green. 2005. "Impact of Humans on the Flux of Terrestrial Sediment to the Global Coastal Ocean." *Science* 308 (5720): 376–80.

Tarui, N., M. Peng, and D. Eversole. 2018. "Economic Impact Analysis of the Potential Erosion of Waikīkī Beach." Update report, University of Hawai'i Sea Grant College Program, University of Hawai'i at Mānoa, Honolulu, HI.

Technical Textiles. 2020. "Pentha Project in India Protected against Cyclones." TechnicalTextile.net, Februrary 25.

Technital. n.d. "Design Services and Works Supervision for the Al Faw Grand Port (Iraq)." Project description, Technital S.p.A., Verona, Italy. https://www.technital.net/projects/ports-and-waterways/industrial-and-commercial-ports/ports-and-waterways-design-services-and-works-supervision-for-the-al-faw-grand-port-iraq.

Tolba, M. K., and N. W. Saab, eds. 2009. *Arab Environment Climate Change: Impact of Climate Change on Arab Countries.* Beirut, Lebanon: Arab Forum for Environment and Development.

Trakadas, A. 2020. "Natural and Anthropogenic Factors Impacting Northern Morocco's Coastal Archaeological Heritage: A Preliminary Assessment." *Journal of Island and Coastal Archaeology:* 1–32.

Tsimplis, M. N., F. M. Calafat, M. Marcos, G. Jordá, D. Gomis, L. Fenoglio-Marc, M. V. Struglia, S. A. Josey, and D. P. Chambers. 2013. "The Effect of the NAO on Sea Level and on Mass Changes in the Mediterranean Sea." *Journal of Geophysical Research: Oceans* 118 (2): 944–52.

Tsimplis, M. N., and T. F. Baker. 2000. "Sea Level Drop in the Mediterranean Sea: An Indicator of Deep Water Salinity and Temperature Changes?" *Geophysical Research Letters* 27 (12): 1731–34.

UNEP (United Nations Environment Programme). 2019. *Sand and Sustainability: Finding New Solutions for Environmental Governance of Global Sand Resources.* Geneva: UNEP.

UNEP (United Nations Environment Programme). 2020. "Status of Signatures and Ratifications of the Barcelona Convention for the Protection of the Marine Environment and the Coastal Region of the Mediterranean and Its Protocols as at 29 October 2020." Contracting Parties documents, UNEP

Coordinating Unit for the Mediterranean Action Plan (MAP), Barcelona Convention Secretariat, Athens. https://wedocs.unep.org/bitstream /handle/20.500.11822/7096/StatusOfSignaturesAndRatifications_20201029 .pdf.

UNWTO (United Nations World Tourism Organization). 2019. *Tourism in the MENA Region.* Madrid: UNWTO.

Uyarra, M. C., I. M. Cote, J. A. Gill, R. R. T. Tinch, D. Viner, and A. R. Watkinson. 2005. "Island-Specific Preferences of Tourists for Environmental Features: Implications of Climate Change for Tourism-Dependent States." *Environmental Conservation* 32 (1): 11–19.

van der Meulen, F., and B. van der Valk. 2019. "Coastal Management Practices." In *Encyclopedia of Coastal Science.* 2nd ed., edited by C. W. Finkl and C. Makowski, 501–10. Cham, Switzerland: Springer

van der Meulen, F., B. van der Valk, K. Vertegaal, and M. Eerden. 2015. "'Building with Nature' at the Dutch Dune Coast: Compensation Target Management in Spanjaards Duin at EU and Regional Policy Levels." *Journal of Coastal Conservation* 19: 707–14.

Whitehead, J. C., C. F. Dumas, J. Herstine, J. Hill, and B. Buerger. 2008. "Valuing Beach Access and Width with Revealed and Stated Preference Data." *Marine Resource Economics* 23 (2): 119–35.

Whitehead, J. C., D. J. Phaneuf, C. F. Dumas, J. Herstine, J. Hill, and B. Buerger. 2010. "Convergent Validity of Revealed and Stated Recreation Behavior with Quality Change: A Comparison of Multiple and Single Site Demands." *Environmental and Resource Economics* 45 (1): 91–112.

Widz, M., and T. Brzezińska-Wójcik. 2020. "Assessment of the Overtourism Phenomenon Risk in Tunisia in Relation to the Tourism Area Life Cycle Concept." *Sustainability* 12 (5): 1–13.

World Bank. 2014. *Turn Down the Heat: Confronting the New Climate Normal.* Washington, DC: World Bank.

World Bank. 2020a. "Implementation Completion and Results Report: India – Integrated Coastal Zone Management Project (P 097985)." Report No. ICR00005155, World Bank, Washington, DC.

World Bank. 2020b. "New World Bank Program to Strengthen Integrated Coastal Zone Management in India." Press release, April 28.

World Bank. 2021. "Preserving Morocco's Coastline." Feature story, January 13. https://www.worldbank.org/en/news/feature/2021/01/13/preserving -moroccos-coastline.